OPTIMAL CONTROL SYSTEMS

Electrical Engineering Textbook Series

Richard C. Dorf, Series Editor
University of California, Davis

Forthcoming and Published Titles

Applied Vector Analysis
Matiur Rahman and Isaac Mulolani

Continuous Signals and Systems with MATLAB
Taan ElAli and Mohammad A. Karim

Discrete Signals and Systems with MATLAB
Taan ElAli

Electromagnetics
Edward J. Rothwell and Michael J. Cloud

Optimal Control Systems
Desineni Subbaram Naidu

OPTIMAL CONTROL SYSTEMS

Desineni Subbaram Naidu
Idaho State University
Pocatello, Idaho, USA

CRC PRESS

Boca Raton   London   New York   Washington, D.C.

Cover photo: Terminal phase (using fuel-optimal control) of the lunar landing of the Apollo 11 mission. Courtesy of NASA.

Library of Congress Cataloging-in-Publication Data

Naidu, D. S. (Desineni S.), 1940-
    Optimal control systems / by Desineni Subbaram Naidu.
      p. cm.— (Electrical engineering textbook series)
    Includes bibliographical references and index.
    ISBN 0-8493-0892-5 (alk. paper)
    1. Automatic control. 2. Control theory. 3. Mathematical optimization. I. Title II. Series.

TJ213 .N2655 2002
629.8—dc21
                    2002067415

Visit the CRC Press Web site at www.crcpress.com

© 2003 by CRC Press LLC

No claim to original U.S. Government works
International Standard Book Number 0-8493-0892-5
Library of Congress Card Number 2002067415
Printed in the United States of America   2 3 4 5 6 7 8 9 0
Printed on acid-free paper

"Because the shape of the whole universe is most per-fect and, in fact, designed by the wisest Creator, nothing in all of the world will occur in which no maximum or minimum rule is somehow shining forth."

Leohard Euler, 1744

Dedication

My deceased parents who shaped my life

Desineni Rama Naidu

Desineni Subbamma

and

My teacher who shaped my education

Buggapati Audi Chetty

Preface

Many systems, physical, chemical, and economical, can be modeled by mathematical relations, such as deterministic and/or stochastic differential and/or difference equations. These systems then change with time or any other independent variable according to the dynamical relations. It is possible to steer these systems from one state to another state by the application of some type of external inputs or controls. If this can be done at all, there may be different ways of doing the same task. If there are different ways of doing the same task, then there may be one way of doing it in the "best" way. This best way can be *minimum* time to go from one state to another state, or *maximum* thrust developed by a rocket engine. The input given to the system corresponding to this best situation is called "optimal" control. The measure of "best" way or performance is called "performance index" or "cost function." Thus, we have an "optimal control system," when a system is controlled in an optimum way satisfying a given performance index. The theory of optimal control systems has enjoyed a flourishing period for nearly two decades after the dawn of the so-called "modern" control theory around the 1960s. The interest in theoretical and practical aspects of the subject has sustained due to its applications to such diverse fields as electrical power, aerospace, chemical plants, economics, medicine, biology, and ecology.

Aim and Scope

In this book we are concerned with essentially the control of physical systems which are "dynamic" and hence described by ordinary differential or difference equations in contrast to "static" systems, which are characterized by algebraic equations. Further, our focus is on "deterministic" systems only.

The development of optimal control theory in the sixties revolved around the "maximum principle" proposed by the Soviet mathematician L. S. Pontryagin and his colleagues whose work was published in English in 1962. Further contributions are due to R. E. Kalman of the United States. Since then, many excellent books on optimal control theory of varying levels of sophistication have been published.

This book is written keeping the "student in mind" and intended to provide the student a simplified treatment of the subject, with an appropriate dose of mathematics. Another feature of this book is to assemble all the topics which can be covered in a one-semester class. A special feature of this book is the presentation of the procedures in the form of a summary table designed in terms of *statement of the problem* and a *step-by-step solution of the problem*. Further, MATLAB© and SIMULINK©[1], including Control System and Symbolic Math Toolboxes, have been incorporated into the book. The book is ideally suited for a one-semester, second level, graduate course in control systems and optimization.

Background and Audience

This is a second level graduate text book and as such the background material required for using this book is a first course on control systems, state space analysis, or linear systems theory. It is suggested that the student review the material in Appendices A and B given at the end of the book. This book is aimed at graduate students in Electrical, Mechanical, Chemical, and Aerospace Engineering and Applied Mathematics. It can also be used by professional scientists and engineers working in a variety of industries and research organizations.

Acknowledgments

This book has grown out of my lecture notes prepared over many years of teaching at the Indian Institute of Technology (IIT), Kharagpur, and Idaho State University (ISU), Pocatello, Idaho. As such, I am indebted to many of my teachers and students. In recent years at ISU, there are many people whom I would like to thank for their encouragement and cooperation. First of all, I would like to thank the *late* Dean Hary Charyulu for his encouragement to graduate work and research which kept me "live" in the area optimal control. Also, I would like to mention a special person, Kevin Moore, whose encouragement and cooperation made my stay at ISU a very pleasant and scholarly productive one for many years during 1990-98. During the last few years, Dean Kunze and Associate Dean Stuffle have been of great help in providing the right atmosphere for teaching and research work.

---

[1]MATLAB and SIMULINK are registered trademarks of The Mathworks, Inc., Natick, MA, USA.

Next, my students over the years were my best critics in providing many helpful suggestions. Among the many, special mention must be made about Martin Murillo, Yoshiko Imura, and Keith Fisher who made several suggestions to my manuscript. In particular, Craig Rieger (of Idaho National Engineering and Environmental Laboratory (INEEL)) deserves special mention for having infinite patience in writing and testing programs in MATLAB$^{©}$ to obtain analytical solutions to matrix Riccati differential and difference equations.

The camera-ready copy of this book was prepared by the author using LATEX of the PCTEX32[2] Version 4.0. The figures were drawn using CorelDRAW[3] and exported into LATEX document.

Several people at the publishing company CRC Press deserve mention. Among them, special mention must be made about Nora Konopka, Acquisition Editor, Electrical Engineering for her interest, understanding and patience with me to see this book to completion. Also, thanks are due to Michael Buso, Michelle Reyes, Helena Redshaw, and Judith Simon Kamin. I would like to make a special mention of Sean Davey who helped me in many issues regarding LATEX. Any corrections and suggestions are welcome via email to *naiduds@isu.edu*

Finally, it is my pleasant duty to thank my wife, Sita and my daughters, Radhika and Kiranmai who have been a great source of encouragement and cooperation throughout my academic life.

Desineni Subbaram Naidu
Pocatello, Idaho
June 2002

---

[2]LATEX is a registered trademark of Personal TEX, Inc., Mill Valley, CA.
[3]CorelDRAW is a registered trademark of Corel Corporation or Corel Corporation Limited.

ACKNOWLEDGMENTS

The permissions given by

1. Prentice Hall for D. E. Kirk, *Optimal Control Theory: An Introduction*, Prentice Hall, Englewood Cliffs, NJ, 1970,

2. John Wiley for F. L. Lewis, *Optimal Control*, John Wiley & Sons, Inc., New York, NY, 1986,

3. McGraw-Hill for M. Athans and P. L. Falb, *Optimal Control: An Introduction to the Theory and Its Applications*, McGraw-Hill Book Company, New York, NY, 1966, and

4. Springer-Verlag for H. H. Goldstine, *A History of the Calculus of Variations*, Springer-Verlag, New York, NY, 1980,

are hereby acknowledged.

AUTHOR'S BIOGRAPHY

Desineni "Subbaram" Naidu received his B.E. degree in Electrical Engineering from Sri Venkateswara University, Tirupati, India, and M.Tech. and Ph.D. degrees in Control Systems Engineering from the Indian Institute of Technology (IIT), Kharagpur, India. He held various positions with the Department of Electrical Engineering at IIT. Dr. Naidu was a recipient of a Senior National Research Council (NRC) Associateship of the National Academy of Sciences, Washington, DC, tenable at NASA Langley Research Center, Hampton, Virginia, during 1985-87 and at the U. S. Air Force Research Laboratory (AFRL) at Wright-Patterson Air Force Base (WPAFB), Ohio, during 1998-99. During 1987-90, he was an adjunct faculty member in the Department of Electrical and Computer Engineering at Old Dominion University, Norfolk, Virginia. Since August 1990, Dr. Naidu has been a professor at Idaho State University. At present he is Director of the Measurement and Control Engineering Research Center; Coordinator, Electrical Engineering program; and Associate Dean of Graduate Studies in the College of Engineering, Idaho State University, Pocatello, Idaho.

Dr. Naidu has over 150 publications including a research monograph, *Singular Perturbation Analysis of Discrete Control Systems*, Lecture Notes in Mathematics, 1985; a book, *Singular Perturbation Methodology in Control Systems*, IEE Control Engineering Series, 1988; and a research monograph entitled, *Aeroassisted Orbital Transfer: Guidance and Control Strategies, Lecture Notes in Control and Information Sciences*, 1994.

Dr. Naidu is (or has been) a member of the Editorial Boards of the IEEE *Transaction on Automatic Control*, (1993-99), the *International Journal of Robust and Nonlinear Control*, (1996-present), the *International Journal of Control-Theory and Advanced Technology* (C-TAT), (1992-1996), and a member of the Editorial Advisory Board of Mechatronics: *The Science of Intelligent Machines, an International Journal*, (1992-present).

Professor Naidu is an elected Fellow of The Institute of Electrical and Electronics Engineers (IEEE), a Fellow of World Innovation Foundation (WIF), an Associate Fellow of the American Institute of Aeronautics and Astronautics (AIAA) and a member of several other organizations such as SIAM, ASEE, etc. Dr. Naidu was a recipient of the Idaho State University Outstanding Researcher Award for 1993-94 and 1994-95 and the Distinguished Researcher Award for 1994-95. Professor Naidu's biography is listed (multiple years) in *Who's Who among America's Teachers*, the Silver Anniversary 25th Edition of *Who's Who in the West*, *Who's Who in Technology*, and *The International Directory of Distinguished Leadership*.

Contents

List of Figures

List of Tables

Chapter 1

Introduction

In this first chapter, we introduce the ideas behind optimization and optimal control and provide a brief history of calculus of variations and optimal control. Also, a brief summary of chapter contents is presented.

1.1 Classical and Modern Control

The *classical* (conventional) control theory concerned with single input and single output (SISO) is mainly based on Laplace transforms theory and its use in system representation in block diagram form. From Figure 1.1, we see that

$$\frac{Y(s)}{R(s)} = \frac{G(s)}{1 + G(s)H(s)} \qquad (1.1.1)$$

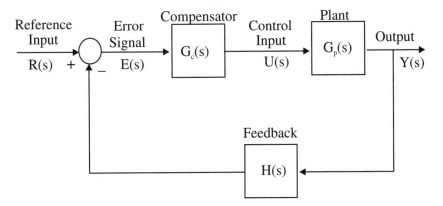

Figure 1.1 Classical Control Configuration

where s is Laplace variable and we used

$$G(s) = G_c(s)G_p(s). \qquad (1.1.2)$$

Note that

1. the input $u(t)$ to the plant is determined by the error $e(t)$ and the compensator, and

2. all the variables are not readily available for feedback. In most cases only one output variable is available for feedback.

The *modern* control theory concerned with multiple inputs and multiple outputs (MIMO) is based on state variable representation in terms of a set of first order differential (or difference) equations. Here, the system (plant) is characterized by state variables, say, in *linear*, time-invariant form as

$$\dot{\mathbf{x}}(t) = \mathbf{A}\mathbf{x}(t) + \mathbf{B}\mathbf{u}(t) \qquad (1.1.3)$$
$$\mathbf{y}(t) = \mathbf{C}\mathbf{x}(t) + \mathbf{D}\mathbf{u}(t) \qquad (1.1.4)$$

where, *dot* denotes differentiation with respect to (w.r.t.) t, $\mathbf{x}(t)$, $\mathbf{u}(t)$, and $\mathbf{y}(t)$ are n, r, and m dimensional *state*, *control*, and *output* vectors respectively, and A is nxn state, B is nxr input, C is mxn output, and D is mxr transfer matrices. Similarly, a *nonlinear* system is characterized by

$$\dot{\mathbf{x}}(t) = \mathbf{f}(\mathbf{x}(t), \mathbf{u}(t), t) \qquad (1.1.5)$$
$$\mathbf{y}(t) = \mathbf{g}(\mathbf{x}(t), \mathbf{u}(t), t). \qquad (1.1.6)$$

The modern theory dictates that all the state variables should be fed back after suitable weighting. We see from Figure 1.2 that in modern control configuration,

1. the input $\mathbf{u}(t)$ is determined by the controller (consisting of error detector and compensator) driven by system states $\mathbf{x}(t)$ and reference signal $\mathbf{r}(t)$,

2. all or most of the state variables are available for control, and

3. it depends on well-established matrix theory, which is amenable for large scale computer simulation.

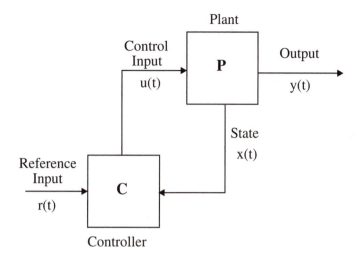

Figure 1.2   Modern Control Configuration

The fact that the state variable representation *uniquely* specifies the transfer function while there are a number of state variable representations for a given transfer function, reveals the fact that state variable representation is a more complete description of a system.

Figure 1.3 shows components of a modern control system. It shows three components of modern control and their important contributors. The first stage of any control system theory is to obtain or formulate the dynamics or *modeling* in terms of dynamical equations such as differential or difference equations. The system dynamics is largely based on the Lagrangian function. Next, the system is *analyzed* for its performance to find out mainly stability of the system and the contributions of Lyapunov to stability theory are well known. Finally, if the system performance is not according to our specifications, we resort to *design* [25, 109]. In optimal control theory, the design is usually with respect to a performance index. We notice that although the concepts such as Lagrange function [85] and V function of Lyapunov [94] are *old*, the techniques using those concepts are *modern*. Again, as the phrase *modern* usually refers to time and what is modern today becomes ancient after a few years, a more appropriate thing is to label them as optimal control, nonlinear control, adaptive control, robust control and so on.

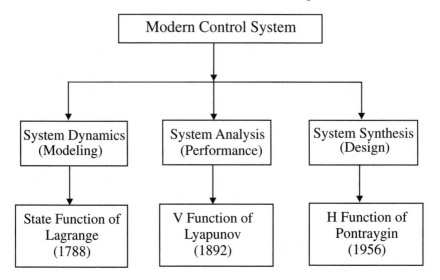

Figure 1.3   Components of a Modern Control System

1.2   *Optimization*

Optimization is a very desirable feature in day-to-day life. We like to work and use our time in an optimum manner, use resources optimally and so on. The subject of optimization is quite general in the sense that it can be viewed in different ways depending on the *approach* (algebraic or geometric), the *interest* (single or multiple), the *nature* of the signals (deterministic or stochastic), and the *stage* (single or multiple) used in optimization. This is shown in Figure 1.4. As we notice that the calculus of variations is one small area of the big picture of the optimization field, and it forms the basis for our study of optimal control systems. Further, optimization can be classified as *static* optimization and *dynamic* optimization.

1. **Static Optimization** is concerned with controlling a plant under *steady state* conditions, i.e., the system variables are not changing with respect to time. The plant is then described by *algebraic* equations. Techniques used are ordinary calculus, Lagrange multipliers, linear and nonlinear programming.

2. **Dynamic Optimization** concerns with the optimal control of plants under *dynamic* conditions, i.e., the system variables are changing with respect to time and thus the time is involved in system description. Then the plant is described by *differential*

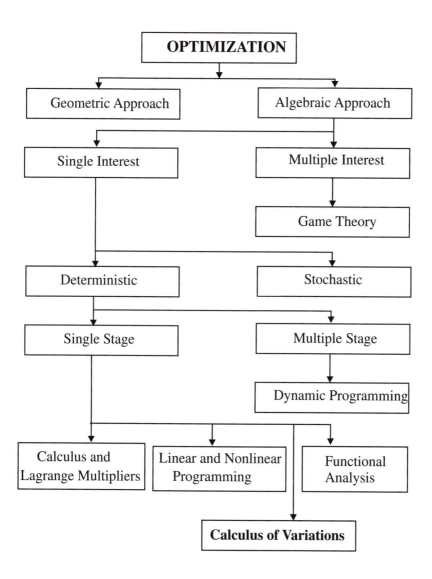

Figure 1.4 Overview of Optimization

(or difference) equations. Techniques used are search techniques, dynamic programming, variational calculus (or calculus of variations) and Pontryagin principle.

1.3   Optimal Control

The main objective of optimal control is to determine control signals that will cause a process (plant) to satisfy some physical constraints and at the same time extremize (maximize or minimize) a chosen performance criterion (performance index or cost function). Referring to Figure 1.2, we are interested in finding the optimal control $\mathbf{u}^*(t)$ (* indicates optimal condition) that will drive the plant P from initial state to final state with some constraints on controls and states and at the same time extremizing the given performance index J.

The formulation of optimal control problem requires

1. a mathematical description (or model) of the process to be controlled (generally in state variable form),

2. a specification of the performance index, and

3. a statement of boundary conditions and the physical constraints on the states and/or controls.

1.3.1   Plant

For the purpose of optimization, we describe a physical plant by a set of linear or nonlinear differential or difference equations. For example, a linear time-invariant system is described by the state and output relations (1.1.3) and (1.1.4) and a nonlinear system by (1.1.5) and (1.1.6).

1.3.2   Performance Index

Classical control design techniques have been successfully applied to linear, time-invariant, single-input, single output (SISO) systems. Typical performance criteria are system time response to step or ramp input characterized by rise time, settling time, peak overshoot, and steady state accuracy; and the frequency response of the system characterized by gain and phase margins, and bandwidth.

In modern control theory, the optimal control problem is to find a control which causes the dynamical system to reach a target or follow a state variable (or trajectory) and at the same time extremize a performance index which may take several forms as described below.

1. **Performance Index for Time-Optimal Control System:** We try to transfer a system from an arbitrary initial state $\mathbf{x}(t_0)$ to a specified final state $\mathbf{x}(t_f)$ in *minimum* time. The corresponding performance index (PI) is

$$J = \int_{t_0}^{t_f} dt = t_f - t_0 = t^*. \qquad (1.3.1)$$

2. **Performance Index for Fuel-Optimal Control System:** Consider a spacecraft problem. Let $u(t)$ be the thrust of a rocket engine and assume that the magnitude $|u(t)|$ of the thrust is proportional to the *rate* of fuel consumption. In order to *minimize* the total expenditure of fuel, we may formulate the performance index as

$$J = \int_{t_0}^{t_f} |u(t)|dt. \qquad (1.3.2)$$

For several controls, we may write it as

$$J = \int_{t_0}^{t_f} \sum_{i=1}^{m} R_i |u_i(t)|dt \qquad (1.3.3)$$

where R is a weighting factor.

3. **Performance Index for Minimum-Energy Control System:** Consider $u_i(t)$ as the current in the ith loop of an electric network. Then $\sum_{i=1}^{m} u_i^2(t)r_i$ (where, r_i is the resistance of the ith loop) is the total power or the total *rate* of energy expenditure of the network. Then, for minimization of the total expended energy, we have a performance criterion as

$$J = \int_{t_0}^{t_f} \sum_{i=1}^{m} u_i^2(t)r_i dt \qquad (1.3.4)$$

or in general,

$$J = \int_{t_0}^{t_f} \mathbf{u}'(t)\mathbf{R}\mathbf{u}(t)dt \qquad (1.3.5)$$

where, $\mathbf{R}$ is a *positive definite* matrix and prime ($'$) denotes transpose here and throughout this book (see Appendix A for more details on *definite* matrices).

Similarly, we can think of minimization of the integral of the squared error of a tracking system. We then have,

$$J = \int_{t_0}^{t_f} \mathbf{x}'(t)\mathbf{Q}\mathbf{x}(t)dt \qquad (1.3.6)$$

where, $\mathbf{x}_d(t)$ is the desired value, $\mathbf{x}_a(t)$ is the actual value, and $\mathbf{x}(t) = \mathbf{x}_a(t) - \mathbf{x}_d(t)$, is the error. Here, $\mathbf{Q}$ is a weighting matrix, which can be *positive semi-definite*.

4. **Performance Index for Terminal Control System:** In a terminal target problem, we are interested in minimizing the error between the desired target position $\mathbf{x}_d(t_f)$ and the actual target position $\mathbf{x}_a(t_f)$ at the end of the maneuver or at the final time t_f. The terminal (final) error is $\mathbf{x}(t_f) = \mathbf{x}_a(t_f) - \mathbf{x}_d(t_f)$. Taking care of positive and negative values of error and weighting factors, we structure the cost function as

$$J = \mathbf{x}'(t_f)\mathbf{F}\mathbf{x}(t_f) \qquad (1.3.7)$$

which is also called the *terminal cost function*. Here, $\mathbf{F}$ is a *positive semi-definite matrix*.

5. **Performance Index for General Optimal Control System:** Combining the above formulations, we have a performance index in general form as

$$J = \mathbf{x}'(t_f)\mathbf{F}\mathbf{x}(t_f) + \int_{t_0}^{t_f} [\mathbf{x}'(t)\mathbf{Q}\mathbf{x}(t) + \mathbf{u}'(t)\mathbf{R}\mathbf{u}(t)]dt \qquad (1.3.8)$$

or,

$$J = S(\mathbf{x}(t_f), t_f) + \int_{t_0}^{t_f} V(\mathbf{x}(t), \mathbf{u}(t), t)dt \qquad (1.3.9)$$

where, $\mathbf{R}$ is a positive definite matrix, and $\mathbf{Q}$ and $\mathbf{F}$ are positive semidefinite matrices, respectively. Note that the matrices $\mathbf{Q}$ and $\mathbf{R}$ may be time varying. The particular form of performance index (1.3.8) is called *quadratic* (in terms of the states and controls) form.

The problems arising in optimal control are classified based on the structure of the performance index J [67]. If the PI (1.3.9) contains the *terminal* cost function $S(\mathbf{x}(t), \mathbf{u}(t), t)$ *only*, it is called the *Mayer* problem, if the PI (1.3.9) has *only* the *integral* cost term, it is called the *Lagrange* problem, and the problem is of the *Bolza* type if the PI contains both the *terminal* cost term and the *integral* cost term as in (1.3.9). There are many other forms of cost functions depending on our performance specifications. However, the above mentioned performance indices (with quadratic forms) lead to some very elegant results in optimal control systems.

1.3.3   Constraints

The control $\mathbf{u}(t)$ and state $\mathbf{x}(t)$ vectors are either *unconstrained* or *constrained* depending upon the physical situation. The unconstrained problem is less involved and gives rise to some elegant results. From the physical considerations, often we have the controls and states, such as currents and voltages in an electrical circuit, speed of a motor, thrust of a rocket, constrained as

$$\mathbf{U}_+ \le \mathbf{u}(t) \le \mathbf{U}_-, \quad \text{and} \quad \mathbf{X}_- \le \mathbf{x}(t) \le \mathbf{X}_+ \qquad (1.3.10)$$

where, $+$, and $-$ indicate the maximum and minimum values the variables can attain.

1.3.4   *Formal Statement of Optimal Control System*

Let us now state formally the optimal control problem even risking repetition of some of the previous equations. The optimal control problem is to find the optimal control $\mathbf{u}^*(t)$ ($*$ indicates extremal or optimal value) which causes the *linear* time-invariant plant (system)

$$\dot{\mathbf{x}}(t) = \mathbf{A}\mathbf{x}(t) + \mathbf{B}\mathbf{u}(t) \qquad (1.3.11)$$

to give the trajectory $\mathbf{x}^*(t)$ that optimizes or extremizes (minimizes or maximizes) a performance index

$$J = \mathbf{x}'(t_f)\mathbf{F}\mathbf{x}(t_f) + \int_{t_0}^{t_f} [\mathbf{x}'(t)\mathbf{Q}\mathbf{x}(t) + \mathbf{u}'(t)\mathbf{R}\mathbf{u}(t)]dt \quad (1.3.12)$$

or which causes the *nonlinear* system

$$\dot{\mathbf{x}}(t) = \mathbf{f}(\mathbf{x}(t), \mathbf{u}(t), t) \qquad (1.3.13)$$

to give the state $\mathbf{x}^*(t)$ that optimizes the general performance index

$$J = S(\mathbf{x}(t_f), t_f) + \int_{t_0}^{t_f} V(\mathbf{x}(t), \mathbf{u}(t), t) dt \qquad (1.3.14)$$

with some constraints on the control variables $\mathbf{u}(t)$ and/or the state variables $\mathbf{x}(t)$ given by (1.3.10). The final time t_f may be *fixed*, or *free*, and the final (target) state may be *fully* or *partially fixed* or *free*. The entire problem statement is also shown pictorially in Figure 1.5. Thus,

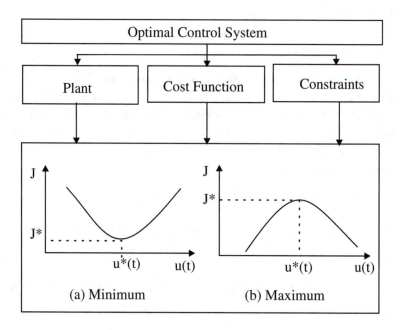

Figure 1.5　Optimal Control Problem

we are basically interested in finding the control $\mathbf{u}^*(t)$ which when applied to the plant described by (1.3.11) or (1.3.13), gives an optimal performance index J^* described by (1.3.12) or (1.3.14).

The optimal control systems are studied in three stages.

1. In the first stage, we just consider the performance index of the form (1.3.14) and use the well-known theory of calculus of variations to obtain optimal functions.

2. In the second stage, we bring in the plant (1.3.11) and try to address the problem of finding optimal control $\mathbf{u}^*(t)$ which will

drive the plant and at the same time optimize the performance index (1.3.12). Next, the above topics are presented in discrete-time domain.

3. Finally, the topic of constraints on the controls and states (1.3.10) is considered along with the plant and performance index to obtain optimal control.

1.4   Historical Tour

We basically consider two stages of the tour: first the development of calculus of variations, and secondly, optimal control theory [134, 58, 99, 28][1].

1.4.1   Calculus of Variations

According to a legend [88], Tyrian princess Dido used a rope made of cowhide in the form of a circular arc to *maximize* the area to be occupied to found Carthage. Although the story of the founding of Carthage is fictitious, it probably inspired a new mathematical discipline, the *calculus of variations* and its extensions such as optimal control theory.

The calculus of variations is that branch of mathematics that deals with finding a function which is an extremum (maximum or minimum) of a functional. A functional is loosely defined as a function of a function. The theory of finding maxima and minima of functions is quite old and can be traced back to the isoperimetric problems considered by Greek mathematicians such as Zenodorus (495-435 B.C.) and by Poppus (c. 300 A.D.). But we will start with the works of Bernoulli. In 1699 Johannes Bernoulli (1667-1748) posed the brachistochrone problem: *the problem of finding the path of quickest descent between two points not in the same horizontal or vertical line.* This problem which was first posed by Galileo (1564-1642) in 1638, was solved by John, his brother Jacob (1654- 1705), by Gottfried Leibniz (1646-1716), and anonymously by Isaac Newton (1642-1727). Leonard Euler (1707-1783) joined John Bernoulli and made some remarkable contributions, which influenced Joseph-Louis Lagrange (1736-1813), who finally gave an el-

---

[1]The permission given by Springer-Verlag for H. H. Goldstine, *A History of the Calculus of Variations*, Springer-Verlag, New York, NY, 1980, is hereby acknowledged.

egant way of solving these types of problems by using the method of (*first*) variations. This led Euler to coin the phrase *calculus of variations*. Later this *necessary* condition for extrema of a functional was called the Euler - the Lagrange equation. Lagrange went on to treat variable end - point problems introducing the multiplier method, which later became one of the most powerful tools-Lagrange (or Euler-Lagrange) multiplier method-in optimization.

The *sufficient* conditions for finding the extrema of functionals in calculus of variations was given by Andrien Marie Legendre (1752-1833) in 1786 by considering additionally the *second* variation. Carl Gustav Jacob Jacobi (1804-1851) in 1836 came up with a more rigorous analysis of the sufficient conditions. This sufficient condition was later on termed as the Legendre-Jacobi condition. At about the same time Sir William Rowan Hamilton (1788-1856) did some remarkable work on mechanics, by showing that the motion of a particle in space, acted upon by various external forces, could be represented by a single function which satisfies *two* first-order partial differential equations. In 1838 Jacobi had some objections to this work and showed the need for only *one* partial differential equation. This equation, called Hamilton-Jacobi equation, later had profound influence on the calculus of variations and dynamic programming, optimal control, and as well as on mechanics.

The distinction between *strong* and *weak* extrema was addressed by Karl Weierstrass (1815-1897) who came up with the idea of the field of extremals and gave the Weierstrass condition, and sufficient conditions for weak and strong extrema. Rudolph Clebsch (1833-1872) and Adolph Mayer proceeded with establishing conditions for the more general class of problems. Clebsch formulated a problem in the calculus of variations by adjoining the constraint conditions in the form of differential equations and provided a condition based on second variation. In 1868 Mayer reconsidered Clebsch's work and gave some elegant results for the general problem in the calculus of variations. Later Mayer described in detail the problems: the problem of Lagrange in 1878, and the problem of Mayer in 1895.

In 1898, Adolf Kneser gave a new approach to the calculus of variations by using the result of Karl Gauss (1777-1855) on geodesics. For variable end-point problems, he established the transversality condition which includes orthogonality as a special case. He along with Oskar Bolza (1857-1942) gave sufficiency proofs for these problems. In 1900, David Hilbert (1862-1943) showed the second variation as a quadratic functional with eigenvalues and eigenfunctions. Between 1908 and 1910, Gilbert Bliss (1876-1951) [23] and Max Mason looked in depth at the results of Kneser. In 1913, Bolza formulated the problem of Bolza as a generalization of the problems of Lagrange and Mayer. Bliss showed that these three problems are equivalent. Other notable contributions to calculus of variations were made by E. J. McShane (1904-1989) [98], M. R. Hestenes [65], H. H. Goldstine and others. There have been a large number of books on the subject of calculus of variations: Bliss (1946) [23], Cicala (1957) [37], Akhiezer (1962) [1], Elsgolts (1962) [47], Gelfand and Fomin (1963) [55], Dreyfus (1966) [45], Forray (1968) [50], Balakrishnan (1969) [8], Young (1969) [146], Elsgolts (1970) [46], Bolza (1973) [26], Smith (1974) [126], Weinstock (1974) [143], Krasnov *et al.* (1975) [81], Leitmann (1981) [88], Ewing (1985) [48], Kamien and Schwartz (1991) [78], Gregory and Lin (1992) [61], Sagan (1992) [118], Pinch (1993) [108], Wan (1994) [141], Giaquinta and Hildebrandt (1995) [56, 57], Troutman (1996) [136], and Milyutin and Osmolovskii (1998) [103].

1.4.2 Optimal Control Theory

The linear quadratic control problem has its origins in the celebrated work of N. Wiener on mean-square filtering for weapon fire control during World War II (1940-45) [144, 145]. Wiener solved the problem of designing filters that minimize a mean-square-error criterion (performance measure) of the form

$$J = E\{e^2(t)\} \tag{1.4.1}$$

where, $e(t)$ is the error, and $E\{x\}$ represents the expected value of the random variable x. For a deterministic case, the above error criterion is generalized as an integral quadratic term as

$$J = \int_0^\infty \mathbf{e}'(t)\mathbf{Q}\mathbf{e}(t)dt \tag{1.4.2}$$

where, $\mathbf{Q}$ is some positive definite matrix. R. Bellman in 1957 [12] introduced the technique of *dynamic programming* to solve discrete-time optimal control problems. But, the most important contribution to optimal control systems was made in 1956 [25] by L. S. Pontryagin (formerly of the United Soviet Socialistic Republic (USSR)) and his associates, in development of his celebrated *maximum principle* described in detail in their book [109]. Also, see a very interesting article on the "discovery of the Maximum Principle" by R. V. Gamkrelidze [52], one of the authors of the original book [109]. At this time in the United States, R. E. Kalman in 1960 [70] provided *linear quadratic regulator (LQR)* and *linear quadratic Gaussian (LQG)* theory to design optimal feedback controls. He went on to present optimal filtering and estimation theory leading to his famous *discrete Kalman filter* [71] and the *continuous Kalman filter* with Bucy [76]. Kalman had a profound effect on optimal control theory and the Kalman filter is one of the most widely used technique in applications of control theory to real world problems in a variety of fields.

At this point we have to mention the *matrix Riccati equation* that appears in all the Kalman filtering techniques and many other fields. C. J. Riccati [114, 22] published his result in 1724 on the solution for some types of nonlinear differential equations, without ever knowing that the Riccati equation would become so famous after more than two centuries!

Thus, optimal control, having its roots in calculus of variations developed during 16th and 17th centuries was really born over 300 years ago [132]. For additional details about the historical perspectives on calculus of variations and optimal control, the reader is referred to some excellent publications [58, 99, 28, 21, 132].

In the so-called *linear quadratic control*, the term "linear" refers to the plant being *linear* and the term "quadratic" refers to the performance index that involves the *square* or *quadratic* of an error, and/or control. Originally, this problem was called the *mean-square* control problem and the term "linear quadratic" did not appear in the literature until the late 1950s.

Basically the *classical* control theory using *frequency* domain deals with single input and single output (SISO) systems, whereas *modern* control theory works with *time* domain for SISO and multi-input and multi-output (MIMO) systems. Although modern control and hence optimal control appeared to be very attractive, it lacked a very important feature of *robustness*. That is, controllers designed based on LQR theory failed to be robust to measurement noise, external disturbances and unmodeled dynamics. On the other hand, frequency domain techniques using the ideas of gain margin and phase margin offer robustness in a natural way. Thus, some researchers [115, 95], especially in the United Kingdom, continued to work on developing frequency domain approaches to MIMO systems.

One important and relevant field that has been developed around the 1980s is the $\mathcal{H}_\infty$-optimal control theory. In this framework, the work developed in the 1960s and 1970s is labeled as $\mathcal{H}_2$-optimal control theory. The seeds for $\mathcal{H}_\infty$-optimal control theory were laid by G. Zames [148], who formulated the optimal $\mathcal{H}_\infty$-sensitivity design problem for SISO systems and solved using optimal Nevanilina-Pick interpolation theory. An important publication in this field came from a group of four active researchers, Doyle, Glover, Khargonekar, and Francis[44], who won the 1991 W. R. G. Baker Award as the best IEEE Transactions paper. There are many other works in the field of $\mathcal{H}_\infty$ control ([51, 96, 43, 128, 7, 60, 131, 150, 39, 34]).

1.5 About This Book

This book, on the subject of optimal control systems, is based on the author's lecture notes used for teaching a graduate level course on this subject. In particular, this author was most influenced by Athans and Falb [6], Schultz and Melsa [121], Sage [119], Kirk [79], Sage and White [120], Anderson and Moore [3] and Lewis and Syrmos [91], and one finds the footprints of these works in the present book.

There were a good number of books on optimal control published during the era of the "glory of modern control," (Leitmann (1964) [87], Tou (1964) [135], Athans and Falb (1966) [6], Dreyfus (1966) [45], Lee and Markus (1967) [86], Petrov (1968) [106], Sage (1968) [119], Citron (1969) [38], Luenberger (1969) [93], Pierre (1969) [107], Pun (1969) [110], Young (1969) [146], Kirk (1970) [79], Boltyanskii [24], Kwakernaak and Sivan (1972) [84], Warga (1972) [142], Berkovitz (1974) [17], Bryson and Ho (1975) [30]), Sage and White (1977) [120], Leitmann (1981) [88]), Ryan (1982) [116]). There has been renewed interest with the second wave of books published during the last few years (Lewis (1986) [89], Stengal (1986) [127], Christensen *et al.* (1987) [36] Anderson and Moore (1990) [3], Hocking (1991) [66], Teo *et al.* (1991) [133], Gregory and Lin (1992) [61], Lewis (1992) [90], Pinch (1993) [108], Dorato *et al.* (1995) [42], Lewis and Syrmos (1995) [91]), Saberi *et al.* (1995) [117], Sima (1996) [124], Siouris [125], Troutman (1996) [136] Bardi and Dolcetta (1997) [9], Vincent and Grantham (1997) [139], Milyutin and Osmolovskii (1998) [103], Bryson (1999) [29], Burl [32], Kolosov (1999) [80], Pytlak (1999) [111], Vinter (2000) [140], Zelikin

(2000) [149], Betts (2001) [20], and Locatelli (2001) [92].

The optimal control theory continues to have a wide variety of applications starting from the traditional electrical power [36] to economics and management [16, 122, 78, 123].

1.6   Chapter Overview

This book is composed of seven chapters. Chapter 2 presents optimal control via calculus of variations. In this chapter, we start with some basic definitions and a simple variational problem of extremizing a functional. We then bring in the plant as a conditional optimization problem and discuss various types of problems based on the boundary conditions. We briefly mention both Lagrangian and Hamiltonian formalisms for optimization. Next, Chapter 3 addresses basically the linear quadratic regulator (LQR) system. Here we discuss the closed-loop optimal control system introducing matrix Riccati differential and algebraic equations. We look at the analytical solution to the Riccati equations and development of MATLAB$^{©}$ routine for the analytical solution. Tracking and other problems of linear quadratic optimal control are discussed in Chapter 4. We also discuss the gain and phase margins of the LQR system.

So far the optimal control of continuous-time systems is described. Next, the optimal control of discrete-time systems is presented in Chapter 5. Here, we start with the basic calculus of variations and then touch upon all the topics discussed above with respect to the continuous-time systems. The Pontryagin Principle and associated topics of dynamic programming and Hamilton-Jacobi-Bellman results are briefly covered in Chapter 6. The optimal control of systems with control and state constraints is described in Chapter 7. Here, we cover topics of control constraints leading to time-optimal, fuel-optimal and energy-optimal control systems and briefly discuss the state constraints problem.

Finally, the Appendices A and B provide summary of results on matrices, vectors, matrix algebra and state space, and Appendix C lists some of the MATLAB$^{©}$ files used in the book.

1.7 Problems

Problem 1.1 A D.C. motor speed control system is described by a second order state equation,

$$\dot{x}_1(t) = 25x_2(t)$$
$$\dot{x}_2(t) = -400x_1(t) - 200x_2(t) + 400u(t),$$

where, $x_1(t)$ = the speed of the motor, and $x_2(t)$ = the current in the armature circuit and the control input $u(t)$ = the voltage input to an amplifier supplying the motor. Formulate a performance index and optimal control problem to keep the speed constant at a particular value.

Problem 1.2 [83] In a liquid-level control system for a storage tank, the valves connecting a reservoir and the tank are controlled by gear train driven by a D. C. motor and an electronic amplifier. The dynamics is described by a third order system

$$\dot{x}_1(t) = -2x_1(t)$$
$$\dot{x}_2(t) = x_3(t)$$
$$\dot{x}_3(t) = -10x_3(t) + 9000u(t)$$

where, $x_1(t)$ = is the height in the tank, $x_2(t)$ = is the angular position of the electric motor driving the valves controlling the liquid from reservoir to tank, $x_3(t)$ = the angular velocity of the motor, and $u(t)$ = is the input to electronic amplifier connected to the input of the motor. Formulate optimal control problem to keep the liquid level constant at a reference value and the system to act only if there is a change in the liquid level.

Problem 1.3 [35] In an inverted pendulum system, it is required to maintain the upright position of the pendulum on a cart. The linearized state equations are

$$\dot{x}_1(t) = x_2(t)$$
$$\dot{x}_2(t) = -x_3(t) + 0.2u(t)$$
$$\dot{x}_3(t) = x_4(t)$$
$$\dot{x}_4(t) = 10x_3(t) - 0.2u(t)$$

where, $x_1(t) =$ is horizontal linear displacement of the cart, $x_2(t) =$ is linear velocity of the cart, $x_3(t) =$ is angular position of the pendulum from vertical line, $x_4(t) =$ is angular velocity, and $u(t) =$ is the horizontal force applied to the cart. Formulate a performance index to keep the pendulum in the vertical position with as little energy as possible.

Problem 1.4 [101] A mechanical system consisting of two masses and two springs, one spring connecting the two masses and the other spring connecting one of the masses to a fixed point. An input is applied to the mass not connected to the fixed point. The displacements $(x_1(t)$ and $x_2(t))$ and the corresponding velocities $(x_3(t)$ and $x_4(t))$ of the two masses provide a fourth-order system described by

$$\dot{x}_1(t) = x_3(t)$$
$$\dot{x}_2(t) = x_4(t)$$
$$\dot{x}_3(t) = -4x_1(t) + 2x_2(t)$$
$$\dot{x}_4(t) = x_1(t) - x_2(t) + u(t)$$

Formulate a performance index to minimize the errors in displacements and velocities and to minimize the control effort.

Problem 1.5 A simplified model of an automobile suspension system is described by

$$m\ddot{x}(t) + kx(t) = bu(t)$$

where, $x(t)$ is the position, $u(t)$ is the input to the suspension system (in the form of an upward force), m is the mass of the suspension system, and k is the spring constant. Formulate the optimal control problem for minimum control energy and passenger comfort. Assume suitable values for all the constants.

Problem 1.6 [112] Consider a continuous stirred tank chemical reactor described by

$$\dot{x}_1(t) = -0.1x_1(t) - 0.12x_2(t)$$
$$\dot{x}_2(t) = -0.3x_1(t) - 0.012x_2(t) - 0.07u(t)$$

where, the normalized deviation state variables of the linearized model are $x_1(t) =$ reaction variable, $x_2(t) =$ temperature and the control variable $u(t) =$ effective cooling rate coefficient. Formulate a suitable performance measure to minimize the deviation errors and to minimize the control effort.

Chapter 2

Calculus of Variations and Optimal Control

Calculus of variations (CoV) or variational calculus deals with finding the *optimum* (maximum or minimum) value of a functional. Variational calculus that originated around 1696 became an independent mathematical discipline after the fundamental discoveries of L. Euler (1709-1783), whom we can claim with good reason as the founder of calculus of variations.

In this chapter, we start with some basic definitions and a simple variational problem of extremizing a functional. We then incorporate the plant as a conditional optimization problem and discuss various types of problems based on the boundary conditions. We briefly mention both the Lagrangian and Hamiltonian formalisms for optimization. It is suggested that the student reviews the material in Appendices A and B given at the end of the book. This chapter is motivated by [47, 79, 46, 143, 81, 48][1].

2.1 Basic Concepts

2.1.1 Function and Functional

We discuss some fundamental concepts associated with *functionals* along side with those of *functions*.

(a) **Function:** A variable x is a *function* of a variable quantity t, (writ-

---

[1]The permission given by Prentice Hall for D. E. Kirk, *Optimal Control Theory: An Introduction*, Prentice Hall, Englewood Cliffs, NJ, 1970, is hereby acknowledged.

ten as $x(t) = f(t)$), if to every value of t over a certain range of t there corresponds a value x; i.e., we have a correspondence: to a number t there corresponds a number x. Note that here t need not be always time but any independent variable.

Example 2.1

Consider

$$x(t) = 2t^2 + 1. \tag{2.1.1}$$

For $t = 1$, $x = 3$, $t = 2$, $x = 9$ and so on. Other functions are $x(t) = 2t$; $x(t_1, t_2) = t_1^2 + t_2^2$.

Next we consider the definition of a *functional* based on that of a function.

(b) Functional: A variable quantity J is a *functional* dependent on a function $f(x)$, written as $J = J(f(x))$, if to each function $f(x)$, there corresponds a value J, i.e., we have a correspondence: to the function $f(x)$ there corresponds a number J. Functional depends on several functions.

Example 2.2

Let $x(t) = 2t^2 + 1$. Then

$$J(x(t)) = \int_0^1 x(t)dt = \int_0^1 (2t^2 + 1)dt = \frac{2}{3} + 1 = \frac{5}{3} \tag{2.1.2}$$

is the area under the curve $x(t)$. If $v(t)$ is the velocity of a vehicle, then

$$J(v(t)) = \int_{t_0}^{t_f} v(t)dt \tag{2.1.3}$$

is the path traversed by the vehicle. Thus, here $x(t)$ and $v(t)$ are functions of t, and J is a functional of $x(t)$ or $v(t)$.

Loosely speaking, a functional can be thought of as a "function of a function."

2.1.2   *Increment*

We consider here *increment* of a function and a functional.

(a) Increment of a Function: In order to consider optimal values of a function, we need the definition of an increment [47, 46, 79].

DEFINITION 2.1   *The increment of the function f, denoted by Δf, is defined as*

$$\Delta f \triangleq f(t + \Delta t) - f(t). \tag{2.1.4}$$

It is easy to see from the definition that Δf depends on both the independent variable t and the increment of the independent variable Δt, and hence strictly speaking, we need to write the increment of a function as $\Delta f(t, \Delta t)$.

Example 2.3

If

$$f(t) = (t_1 + t_2)^2 \tag{2.1.5}$$

find the increment of the function $f(t)$.

Solution: The increment Δf becomes

$$
\begin{aligned}
\Delta f &\triangleq f(t + \Delta t) - f(t) \\
&= (t_1 + \Delta t_1 + t_2 + \Delta t_2)^2 - (t_1 + t_2)^2 \\
&= (t_1 + \Delta t_1)^2 + (t_2 + \Delta t_2)^2 + 2(t_1 + \Delta t_1)(t_2 + \Delta t_2) - \\
&\quad (t_1^2 + t_2^2 + 2t_1 t_2) \\
&= 2(t_1 + t_2)\Delta t_1 + 2(t_1 + t_2)\Delta t_2 + (\Delta t_1)^2 + (\Delta t_2)^2 \\
&\quad + 2\Delta t_1 \Delta t_2. \tag{2.1.6}
\end{aligned}
$$

(b) Increment of a Functional: Now we are ready to define the increment of a functional.

DEFINITION 2.2   *The increment of the functional J, denoted by ΔJ, is defined as*

$$\boxed{\Delta J \triangleq J(x(t) + \delta x(t)) - J(x(t)).} \tag{2.1.7}$$

Here $\delta x(t)$ is called the *variation* of the function $x(t)$. Since the increment of a functional is dependent upon the function $x(t)$ and its variation $\delta x(t)$, strictly speaking, we need to write the increment as $\Delta J(x(t), \delta x(t))$.

Example 2.4

Find the increment of the functional

$$J = \int_{t_0}^{t_f} \left[2x^2(t) + 1 \right] dt. \qquad (2.1.8)$$

Solution: The increment of J is given by

$$\Delta J \triangleq J(x(t) + \delta x(t)) - J(x(t)),$$

$$= \int_{t_0}^{t_f} \left[2(x(t) + \delta x(t))^2 + 1 \right] dt - \int_{t_0}^{t_f} \left[2x^2(t) + 1 \right] dt,$$

$$= \int_{t_0}^{t_f} \left[4x(t)\delta x(t) + 2(\delta x(t))^2 \right] dt. \qquad (2.1.9)$$

2.1.3   *Differential and Variation*

Here, we consider the *differential* of a function and the *variation* of a functional.

(a) Differential of a Function: Let us define at a point t^* the increment of the function f as

$$\Delta f \triangleq f(t^* + \Delta t) - f(t^*). \qquad (2.1.10)$$

By expanding $f(t^* + \Delta t)$ in a Taylor series about t^*, we get

$$\Delta f = f(t^*) + \left(\frac{df}{dt}\right)_* \Delta t + \frac{1}{2!}\left(\frac{d^2 f}{dt^2}\right)_* (\Delta t)^2 + \cdots - f(t^*). \qquad (2.1.11)$$

Neglecting the higher order terms in Δt,

$$\Delta f = \left(\frac{df}{dt}\right)_* \Delta t = \dot{f}(t^*)\Delta t = df. \qquad (2.1.12)$$

Here, df is called the *differential* of f at the point t^*. $\dot{f}(t^*)$ is the *derivative* or slope of f at t^*. In other words, the differential df is the first order approximation to increment Δt. Figure 2.1 shows the relation between increment, differential and derivative.

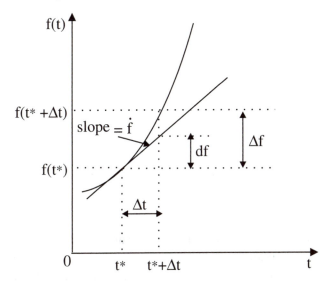

Figure 2.1   Increment Δf, Differential df, and Derivative $\dot{f}$ of a Function $f(t)$

Example 2.5

Let $f(t) = t^2 + 2t$. Find the increment and the derivative of the function $f(t)$.

Solution: By definition, the increment Δf is

$$
\begin{aligned}
\Delta f &\triangleq f(t + \Delta t) - f(t), \\
&= (t + \Delta t)^2 + 2(t + \Delta t) - (t^2 + 2t), \\
&= 2t\Delta t + 2\Delta t + \cdots + \text{higher order terms}, \\
&= 2(t + 1)\Delta t, \\
&= \dot{f}(t)\Delta t. \qquad\qquad\qquad\qquad\qquad\qquad (2.1.13)
\end{aligned}
$$

Here, $\dot{f}(t) = 2(t + 1)$.

(b) Variation of a Functional: Consider the increment of a functional

$$
\Delta J \triangleq J(x(t) + \delta x(t)) - J(x(t)). \qquad\qquad (2.1.14)
$$

Expanding $J(x(t) + \delta x(t))$ in a Taylor series, we get

$$\Delta J = J(x(t)) + \frac{\partial J}{\partial x}\delta x(t) + \frac{1}{2!}\frac{\partial^2 J}{\partial x^2}(\delta x(t))^2 + \cdots - J(x(t))$$

$$= \frac{\partial J}{\partial x}\delta x(t) + \frac{1}{2!}\frac{\partial^2 J}{\partial x^2}(\delta x(t))^2 + \cdots$$

$$= \delta J + \delta^2 J + \cdots, \qquad (2.1.15)$$

where,

$$\delta J = \frac{\partial J}{\partial x}\delta x(t) \quad \text{and} \quad \delta^2 J = \frac{1}{2!}\frac{\partial^2 J}{\partial x^2}(\delta x(t))^2 \qquad (2.1.16)$$

are called the *first variation* (or simply the *variation*) and the *second variation* of the functional J, respectively. The variation δJ of a functional J is the *linear* (or first order approximate) part (in $\delta x(t)$) of the increment ΔJ. Figure 2.2 shows the relation between increment and the first variation of a functional.

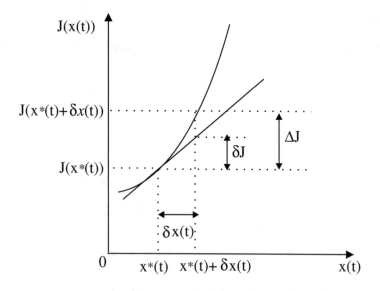

Figure 2.2   Increment ΔJ and the First Variation δJ of the Functional J

Example 2.6

Given the functional

$$J(x(t)) = \int_{t_0}^{t_f} [2x^2(t) + 3x(t) + 4]dt, \qquad (2.1.17)$$

evaluate the variation of the functional.

Solution: First, we form the increment and then extract the variation as the first order approximation. Thus

$$\Delta J \triangleq J(x(t) + \delta x(t)) - J(x(t)),$$

$$= \int_{t_0}^{t_f} \Big[2(x(t) + \delta x(t))^2 + 3(x(t) + \delta x(t)) + 4$$

$$- (2x^2(t) + 3x(t) + 4) \Big] dt,$$

$$= \int_{t_0}^{t_f} \Big[4x(t)\delta x(t) + 2(\delta x(t))^2 + 3\delta x(t) \Big] dt. \qquad (2.1.18)$$

Considering only the first order terms, we get the (first) variation as

$$\delta J(x(t), \delta x(t)) = \int_{t_0}^{t_f} (4x(t) + 3)\delta x(t)dt. \qquad (2.1.19)$$

2.2 *Optimum of a Function and a Functional*

We give some definitions for optimum or extremum (maximum or minimum) of a function and a functional [47, 46, 79]. The *variation* plays the same role in determining optimal value of a functional as the *differential* does in finding extremal or optimal value of a function.

DEFINITION 2.3 **Optimum of a Function:** *A function $f(t)$ is said to have a relative optimum at the point t^* if there is a positive parameter ϵ such that for all points t in a domain $\mathcal{D}$ that satisfy $|t - t^*| < \epsilon$, the increment of $f(t)$ has the same sign (positive or negative).*

In other words, if

$$\Delta f = f(t) - f(t^*) \geq 0, \qquad (2.2.1)$$

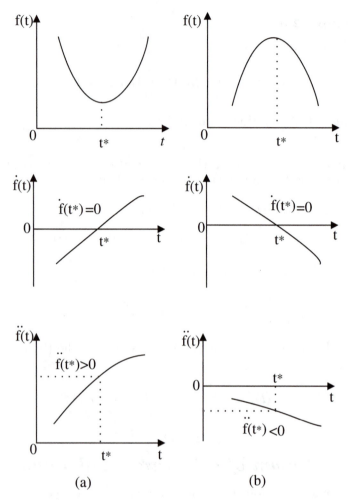

Figure 2.3 (a) Minimum and (b) Maximum of a Function $f(t)$

then, $f(t^*)$ is a relative local *minimum*. On the other hand, if

$$\Delta f = f(t) - f(t^*) \leq 0, \tag{2.2.2}$$

then, $f(t^*)$ is a relative local *maximum*. If the previous relations are valid for arbitrarily large ϵ, then, $f(t^*)$ is said to have a *global* absolute optimum. Figure 2.3 illustrates the (a) minimum and (b) maximum of a function.

It is well known that the *necessary* condition for optimum of a function is that the (first) differential vanishes, i.e., $df = 0$. The *sufficient* condition

1. for *minimum* is that the second differential is positive,
   i.e., $d^2 f > 0$, and

2. for *maximum* is that the second differential is negative,
   i.e., $d^2 f < 0$.

If $d^2 f = 0$, it corresponds to a *stationary* (or inflection) point.

DEFINITION 2.4   Optimum of a Functional: *A functional J is said to have a relative optimum at* x^* *if there is a positive* ϵ *such that for all functions x in a domain* Ω *which satisfy* $|x - x^*| < \epsilon$, *the increment of J has the same sign.*

In other words, if

$$\Delta J = J(x) - J(x^*) \geq 0, \tag{2.2.3}$$

then $J(x^*)$ is a relative *minimum*. On the other hand, if

$$\Delta J = J(x) - J(x^*) \leq 0, \tag{2.2.4}$$

then, $J(x^*)$ is a relative *maximum*. If the above relations are satisfied for arbitrarily large ϵ, then, $J(x^*)$ is a *global* absolute optimum.

Analogous to finding extremum or optimal values for *functions*, in variational problems concerning *functionals*, the result is that the variation must be zero on an optimal curve. Let us now state the result in the form of a theorem, known as *fundamental theorem of the calculus of variations*, the proof of which can be found in any book on calculus of variations [47, 46, 79].

THEOREM 2.1
For $x^*(t)$ to be a candidate for an optimum, the (first) variation of J must be zero on $x^*(t)$, i.e., $\delta J(x^*(t), \delta x(t)) = 0$ for all admissible values of $\delta x(t)$. This is a necessary condition. As a sufficient condition for minimum, the second variation $\delta^2 J > 0$, and for maximum $\delta^2 J < 0$.

2.3   The Basic Variational Problem

2.3.1   Fixed-End Time and Fixed-End State System

We address a fixed-end time and fixed-end state problem, where both the *initial* time and state and the *final* time and state are fixed or given *a priori.* Let $x(t)$ be a scalar function with continuous first derivatives and the vector case can be similarly dealt with. The problem is to find the *optimal* function $x^*(t)$ for which the functional

$$J(x(t)) = \int_{t_0}^{t_f} V(x(t), \dot{x}(t), t)dt \qquad (2.3.1)$$

has a relative *optimum*. It is assumed that the integrand V has continuous first and second partial derivatives w.r.t. all its arguments; t_0 and t_f are fixed (or given a priori) and the end points are fixed, i.e.,

$$x(t = t_0) = x_0; \quad x(t = t_f) = x_f. \qquad (2.3.2)$$

We already know from Theorem 2.1 that the necessary condition for an optimum is that the *variation of a functional vanishes*. Hence, in our attempt to find the optimum of $x(t)$, we first define the increment for J, obtain its variation and finally apply the fundamental theorem of the calculus of variations (Theorem 2.1).

Thus, the various steps involved in finding the optimal solution to the fixed-end time and fixed-end state system are first listed and then discussed in detail.

- **Step 1:** *Assumption of an Optimum*

- **Step 2:** *Variations and Increment*

- **Step 3:** *First Variation*

- **Step 4:** *Fundamental Theorem*

- **Step 5:** *Fundamental Lemma*

- **Step 6:** *Euler-Lagrange Equation*

- **Step 1:** *Assumption of an Optimum:* Let us assume that $x^*(t)$ is the optimum attained for the function $x(t)$. Take some admissible function $x_a(t) = x^*(t) + \delta x(t)$ close to $x^*(t)$, where $\delta x(t)$ is the variation of $x^*(t)$ as shown in Figure 2.4. The function $x_a(t)$ should also satisfy the boundary conditions (2.3.2) and hence it is necessary that

$$\delta x(t_0) = \delta x(t_f) = 0. \qquad (2.3.3)$$

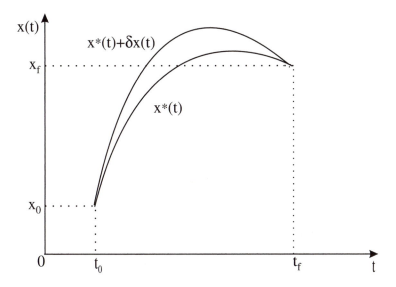

Figure 2.4 Fixed-End Time and Fixed-End State System

- **Step 2:** *Variations and Increment:* Let us first define the increment as

$$\Delta J(x^*(t), \delta x(t)) \triangleq J(x^*(t) + \delta x(t), \dot{x}^*(t) + \delta \dot{x}(t), t)$$
$$-J(x^*(t), \dot{x}^*(t), t)$$
$$= \int_{t_0}^{t_f} V(x^*(t) + \delta x(t), \dot{x}^*(t) + \delta \dot{x}(t), t) \, dt$$
$$-\int_{t_0}^{t_f} V(x^*(t), \dot{x}^*(t), t) dt. \qquad (2.3.4)$$

which by combining the integrals can be written as

$$\Delta J(x^*(t), \delta x(t)) = \int_{t_0}^{t_f} [V(x^*(t) + \delta x(t), \dot{x}^*(t) + \delta \dot{x}(t), t)$$
$$-V(x^*(t), \dot{x}^*(t), t)] \, dt. \qquad (2.3.5)$$

where,

$$\dot{x}(t) = \frac{dx(t)}{dt} \quad \text{and} \quad \delta \dot{x}(t) = \frac{d}{dt} \{\delta x(t)\} \qquad (2.3.6)$$

Expanding V in the increment (2.3.5) in a Taylor series about the point $x^*(t)$ and $\dot{x}^*(t)$, the increment ΔJ becomes (note the cancelation of $V(x^*(t), \dot{x}^*(t), t))$

$$
\begin{aligned}
\Delta J &= \Delta J(x^*(t), \delta x(t)) \\
&= \int_{t_0}^{t_f} \left[\frac{\partial V(x^*(t), \dot{x}^*(t), t)}{\partial x} \delta x(t) + \frac{\partial V(x^*(t), \dot{x}^*(t), t)}{\partial \dot{x}} \delta \dot{x}(t) \right. \\
&\quad + \frac{1}{2!} \left\{ \frac{\partial^2 V(\ldots)}{\partial x^2} (\delta x(t))^2 + \frac{\partial^2 V(\ldots)}{\partial \dot{x}^2} (\delta \dot{x}(t))^2 + \right. \\
&\quad \left. \left. + 2\frac{\partial^2 V(\ldots)}{\partial x \partial \dot{x}} \delta x(t) \delta \dot{x}(t) \right\} + \cdots \right] dt. \quad (2.3.7)
\end{aligned}
$$

Here, the partial derivatives are w.r.t. $x(t)$ and $\dot{x}(t)$ at the optimal condition (*) and * is omitted for simplicity.

- **Step 3:** *First Variation:* Now, we obtain the variation by retaining the terms that are *linear* in $\delta x(t)$ and $\delta \dot{x}(t)$ as

$$
\begin{aligned}
\delta J(x^*(t), \delta x(t)) &= \int_{t_0}^{t_f} \left[\frac{\partial V(x^*(t), \dot{x}^*(t), t)}{\partial x} \delta x(t) \right. \\
&\quad \left. + \frac{\partial V(x^*(t), \dot{x}^*(t), t)}{\partial \dot{x}} \delta \dot{x}(t) \right] dt. \quad (2.3.8)
\end{aligned}
$$

To express the relation for the first variation (2.3.8) entirely in terms containing $\delta x(t)$ (since $\delta \dot{x}(t)$ is dependent on $\delta x(t)$), we integrate by parts the term involving $\delta \dot{x}(t)$ as (omitting the arguments in V for simplicity)

$$
\begin{aligned}
\int_{t_0}^{t_f} \left(\frac{\partial V}{\partial \dot{x}} \right)_* \delta \dot{x}(t) dt &= \int_{t_0}^{t_f} \left(\frac{\partial V}{\partial \dot{x}} \right)_* \frac{d}{dt} (\delta x(t)) dt \\
&= \int_{t_0}^{t_f} \left(\frac{\partial V}{\partial \dot{x}} \right)_* d(\delta x(t)), \\
&= \left[\left(\frac{\partial V}{\partial \dot{x}} \right)_* \delta x(t) \right]_{t_0}^{t_f} \\
&\quad - \int_{t_0}^{t_f} \delta x(t) \frac{d}{dt} \left(\frac{\partial V}{\partial \dot{x}} \right)_* dt. \quad (2.3.9)
\end{aligned}
$$

In the above, we used the well-known integration formula $\int u\, dv = uv - \int v\, du$ where $u = \partial V/\partial \dot{x}$ and $v = \delta x(t)$). Using (2.3.9), the relation (2.3.8) for first variation becomes

$$\delta J(x^*(t), \delta x(t)) = \int_{t_0}^{t_f} \left(\frac{\partial V}{\partial x}\right)_* \delta x(t) dt + \left[\left(\frac{\partial V}{\partial \dot{x}}\right)_* \delta x(t)\right]_{t_0}^{t_f}$$

$$- \int_{t_0}^{t_f} \frac{d}{dt}\left(\frac{\partial V}{\partial \dot{x}}\right)_* \delta x(t) dt,$$

$$= \int_{t_0}^{t_f} \left[\left(\frac{\partial V}{\partial x}\right)_* - \frac{d}{dt}\left(\frac{\partial V}{\partial \dot{x}}\right)_*\right] \delta x(t) dt$$

$$+ \left[\left(\frac{\partial V}{\partial \dot{x}}\right)_* \delta x(t)\right]_{t_0}^{t_f}. \tag{2.3.10}$$

Using the relation (2.3.3) for boundary variations in (2.3.10), we get

$$\delta J(x^*(t), \delta x(t)) = \int_{t_0}^{t_f} \left[\left(\frac{\partial V}{\partial x}\right)_* - \frac{d}{dt}\left(\frac{\partial V}{\partial \dot{x}}\right)_*\right] \delta x(t) dt. \tag{2.3.11}$$

- **Step 4:** *Fundamental Theorem:* We now apply the *fundamental theorem of the calculus of variations* (Theorem 2.1), i.e., the variation of J must vanish for an optimum. That is, for the optimum $x^*(t)$ to exist, $\delta J(x^*(t), \delta x(t)) = 0$. Thus the relation (2.3.11) becomes

$$\int_{t_0}^{t_f} \left[\left(\frac{\partial V}{\partial x}\right)_* - \frac{d}{dt}\left(\frac{\partial V}{\partial \dot{x}}\right)_*\right] \delta x(t) dt = 0. \tag{2.3.12}$$

Note that the function $\delta x(t)$ must be zero at t_0 and t_f, but for this, it is completely arbitrary.

- **Step 5:** *Fundamental Lemma:* To simplify the condition obtained in the equation (2.3.12), let us take advantage of the following lemma called the *fundamental lemma of the calculus of variations* [47, 46, 79].

LEMMA 2.1

If for every function $g(t)$ which is continuous,

$$\int_{t_0}^{t_f} g(t)\delta x(t) dt = 0 \tag{2.3.13}$$

where the function $\delta x(t)$ is continuous in the interval $[t_0, t_f]$, then the function $g(t)$ must be zero everywhere throughout the interval $[t_0, t_f]$. (see Figure 2.5.)

Proof: We prove this by contradiction. Let us assume that $g(t)$ is nonzero (positive or negative) during a short interval $[t_a, t_b]$. Next, let us select $\delta x(t)$, which is arbitrary, to be positive (or negative) throughout the interval where $g(t)$ has a nonzero value. By this selection of $\delta x(t)$, the value of the integral in (2.3.13) will be nonzero. This contradicts our assumption that $g(t)$ is non-zero during the interval. Thus $g(t)$ must be identically zero everywhere during the entire interval $[t_0, t_f]$ in (2.3.13). Hence the lemma.

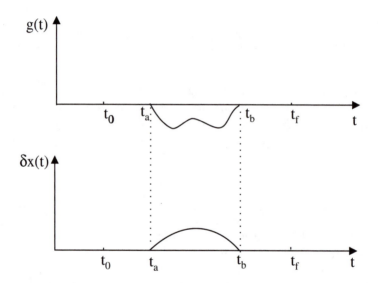

Figure 2.5   *A Nonzero $g(t)$ and an Arbitrary $\delta x(t)$*

- **Step 6:** *Euler-Lagrange Equation:* Applying the previous lemma to (2.3.12), a *necessary* condition for $x^*(t)$ to be an optimal of the functional J given by (2.3.1) is

$$\left(\frac{\partial V(x^*(t), \dot{x}^*(t), t)}{\partial x} \right)_* - \frac{d}{dt} \left(\frac{\partial V(x^*(t), \dot{x}^*(t), t)}{\partial \dot{x}} \right)_* = 0 \quad (2.3.14)$$

or in simplified notation omitting the arguments in V,

$$\left(\frac{\partial V}{\partial x}\right)_* - \frac{d}{dt}\left(\frac{\partial V}{\partial \dot{x}}\right)_* = 0 \qquad (2.3.15)$$

for all $t \in [t_0, t_f]$. This equation is called Euler equation, first published in 1741 [126].

A historical note is worthy of mention.

Euler obtained the equation (2.3.14) in 1741 using an elaborate and cumbersome procedure. Lagrange studied Euler's results and wrote a letter to Euler in 1755 in which he obtained the previous equation by a more elegant method of "variations" as described above. Euler recognized the simplicity and generality of the method of Lagrange and introduced the name **calculus of variations***. The all important fundamental equation (2.3.14) is now generally known as Euler-Lagrange (E.-L.) equation after these two great mathematicians of the 18th century. Lagrange worked further on optimization and came up with the well-known Lagrange multiplier rule or method.*

2.3.2   Discussion on Euler-Lagrange Equation

We provide some comments on the Euler-Lagrange equation [47, 46].

1. The Euler-Lagrange equation (2.3.14) can be written in many different forms. Thus (2.3.14) becomes

$$V_x - \frac{d}{dt}(V_{\dot{x}}) = 0 \qquad (2.3.16)$$

where,

$$V_x = \frac{\partial V}{\partial x} = V_x(x^*(t), \dot{x}^*(t), t); \qquad V_{\dot{x}} = \frac{\partial V}{\partial \dot{x}} = V_{\dot{x}}(x^*(t), \dot{x}^*(t), t). \qquad (2.3.17)$$

Since V is a function of three arguments $x^*(t)$, $\dot{x}^*(t)$, and t, and that $x^*(t)$ and $\dot{x}^*(t)$ are in turn functions of t, we get

$$\frac{d}{dt}\left(\frac{\partial V}{\partial \dot{x}}\right)_* = \frac{d}{dt}\left(\frac{\partial V(x^*(t), \dot{x}^*(t), t)}{\partial \dot{x}}\right)_*,$$

$$= \frac{d}{dt}\left(\frac{\partial^2 V}{\partial x \partial \dot{x}}dx + \frac{\partial^2 V}{\partial \dot{x}\partial \dot{x}}d\dot{x} + \frac{\partial^2 V}{\partial t \partial \dot{x}}dt\right)_*,$$

$$= \left(\frac{\partial^2 V}{\partial x \partial \dot{x}}\right)_*\left(\frac{dx}{dt}\right)_* + \left(\frac{\partial^2 V}{\partial \dot{x}\partial \dot{x}}\right)_*\left(\frac{d^2 x}{dt^2}\right)_* + \left(\frac{\partial^2 V}{\partial t \partial \dot{x}}\right)_*$$

$$= V_{x\dot{x}}\dot{x}^*(t) + V_{\dot{x}\dot{x}}\ddot{x}^*(t) + V_{t\dot{x}}. \qquad (2.3.18)$$

Combining (2.3.16) and (2.3.18), we get an alternate form for the EL equation as

$$\boxed{V_x - V_{t\dot{x}} - V_{x\dot{x}}\dot{x}^*(t) - V_{\dot{x}\dot{x}}\ddot{x}^*(t) = 0.} \qquad (2.3.19)$$

2. The presence of $\frac{d}{dt}$ and/or $\dot{x}^*(t)$ in the EL equation (2.3.14) means that it is a *differential* equation.

3. In the EL equation (2.3.14), the term $\frac{\partial V(x^*(t), \dot{x}^*(t), t)}{\partial \dot{x}}$ is in general a function of $x^*(t)$, $\dot{x}^*(t)$, and t. Thus when this function is differentiated w.r.t. t, $\ddot{x}^*(t)$ may be present. This means that the differential equation (2.3.14) is in general of *second* order. This is also evident from the alternate form (2.3.19) for the EL equation.

4. There may also be terms involving products or powers of $\ddot{x}^*(t)$, $\dot{x}^*(t)$, and $x^*(t)$, in which case, the differential equation becomes *nonlinear*.

5. The explicit presence of t in the arguments indicates that the coefficients may be *time-varying*.

6. The conditions at initial point $t = t_0$ and final point $t = t_f$ leads to a *boundary value* problem.

7. Thus, the Euler-Lagrange equation (2.3.14) is, in general, a non-linear, time-varying, two-point boundary value, second order, or-dinary differential equation. Thus, we often have a nonlinear *two-point boundary value problem* (TPBVP). The solution of the nonlinear TPBVP is quite a formidable task and often done us-ing numerical techniques. This is the price we pay for demanding optimal performance!

8. Compliance with the Euler-Lagrange equation is only a *necessary* condition for the optimum. Optimal may sometimes not yield either a maximum or a minimum; just as inflection points where the derivative vanishes in differential calculus. However, if the Euler-Lagrange equation is not satisfied for any function, this indicates that the optimum does not exist for that functional.

2.3.3   Different Cases for Euler-Lagrange Equation

We now discuss various cases of the EL equation.

Case 1: V is dependent of $\dot{x}(t)$, and t. That is, $V = V(\dot{x}(t), t)$. Then $V_x = 0$. The Euler-Lagrange equation (2.3.16) becomes

$$\frac{d}{dt}(V_{\dot{x}}) = 0. \tag{2.3.20}$$

This leads us to

$$V_{\dot{x}} = \frac{\partial V(\dot{x}^*(t), t)}{\partial \dot{x}} = C \tag{2.3.21}$$

where, C is a constant of integration.

Case 2: V is dependent of $\dot{x}(t)$ only. That is, $V = V(\dot{x}(t))$. Then $V_x = 0$. The Euler-Lagrange equation (2.3.16) becomes

$$\frac{d}{dt}(V_{\dot{x}}) = 0 \longrightarrow V_{\dot{x}} = C. \tag{2.3.22}$$

In general, the solution of either (2.3.21) or (2.3.22) becomes

$$\dot{x}^*(t) = C_1 \longrightarrow x^*(t) = C_1 t + C_2. \tag{2.3.23}$$

This is simply an equation of a straight line.

Case 3: V is dependent of $x(t)$ and $\dot{x}(t)$. That is, $V = V(x(t), \dot{x}(t))$. Then $V_{t\dot{x}} = 0$. Using the other form of the Euler-Lagrange equation (2.3.19), we get

$$V_x - V_{x\dot{x}}\dot{x}^*(t) - V_{\dot{x}\dot{x}}\ddot{x}^*(t) = 0. \tag{2.3.24}$$

Multiplying the previous equation by $\dot{x}^*(t)$, we have

$$\dot{x}^*(t)\left[V_x - V_{x\dot{x}}\dot{x}^*(t) - V_{\dot{x}\dot{x}}\ddot{x}^*(t)\right] = 0. \tag{2.3.25}$$

This can be rewritten as

$$\frac{d}{dt}(V - \dot{x}^*(t)V_{\dot{x}}) = 0 \longrightarrow V - \dot{x}^*(t)V_{\dot{x}} = C. \tag{2.3.26}$$

The previous equation can be solved using any of the techniques such as, separation of variables.

Case 4: V is dependent of $x(t)$, and t, i.e., $V = V(x(t), t)$. Then, $V_{\dot{x}} = 0$ and the Euler-Lagrange equation (2.3.16) becomes

$$\frac{\partial V(x^*(t), t)}{\partial x} = 0. \qquad (2.3.27)$$

The solution of this equation does not contain any arbitrary constants and therefore generally speaking does not satisfy the boundary conditions $x(t_0)$ and $x(t_f)$. Hence, in general, no solution exists for this variational problem. Only in rare cases, when the function $x(t)$ satisfies the given boundary conditions $x(t_0)$ and $x(t_f)$, it becomes an optimal function.

Let us now illustrate the application of the EL equation with a very simple classic example of finding the shortest distance between two points. Often, we omit the $*$ (which indicates an optimal or extremal value) during the working of a problem and attach the same to the final solution.

Example 2.7

Find the minimum length between any two points.

Solution: It is well known that the solution to this problem is a straight line. However, we like to illustrate the application of Euler-Lagrange equation for this simple case. Consider the arc between two points A and B as shown in Figure 2.6. Let ds be the small arc length, and dx and dt are the small rectangular coordinate values. Note that t is the independent variable representing distance and not time. Then,

$$(ds)^2 = (dx)^2 + (dt)^2. \qquad (2.3.28)$$

Rewriting

$$ds = \sqrt{1 + \dot{x}^2(t)}dt, \quad \text{where} \quad \dot{x}(t) = \frac{dx}{dt}. \qquad (2.3.29)$$

Now the total arc length S between two points $x(t = t_0)$ and $x(t = t_f)$ is the performance index J to be minimized. Thus,

$$S = J = \int ds = \int_{t_0}^{t_f} \sqrt{1 + \dot{x}^2(t)}dt = \int_{t_0}^{t_f} V(\dot{x}(t))dt \qquad (2.3.30)$$

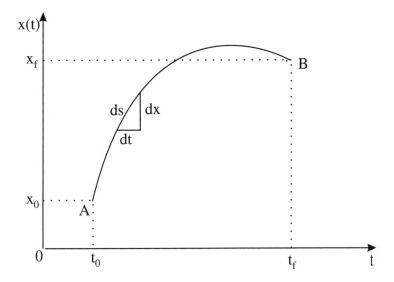

Figure 2.6   Arc Length

where, $V(\dot{x}(t)) = \sqrt{1 + \dot{x}^2(t)}$. Note that V is a function of $\dot{x}(t)$ only. Applying the Euler-Lagrange equation (2.3.22) to the performance index (2.3.30), we get

$$\frac{\dot{x}^*(t)}{\sqrt{1 + \dot{x}^{*2}(t)}} = C. \tag{2.3.31}$$

Solving this equation, we get the optimal solution as

$$x^*(t) = C_1 t + C_2. \tag{2.3.32}$$

This is evidently an equation for a straight line and the constants C_1 and C_2 are evaluated from the given boundary conditions. For example, if $x(0) = 1$ and $x(2) = 5$, $C_1 = 2$ and $C_2 = 1$ the straight line is $x^*(t) = 2t + 1$.

Although the previous example is a simple one,

1. it illustrates the formulation of a performance index from a given simple specification or a statement, and

2. the solution is well known *a priori* so that we can easily verify the application of the Euler-Lagrange equation.

In the previous example, we notice that the integrand V in the functional (2.3.30), is a function of $\dot{x}(t)$ only. Next, we take an example, where, V is a function of $x(t)$, $\dot{x}(t)$ and t.

Example 2.8

Find the optimum of

$$J = \int_0^2 \left[\dot{x}^2(t) - 2tx(t) \right] dt \qquad (2.3.33)$$

that satisfy the boundary (initial and final) conditions

$$x(0) = 1 \quad \text{and} \quad x(2) = 5. \qquad (2.3.34)$$

Solution: In the EL equation (2.3.19), we first identify that $V = \dot{x}^2(t) - 2tx(t)$. Then applying the EL equation (2.3.15) to the performance index (2.3.33) we get

$$\frac{\partial V}{\partial x} - \frac{d}{dt}\left(\frac{\partial V}{\partial \dot{x}} \right) = 0 \longrightarrow -2t - \frac{d}{dt}\left(2\dot{x}(t) \right) = 0$$

$$\longrightarrow \ddot{x}(t) = t. \qquad (2.3.35)$$

Solving the previous simple differential equation, we have

$$x^*(t) = \frac{t^3}{6} + C_1 t + C_2 \qquad (2.3.36)$$

where, C_1 and C_2 are constants of integration. Using the given boundary conditions (2.3.19) in (2.3.36), we have

$$x(0) = 1 \longrightarrow C_2 = 1, \quad x(2) = 5 \longrightarrow C_1 = \frac{4}{3}. \qquad (2.3.37)$$

With these values for the constants, we finally have the optimal function as

$$x^*(t) = \frac{t^3}{6} + \frac{4}{3}t + 1. \qquad (2.3.38)$$

Another classical example in the calculus of variations is the *brachistochrone* (from *brachisto*, the shortest, and *chrones*, time) problem and this problem is dealt with in almost all books on calculus of variations [126].

Further, note that we have considered here only the so-called fixedend point problem where both (initial and final) ends are fixed or given in advance. Other types of problems such as free-end point problems are not presented here but can be found in most of the books on the calculus of variations [79, 46, 81, 48]. However, these free-end point problems are better considered later in this chapter when we discuss the optimal control problem consisting of a performance index and a physical plant.

2.4   The Second Variation

In the study of extrema of functionals, we have so far considered only the *necessary* condition for a functional to have a relative or *weak extremum*, i.e., the condition that the first variation vanish leading to the classic *Euler-Lagrange equation*. To establish the nature of optimum (maximum or minimum), it is required to examine the *second variation*. In the relation (2.3.7) for the increment consider the terms corresponding to the second variation [120],

$$\delta^2 J = \int_{t_0}^{t_f} \frac{1}{2!} \left[\left(\frac{\partial^2 V}{\partial x^2} \right)_* (\delta x(t))^2 + \left(\frac{\partial^2 V}{\partial \dot{x}^2} \right)_* (\delta \dot{x}(t))^2 \right.$$
$$\left. + 2 \left(\frac{\partial^2 V}{\partial x \partial \dot{x}} \right)_* \delta x(t) \delta \dot{x}(t) \right] dt. \qquad (2.4.1)$$

Consider the last term in the previous equation and rewrite it in terms of $\delta x(t)$ only using integration by parts ($\int u\,dv = uv - \int v\,du$ where, $u = \frac{\partial^2 V}{\partial x \partial \dot{x}} \delta x(t)$ and $v = \delta x(t)$). Then using $\delta x(t_0) = \delta x(t_f) = 0$ for fixed-end conditions, we get

$$\delta^2 J = \frac{1}{2} \int_{t_0}^{t_f} \left[\left\{ \left(\frac{\partial^2 V}{\partial x^2} \right)_* - \frac{d}{dt} \left(\frac{\partial^2 V}{\partial x \partial \dot{x}} \right)_* \right\} (\delta x(t))^2 \right.$$
$$\left. + \left(\frac{\partial^2 V}{\partial \dot{x}^2} \right)_* (\delta \dot{x}(t))^2 \right] dt. \qquad (2.4.2)$$

According to Theorem 2.1, the fundamental theorem of the calculus of variations, the sufficient condition for a *minimum* is $\delta^2 J > 0$. This, for arbitrary values of $\delta x(t)$ and $\delta \dot{x}(t)$, means that

$$\left(\frac{\partial^2 V}{\partial x^2} \right)_* - \frac{d}{dt} \left(\frac{\partial^2 V}{\partial x \partial \dot{x}} \right)_* > 0, \qquad (2.4.3)$$

$$\left(\frac{\partial^2 V}{\partial \dot{x}^2} \right)_* > 0. \qquad (2.4.4)$$

For *maximum*, the signs of the previous conditions are reversed. Alternatively, we can rewrite the second variation (2.4.1) in matrix form as

$$\delta^2 J = \frac{1}{2} \int_{t_0}^{t_f} [\delta x(t) \ \delta \dot{x}(t)] \begin{bmatrix} \frac{\partial^2 V}{\partial x^2} & \frac{\partial^2 V}{\partial x \partial \dot{x}} \\ \frac{\partial^2 V}{\partial x \partial \dot{x}} & \frac{\partial^2 V}{\partial \dot{x}^2} \end{bmatrix}_* \begin{bmatrix} \delta x(t) \\ \delta \dot{x}(t) \end{bmatrix} dt$$
$$= \frac{1}{2} \int_{t_0}^{t_f} [\delta x(t) \ \delta \dot{x}(t)] \Pi \begin{bmatrix} \delta x(t) \\ \delta \dot{x}(t) \end{bmatrix} dt \qquad (2.4.5)$$

where,

$$\Pi = \begin{bmatrix} \dfrac{\partial^2 V}{\partial x^2} & \dfrac{\partial^2 V}{\partial x \partial \dot{x}} \\[2ex] \dfrac{\partial^2 V}{\partial x \partial \dot{x}} & \dfrac{\partial^2 V}{\partial \dot{x}^2} \end{bmatrix}_*. \tag{2.4.6}$$

If the matrix Π in the previous equation is positive (negative) definite, we establish a minimum (maximum). In many cases since $\delta x(t)$ is arbitrary, the coefficient of $(\delta \dot{x}(t))^2$, i.e., $\partial^2 V / \partial \dot{x}^2$ determines the sign of $\delta^2 J$. That is, the sign of second variation agrees with the sign of $\partial^2 V / \partial \dot{x}^2$. Thus, for *minimization* requirement

$$\boxed{\left(\dfrac{\partial^2 V}{\partial \dot{x}^2} \right)_* > 0.} \tag{2.4.7}$$

For *maximization*, the sign of the previous equation reverses. In the literature, this condition is called *Legendre condition* [126].

> *In 1786, Legendre obtained this result of deciding whether a given optimum is maximum or minimum by examining the second variation. The second variation technique was further generalized by Jacobi in 1836 and hence this condition is usually called Legendre-Jacobi condition.*

Example 2.9

Verify that the straight line represents the minimum distance between two points.

Solution: This is an obvious solution, however, we illustrate the second variation. Earlier in Example 2.7, we have formulated a functional for the distance between two points as

$$J = \int_{t_0}^{t_f} \sqrt{1 + \dot{x}^2(t)} dt = \int_{t_0}^{t_f} V(\dot{x}(t)) dt \tag{2.4.8}$$

and found that the optimum is a straight line $x^*(t) = C_1 t + C_2$. To satisfy the sufficiency condition (2.4.7), we find

$$\left(\dfrac{\partial V}{\partial \dot{x}} \right)_* = \dfrac{\dot{x}^*(t)}{\sqrt{1 + \dot{x}^{*2}(t)}} \quad \text{and} \quad \left(\dfrac{\partial^2 V}{\partial \dot{x}^2} \right)_* = \dfrac{1}{[1 + \dot{x}^{*2}(t)]^{3/2}}. \tag{2.4.9}$$

Since $\dot{x}^*(t)$ is a constant (+ve or -ve), the previous equation satisfies the condition (2.4.7). Hence, the distance between two points as given by $x^*(t)$ (straight line) is minimum.

Next, we begin the second stage of optimal control. We consider optimization (or extremization) of a *functional* with a plant, which is considered as a constraint or a condition along with the functional. In other words, we address the extremization of a functional with some condition, which is in the form of a plant equation. The plant takes the form of state equation leading us to optimal control of dynamic systems. This section is motivated by [6, 79, 120, 108].

2.5   *Extrema of Functions with Conditions*

We begin with an example of finding the extrema of a function under a condition (or constraint). We solve this example with two methods, first by *direct* method and then by Lagrange *multiplier* method. Let us note that we consider this simple example only to illustrate some basic concepts associated with conditional extremization [120].

Example 2.10

A manufacturer wants to maximize the volume of the material stored in a circular tank subject to the condition that the material used for the tank is limited (constant). Thus, for a constant thickness of the material, the manufacturer wants to minimize the volume of the material used and hence part of the cost for the tank.

Solution: If a fixed metal thickness is assumed, this condition implies that the cross-sectional area of the tank material is constant. Let d and h be the diameter and the height of the circular tank. Then the volume contained by the tank is

$$V(d, h) = \pi d^2 h/4 \qquad (2.5.1)$$

and the cross-sectional surface area (upper, lower and side) of the tank is

$$A(d, h) = 2\pi d^2/4 + \pi dh = A_0. \qquad (2.5.2)$$

Our intent is to maximize $V(d, h)$ keeping $A(d, h) = A_0$, where A_0 is a given constant. We discuss two methods: first one is called the *Direct Method* using simple calculus and the second one is called *Lagrange Multiplier Method* using the Lagrange multiplier method.

1 Direct Method: In solving for the optimum value directly, we eliminate one of the variables, say h, from the volume relation (2.5.1) using the area relation (2.5.2). By doing so, the condition is *embedded* in the original function to be extremized. From (2.5.2),

$$h = \frac{A_0 - \pi d^2/2}{\pi d}. \qquad (2.5.3)$$

Using the relation (2.5.3) for height in the relation (2.5.1) for volume

$$V(d) = A_0 d/4 - \pi d^3/8.\qquad(2.5.4)$$

Now, to find the extrema of this simple calculus problem, we differentiate (2.5.4) w.r.t. d and set it to zero to get

$$\frac{A_0}{4} - \frac{3}{8}\pi d^2 = 0.\qquad(2.5.5)$$

Solving, we get the optimal value of d as

$$d^* = \sqrt{\frac{2A_0}{3\pi}}.\qquad(2.5.6)$$

By demanding that as per the Definition 2.3 for optimum of a function, the second derivative of V w.r.t. d in (2.5.4) be *negative* for *maximum*, we can easily see that the positive value of the square root function corresponds to the maximum value of the function. Substituting the optimal value of the diameter (2.5.6) in the original cross-sectional area given by (2.5.2), and solving for the optimum h^*, we get

$$h^* = \sqrt{\frac{2A_0}{3\pi}}.\qquad(2.5.7)$$

Thus, we see from (2.5.6) and (2.5.7) that the volume stored by a tank is maximized if the height of the tank is made equal to its diameter.

2 Lagrange Multiplier Method: Now we solve the above problem by applying Lagrange multiplier method. We form a new function to be extremized by *adjoining* a given condition to the original function. The new adjoined function is extremized in the normal way by taking the partial derivatives w.r.t. all its variables, making them equal to zero, and solving for these variables which are extremals. Let the original volume relation (2.5.1) to be extremized be rewritten as

$$f(d, h) = \pi d^2 h/4\qquad(2.5.8)$$

and the condition (2.5.2) to be satisfied as

$$g(d, h) = 2\pi d^2/4 + \pi dh - A_0 = 0.\qquad(2.5.9)$$

Then a new adjoint function $\mathcal{L}$ (called Lagrangian) is formed as

$$\begin{aligned}\mathcal{L}(d, h, \lambda) &= f(d, h) + \lambda g(d, h)\\&= \pi d^2 h/4 + \lambda(2\pi d^2/4 + \pi dh - A_0)\end{aligned}\qquad(2.5.10)$$

where, λ, a parameter yet to be determined, is called the *Lagrange multiplier*. Now, since the Lagrangian $\mathcal{L}$ is a function of three optimal variables d, h, and λ, we take the partial derivatives of $\mathcal{L}(d, h, \lambda)$ w.r.t. each of the variables d, h and λ and set them to zero. Thus,

$$\frac{\partial \mathcal{L}}{\partial d} = \pi dh/2 + \lambda(\pi d + \pi h) = 0 \qquad (2.5.11)$$

$$\frac{\partial \mathcal{L}}{\partial h} = \pi d^2/4 + \lambda(\pi d) = 0 \qquad (2.5.12)$$

$$\frac{\partial \mathcal{L}}{\partial \lambda} = 2\pi d^2/4 + \pi dh - A_0 = 0. \qquad (2.5.13)$$

Now, solving the previous three relations (2.5.11) to (2.5.13) for the three variables d^*, h^*, and λ^*, we get

$$d^* = \sqrt{\frac{2A_0}{3\pi}}; \quad h^* = \sqrt{\frac{2A_0}{3\pi}}; \quad \lambda^* = -\sqrt{\frac{A_0}{24\pi}}. \qquad (2.5.14)$$

Once again, to maximize the volume of a cylindrical tank, we need to have the height (h^*) equal to the diameter (d^*) of the tank. Note that we need to take the negative value of the square root function for λ in (2.5.14) in order to satisfy the physical requirement that the diameter d obtained from (2.5.12) as

$$d = -4\lambda \qquad (2.5.15)$$

is a positive value.

Now, we generalize the previous two methods.

2.5.1   Direct Method

Now we generalize the preceding method of elimination using differential calculus. Consider the extrema of a function $f(x_1, x_2)$ with two *interdependent* variables x_1 and x_2, subject to the condition

$$g(x_1, x_2) = 0. \qquad (2.5.16)$$

As a necessary condition for extrema, we have

$$df = \frac{\partial f}{\partial x_1} dx_1 + \frac{\partial f}{\partial x_2} dx_2 = 0. \qquad (2.5.17)$$

However, since dx_1 and dx_2 are not *arbitrary*, but related by the condition

$$dg = \frac{\partial g}{\partial x_1} dx_1 + \frac{\partial g}{\partial x_2} dx_2 = 0, \qquad (2.5.18)$$

it is not possible to conclude as in the case of extremization of functions without conditions that

$$\frac{\partial f}{\partial x_1} = 0 \quad \text{and} \quad \frac{\partial f}{\partial x_2} = 0 \tag{2.5.19}$$

in the necessary condition (2.5.17). This is easily seen, since if the set of extrema conditions (2.5.19) is solved for optimal values x_1^* and x_2^*, there is no guarantee that these optimal values, would, in general satisfy the given condition (2.5.16).

In order to find optimal values that satisfy both the condition (2.5.16) and that of the extrema conditions (2.5.17), we arbitrarily choose one of the variables, say x_1, as the *independent* variable. Then x_2 becomes a *dependent* variable as per the condition (2.5.16). Now, assuming that $\partial g/\partial x_2 \neq 0$, (2.5.18) becomes

$$dx_2 = - \left\{ \frac{\partial g/\partial x_1}{\partial g/\partial x_2} \right\} dx_1 \tag{2.5.20}$$

and using (2.5.20) in the necessary condition (2.5.17), we have

$$df = \left[\frac{\partial f}{\partial x_1} - \frac{\partial f}{\partial x_2} \left\{ \frac{\partial g/\partial x_1}{\partial g/\partial x_2} \right\} \right] dx_1 = 0. \tag{2.5.21}$$

As we have chosen dx_1 to be the *independent*, we now can consider it to be *arbitrary*, and conclude that in order to satisfy (2.5.21), we have the coefficient of dx_1 to be zero. That is

$$\left(\frac{\partial f}{\partial x_1} \right) \left(\frac{\partial g}{\partial x_2} \right) - \left(\frac{\partial f}{\partial x_2} \right) \left(\frac{\partial g}{\partial x_1} \right) = 0. \tag{2.5.22}$$

Now, the relation (2.5.22) and the condition (2.5.16) are solved simultaneously for the optimal solutions x_1^* and x_2^*. Equation (2.5.22) can be rewritten as

$$\begin{vmatrix} \dfrac{\partial f}{\partial x_1} & \dfrac{\partial f}{\partial x_2} \\[2mm] \dfrac{\partial g}{\partial x_1} & \dfrac{\partial g}{\partial x_2} \end{vmatrix} = 0. \tag{2.5.23}$$

This is also, as we know, the Jacobian of f and g w.r.t. x_1 and x_2. This method of elimination of the dependent variables is quite tedious for higher order problems.

2.5.2   *Lagrange Multiplier Method*

We now generalize the second method of solving the same problem of extrema of functions with conditions. Consider again the extrema of the function $f(x_1, x_2)$ subject to the condition

$$g(x_1, x_2) = 0. \qquad (2.5.24)$$

In this method, we form an *augmented* Lagrangian function

$$\mathcal{L}(x_1, x_2, \lambda) = f(x_1, x_2) + \lambda g(x_1, x_2) \qquad (2.5.25)$$

where, λ, a parameter (multiplier) yet to be determined, is the Lagrange multiplier. Let us note that using the given condition (2.5.24) in the Lagrangian (2.5.25), we have

$$\mathcal{L}(x_1, x_2) = f(x_1, x_2) \qquad (2.5.26)$$

and therefore a *necessary* condition for extrema is that

$$df = d\mathcal{L} = 0. \qquad (2.5.27)$$

Accepting the idea that the Lagrangian (2.5.25) is a better representation of the entire problem than the equation (2.5.26) in finding the extrema, we have from the Lagrangian relation (2.5.25)

$$d\mathcal{L} = df + \lambda dg = 0. \qquad (2.5.28)$$

Using (2.5.17) and (2.5.18) in (2.5.28), and rearranging

$$\left[\frac{\partial f}{\partial x_1} + \lambda \frac{\partial g}{\partial x_1} \right] dx_1 + \left[\frac{\partial f}{\partial x_2} + \lambda \frac{\partial g}{\partial x_2} \right] dx_2 = 0. \qquad (2.5.29)$$

Now dx_1 and dx_2 are *both* not independent and hence cannot immediately conclude that each of the coefficients of dx_1 and dx_2 in (2.5.29) must be zero. Let us choose dx_1 to be *independent* differential and then dx_2 becomes a *dependent* differential as per (2.5.18). Further, let us choose the multiplier λ, which has been introduced by us and is at our disposal, to make one of the coefficients of dx_1 or dx_2 in (2.5.29) zero. For example, let λ take on the value λ^* that makes the coefficient of the *dependent* differential dx_2 equal zero, that is

$$\frac{\partial f}{\partial x_2} + \lambda^* \frac{\partial g}{\partial x_2} = 0. \qquad (2.5.30)$$

With (2.5.30), the equation (2.5.29) reduces to

$$\left[\frac{\partial f}{\partial x_1} + \lambda \frac{\partial g}{\partial x_1}\right] dx_1 = 0. \tag{2.5.31}$$

Since, dx_1 is the *independent* differential, it can be varied *arbitrarily*. Hence, for (2.5.31) to be satisfied for all dx_1, the coefficient of dx_1 must be zero. That is

$$\frac{\partial f}{\partial x_1} + \lambda^* \frac{\partial g}{\partial x_1} = 0. \tag{2.5.32}$$

Now from (2.5.25), note that

$$\frac{\partial \mathcal{L}}{\partial \lambda} = 0 \tag{2.5.33}$$

yields the constraint relation (2.5.16). Combining the results from (2.5.32), (2.5.30), and (2.5.33), we have

$$\frac{\partial \mathcal{L}}{\partial x_1} = \frac{\partial f}{\partial x_1} + \lambda^* \frac{\partial g}{\partial x_1} = 0 \tag{2.5.34}$$

$$\frac{\partial \mathcal{L}}{\partial x_2} = \frac{\partial f}{\partial x_2} + \lambda^* \frac{\partial g}{\partial x_2} = 0 \tag{2.5.35}$$

$$\frac{\partial \mathcal{L}}{\partial \lambda} = g(x_1^*, x_2^*) = 0. \tag{2.5.36}$$

The preceding three equations are to be solved simultaneously to obtain x_1^*, x_2^*, and λ^*. By eliminating λ^* between (2.5.34) and (2.5.35)

$$\left(\frac{\partial f}{\partial x_1}\right)\left(\frac{\partial g}{\partial x_2}\right) - \left(\frac{\partial f}{\partial x_2}\right)\left(\frac{\partial g}{\partial x_1}\right) = 0 \tag{2.5.37}$$

which is the same condition as (2.5.22) obtained by the direct method, thus indicating that we have the same result by Lagrange multiplier method.

Let us note that the necessary conditions (2.5.34) and (2.5.35) are just the same equations which would have been obtained from considering the differentials dx_1 and dx_2 as though they were *independent* in (2.5.29). Introduction of the multiplier λ has allowed us to treat all the variables in the augmented function $\mathcal{L}(x_1, x_2, \lambda)$ as though each variable is *independent*. Thus, the multiplier λ has acted like a *catalyst*, appearing in the intermediate stage only.

Summarizing, the extrema of a function $f(x_1, x_2)$ subject to the condition (or constraint) $g(x_1, x_2) = 0$ is equivalent to extrema of a single augmented function $\mathcal{L}(x_1, x_2, \lambda) = f(x_1, x_2) + \lambda g(x_1, x_2)$ as though x_1, x_2 and λ are *independent*. We now generalize this result.

THEOREM 2.2

Consider the extrema of a continuous, real-valued function $f(\mathbf{x}) = f(x_1, x_2, \cdots, x_n)$ subject to the conditions

$$g_1(\mathbf{x}) = g_1(x_1, x_2, \cdots, x_n) = 0$$
$$g_2(\mathbf{x}) = g_2(x_1, x_2, \cdots, x_n) = 0$$
$$\cdots$$
$$g_m(\mathbf{x}) = g_m(x_1, x_2, \cdots, x_n) = 0 \qquad (2.5.38)$$

where, f and $\mathbf{g}$ have continuous partial derivatives, and $m < n$. Let $\lambda_1, \lambda_2, \cdots, \lambda_m$ be the Lagrange multipliers corresponding to m conditions, such that the augmented Lagrangian function is formed as

$$\mathcal{L}(\mathbf{x}, \boldsymbol{\lambda}) = f(\mathbf{x}) + \boldsymbol{\lambda}'\mathbf{g}(\mathbf{x}), \qquad (2.5.39)$$

where, $\boldsymbol{\lambda}'$ is the transpose of $\boldsymbol{\lambda}$. Then, the optimal values $\mathbf{x}^*$ and $\boldsymbol{\lambda}^*$ are the solutions of the following $n + m$ equations

$$\frac{\partial \mathcal{L}}{\partial \mathbf{x}} = \frac{\partial f}{\partial \mathbf{x}} + \boldsymbol{\lambda}'\frac{\partial \mathbf{g}}{\partial \mathbf{x}} = 0 \qquad (2.5.40)$$

$$\frac{\partial \mathcal{L}}{\partial \boldsymbol{\lambda}} = \mathbf{g}(\mathbf{x}) = 0. \qquad (2.5.41)$$

Features of Lagrange Multiplier: The Lagrange multiplier method is a powerful one in finding the extrema of functions subject to conditions. It has the following attractive features:

1. The importance of the Lagrange multiplier technique lies on the fact that the problem of determining the extrema of the function $f(\mathbf{x})$ subject to the conditions $\mathbf{g}(\mathbf{x}) = 0$ is *embedded* within the simple problem of determining the extrema of the simple *augmented* function $\mathcal{L}(\mathbf{x}, \boldsymbol{\lambda}) = f(\mathbf{x}) + \boldsymbol{\lambda}'\mathbf{g}(\mathbf{x})$.

2. Introduction of Lagrange multiplier allows us to treat all the variables $\mathbf{x}$ and $\boldsymbol{\lambda}$ in the augmented function $\mathcal{L}(\mathbf{x}, \boldsymbol{\lambda})$ as though each were *independent*.

3. The multiplier λ acts like a *catalyst* in the sense that it is introduced to perform a certain duty as given by item 2.

4. The increased dimensionality $(n + m)$ which is characteristic of the Lagrange multiplier method, is generally more than compensated by the relative *simplicity* and *systematic* procedure of the technique.

   The multiplier method was given by Lagrange in 1788.

2.6   Extrema of Functionals with Conditions

In this section, we extend our ideas to functionals based on those developed in the last section for functions. First, we consider a functional with two variables, use the results of the previous section on the CoV, derive the necessary conditions and then extend the same for a general nth order vector case. Consider the extremization of the performance index in the form of a functional

$$J(x_1(t), x_2(t), t) = J = \int_{t_0}^{t_f} V(x_1(t), x_2(t), \dot{x}_1(t), \dot{x}_2(t), t)dt \quad (2.6.1)$$

subject to the condition (plant or system equation)

$$g(x_1(t), x_2(t), \dot{x}_1(t), \dot{x}_2(t)) = 0 \quad (2.6.2)$$

with fixed-end-point conditions

$$\begin{aligned} x_1(t_0) &= x_{10}; \quad x_2(t_0) = x_{20} \\ x_1(t_f) &= x_{1f}; \quad x_2(t_f) = x_{2f}. \end{aligned} \quad (2.6.3)$$

Now we address this problem under the following steps.

- **Step 1:** *Lagrangian*

- **Step 2:** *Variations and Increment*

- **Step 3:** *First Variation*

- **Step 4:** *Fundamental Theorem*

- **Step 5:** *Fundamental Lemma*

- **Step 6:** *Euler-Lagrange Equation*

- **Step 1:** *Lagrangian:* We form an *augmented* functional

$$J_a = \int_{t_0}^{t_f} \mathcal{L}(x_1(t), x_2(t), \dot{x}_1(t), \dot{x}_2(t), \lambda(t), t)dt \qquad (2.6.4)$$

where, $\lambda(t)$ is the Lagrange multiplier, and the Lagrangian $\mathcal{L}$ is defined as

$$\begin{aligned}
\mathcal{L} &= \mathcal{L}(x_1(t), x_2(t), \dot{x}_1(t), \dot{x}_2(t), \lambda(t), t) \\
&= V(x_1(t), x_2(t), \dot{x}_1(t), \dot{x}_2(t), t) \\
&\quad + \lambda(t)g(x_1(t), x_2(t), \dot{x}_1(t), \dot{x}_2(t)) \qquad (2.6.5)
\end{aligned}$$

Note from the performance index (2.6.1) and the augmented performance index (2.6.4) that $J_a = J$ if the condition (2.6.2) is satisfied for any $\lambda(t)$.

- **Step 2:** *Variations and Increment:* Next, assume optimal values and then consider the *variations* and *increment* as

$$x_i(t) = x_i^*(t) + \delta x_i(t), \qquad \dot{x}_i(t) = \dot{x}_i^*(t) + \delta\dot{x}_i(t), \quad i = 1, 2$$
$$\Delta J_a = J_a(x_i^*(t) + \delta x_i(t), \dot{x}_i^*(t) + \delta\dot{x}_i(t), t) - J_a(x_i^*(t), \dot{x}_i^*(t), t), \qquad (2.6.6)$$

for $i = 1, 2$.

- **Step 3:** *First Variation:* Then using the Taylor series expansion and retaining linear terms only, the *first* variation of the functional J_a becomes

$$\begin{aligned}
\delta J_a = \int_{t_0}^{t_f} \Bigg[&\left(\frac{\partial\mathcal{L}}{\partial x_1}\right)_* \delta x_1(t) + \left(\frac{\partial\mathcal{L}}{\partial x_2}\right)_* \delta x_2(t) \\
&+ \left(\frac{\partial\mathcal{L}}{\partial\dot{x}_1}\right)_* \delta\dot{x}_1(t) + \left(\frac{\partial\mathcal{L}}{\partial\dot{x}_2}\right)_* \delta\dot{x}_2(t) \Bigg] dt. \qquad (2.6.7)
\end{aligned}$$

As before in the section on CoV, we rewrite the terms containing $\delta\dot{x}_1(t)$ and $\delta\dot{x}_2(t)$ in terms of those containing $\delta x_1(t)$ and $\delta x_2(t)$

only (using integration by parts, $\int u\,dv = uv - \int v\,du$). Thus

$$\int_{t_0}^{t_f} \left(\frac{\partial \mathcal{L}}{\partial x_1}\right)_* \delta \dot{x}_1(t)dt = \int_{t_0}^{t_f} \left(\frac{\partial \mathcal{L}}{\partial \dot{x}_1}\right)_* \frac{d}{dt}(\delta x_1(t))dt$$

$$= \int_{t_0}^{t_f} \left(\frac{\partial \mathcal{L}}{\partial \dot{x}_1}\right)_* d(\delta x_1(t))$$

$$= \left[\left(\frac{\partial \mathcal{L}}{\partial \dot{x}_1}\right)_* \delta x_1(t)\right]\Big|_{t_0}^{t_f}$$

$$- \int_{t_0}^{t_f} \frac{d}{dt}\left(\frac{\partial \mathcal{L}}{\partial \dot{x}_1}\right)_* \delta x_1(t)dt.$$

$$(2.6.8)$$

Using the above, we have the first variation (2.6.7) as

$$\delta J_a = \int_{t_0}^{t_f} \left[\left(\frac{\partial \mathcal{L}}{\partial x_1}\right)_* \delta x_1(t) + \left(\frac{\partial \mathcal{L}}{\partial x_2}\right)_* \delta x_2(t)\right] dt$$

$$+ \left[\left(\frac{\partial \mathcal{L}}{\partial \dot{x}_1}\right)_* \delta x_1(t)\right]\Big|_{t_0}^{t_f} + \left[\left(\frac{\partial \mathcal{L}}{\partial \dot{x}_2}\right)_* \delta x_2(t)\right]\Big|_{t_0}^{t_f}$$

$$- \int_{t_0}^{t_f} \frac{d}{dt}\left(\frac{\partial \mathcal{L}}{\partial \dot{x}_1}\right)_* \delta x_1(t)dt - \int_{t_0}^{t_f} \frac{d}{dt}\left(\frac{\partial \mathcal{L}}{\partial \dot{x}_2}\right)_* \delta x_2(t)dt.$$

$$(2.6.9)$$

Since this is a fixed-final time and fixed-final state problem as given by (2.6.3), no variations are allowed at the final point. This means

$$\delta x_1(t_0) = \delta x_2(t_0) = \delta x_1(t_f) = \delta x_2(t_f) = 0. \qquad (2.6.10)$$

Using the boundary variations (2.6.10) in the augmented first variation (2.6.9), we have

$$\delta J_a = \int_{t_0}^{t_f} \left[\left(\frac{\partial \mathcal{L}}{\partial x_1}\right)_* - \frac{d}{dt}\left(\frac{\partial \mathcal{L}}{\partial \dot{x}_1}\right)_*\right] \delta x_1(t)dt$$

$$+ \int_{t_0}^{t_f} \left[\left(\frac{\partial \mathcal{L}}{\partial x_2}\right)_* - \frac{d}{dt}\left(\frac{\partial \mathcal{L}}{\partial \dot{x}_2}\right)_*\right] \delta x_2(t)dt. \quad (2.6.11)$$

- **Step 4:** *Fundamental Theorem:* Now, we proceed as follows.

    1. We invoke the fundamental theorem of the calculus of variations (Theorem 2.1) and make the first variation (2.6.11) equal to zero.

2. Remembering that *both* $\delta x_1(t)$ and $\delta x_2(t)$ are not independent, because $x_1(t)$ and $x_2(t)$ are related by the condition
   (2.6.2), we choose $\delta x_2(t)$ as the *independent* variation and
   $\delta x_1(t)$ as the *dependent* variation.

3. Let us choose the multiplier $\lambda^*(t)$ which is arbitrarily introduced and is at our disposal, in such a way that the coefficient of the *dependent* variation $\delta x_1(t)$ in (2.6.11) vanish.
   That is

$$\left(\frac{\partial \mathcal{L}}{\partial x_1}\right)_* - \frac{d}{dt}\left(\frac{\partial \mathcal{L}}{\partial \dot{x}_1}\right)_* = 0. \tag{2.6.12}$$

With these choices, the first variation (2.6.11) becomes

$$\int_{t_0}^{t_f}\left[\left(\frac{\partial \mathcal{L}}{\partial x_2}\right)_* - \frac{d}{dt}\left(\frac{\partial \mathcal{L}}{\partial \dot{x}_2}\right)_*\right]\delta x_2(t)dt = 0. \tag{2.6.13}$$

- **Step 5:** *Fundamental Lemma:* Using the fundamental lemma of
  CoV (Lemma 2.1) and noting that since $\delta x_2(t)$ has been chosen
  to be *independent* variation and hence *arbitrary*, the only way
  (2.6.13) can be satisfied, in general, is that the coefficient of $\delta x_1(t)$
  also vanish. That is

$$\left(\frac{\partial \mathcal{L}}{\partial x_2}\right)_* - \frac{d}{dt}\left(\frac{\partial \mathcal{L}}{\partial \dot{x}_2}\right)_* = 0. \tag{2.6.14}$$

Also, from the Lagrangian(2.6.5) note that

$$\left(\frac{\partial \mathcal{L}}{\partial \lambda}\right)_* = 0 \tag{2.6.15}$$

yields the constraint relation (2.6.2).

- **Step 6:** *Euler-Lagrange Equation:* Combining the various relations (2.6.12), (2.6.14), and (2.6.15), the *necessary* conditions for
  extremization of the functional (2.6.1) subject to the condition
  (2.6.2) (according to Euler-Lagrange equation) are

$$\left(\frac{\partial \mathcal{L}}{\partial x_1}\right)_* - \frac{d}{dt}\left(\frac{\partial \mathcal{L}}{\partial \dot{x}_1}\right)_* = 0 \tag{2.6.16}$$

$$\left(\frac{\partial \mathcal{L}}{\partial x_2}\right)_* - \frac{d}{dt}\left(\frac{\partial \mathcal{L}}{\partial \dot{x}_2}\right)_* = 0 \tag{2.6.17}$$

$$\left(\frac{\partial \mathcal{L}}{\partial \lambda}\right)_* - \frac{d}{dt}\left(\frac{\partial \mathcal{L}}{\partial \dot{\lambda}}\right)_* = 0. \tag{2.6.18}$$

Let us note that these conditions are just the ones that would have been obtained from the Lagrangian (2.6.5), as if both $\delta x_1(t)$ and $\delta x_2(t)$ had been *independent*. Also, in (2.6.18), the Lagrangian $\mathcal{L}$ is independent of $\lambda(t)$ and hence the condition (2.6.18) is really the given plant equation (2.6.2).

Thus, the introduction of the Lagrange multiplier $\lambda(t)$ has enabled us to treat the variables $x_1(t)$ and $x_2(t)$ as though they were *independent*, in spite of the fact that they are related by the condition (2.6.2). The solution of the two, second-order differential equations (2.6.16) and (2.6.17) and the condition relation (2.6.2) or (2.6.18) along with the boundary conditions (2.6.3) give the optimal solutions $x_1^*(t)$, $x_2^*(t)$, and $\lambda^*(t)$.

Now, we generalize the preceding procedure for an nth order system. Consider the extremization of a functional

$$J = \int_{t_0}^{t_f} V(\mathbf{x}(t), \dot{\mathbf{x}}(t), t)dt \qquad (2.6.19)$$

where, $\mathbf{x}(t)$ is an nth order state vector, subject to the plant equation (condition)

$$g_i(\mathbf{x}(t), \dot{\mathbf{x}}(t), t) = 0; \quad i = 1, 2, \ldots, m \qquad (2.6.20)$$

and boundary conditions, $\mathbf{x}(0)$ and $\mathbf{x}(t_f)$. We form an augmented functional

$$J_a = \int_{t_0}^{t_f} \mathcal{L}(\mathbf{x}(t), \dot{\mathbf{x}}(t), \boldsymbol{\lambda}(t), t)dt \qquad (2.6.21)$$

where, the Lagrangian $\mathcal{L}$ is given by

$$\boxed{\mathcal{L}(\mathbf{x}(t), \dot{\mathbf{x}}(t), \boldsymbol{\lambda}(t), t) = V(\mathbf{x}(t), \dot{\mathbf{x}}(t), t) + \boldsymbol{\lambda}'(t)g_i(\mathbf{x}(t), \dot{\mathbf{x}}(t), t)} \qquad (2.6.22)$$

and the Lagrange multiplier $\boldsymbol{\lambda}(t) = [\lambda_1(t), \lambda_2(t), \ldots, \lambda_m(t)]'$. We now apply the Euler-Lagrange equation on J_a to yield

$$\boxed{\left(\frac{\partial \mathcal{L}}{\partial \mathbf{x}}\right)_* - \frac{d}{dt}\left(\frac{\partial \mathcal{L}}{\partial \dot{\mathbf{x}}}\right)_* = 0,} \qquad (2.6.23)$$

$$\boxed{\left(\frac{\partial \mathcal{L}}{\partial \boldsymbol{\lambda}}\right)_* - \frac{d}{dt}\left(\frac{\partial \mathcal{L}}{\partial \dot{\boldsymbol{\lambda}}}\right)_* = 0 \longrightarrow g_i(\mathbf{x}(t), \dot{\mathbf{x}}(t), t) = 0.} \qquad (2.6.24)$$

Note that from (2.6.22), the Lagrangian $\mathcal{L}$ is independent of $\dot{\lambda}(t)$ and hence the Euler-Lagrange equation (2.6.24) is nothing but the given relation regarding the plant or the system (2.6.20). Thus, we solve the Euler-Lagrange equation (2.6.23) along with the given boundary conditions. Let us now illustrate the preceding method by a simple example.

Example 2.11

Minimize the performance index

$$J = \int_0^1 \left[x^2(t) + u^2(t) \right] dt \qquad (2.6.25)$$

with boundary conditions

$$x(0) = 1; \quad x(1) = 0 \qquad (2.6.26)$$

subject to the condition (plant equation)

$$\dot{x}(t) = -x(t) + u(t). \qquad (2.6.27)$$

Solution: Let us solve this problem by the two methods, i.e., the direct method and the Lagrange multiplier method.

1 Direct Method: Here, we eliminate $u(t)$ between the performance index (2.6.25) and the plant (2.6.27) to get the functional as

$$J = \int_0^1 [x^2(t) + (\dot{x}(t) + x(t))^2] dt$$

$$= \int_0^1 [2x^2(t) + \dot{x}^2(t) + 2x(t)\dot{x}(t)] dt. \qquad (2.6.28)$$

Now, we notice that the functional (2.6.28) absorbed the condition (2.6.27) within itself, and we need to consider it as a straight forward extremization of a functional as given earlier. Thus, applying the Euler-Lagrange equation

$$\left(\frac{\partial V}{\partial x} \right)_* - \frac{d}{dt} \left(\frac{\partial V}{\partial \dot{x}} \right)_* = 0 \qquad (2.6.29)$$

to the functional (2.6.28), where,

$$V = 2x^2(t) + \dot{x}^2(t) + 2x(t)\dot{x}(t), \qquad (2.6.30)$$

we get

$$4x^*(t) + 2\dot{x}^*(t) - \frac{d}{dt}(2\dot{x}^*(t) + 2x^*(t)) = 0. \tag{2.6.31}$$

Simplifying the above

$$\ddot{x}^*(t) - 2x^*(t) = 0 \tag{2.6.32}$$

the solution (see later for use of MATLAB©) of which gives the optimal as

$$x^*(t) = C_1 e^{-\sqrt{2}t} + C_2 e^{\sqrt{2}t} \tag{2.6.33}$$

where, the constants C_1 and C_2, evaluated using the given boundary conditions (2.6.26), are found to be

$$C_1 = 1/(1 - e^{-2\sqrt{2}}); \quad C_2 = 1/(1 - e^{2\sqrt{2}}). \tag{2.6.34}$$

Finally, knowing the optimal $x^*(t)$, the optimal control $u^*(t)$ is found from the plant (2.6.27) to be

$$u^*(t) = \dot{x}^*(t) + x^*(t)$$
$$= C_1(1 - \sqrt{2})e^{-\sqrt{2}t} + C_2(1 + \sqrt{2})e^{\sqrt{2}t}. \tag{2.6.35}$$

Although the method appears to be simple, let us note that it is not always possible to eliminate $u(t)$ from (2.6.25) and (2.6.27) especially for higher-order systems.

2 Lagrange Multiplier Method: Here, we use the ideas developed in the previous section on the extremization of functions with conditions. Consider the optimization of the functional (2.6.25) with the boundary conditions (2.6.26) under the condition describing the plant (2.6.27). First we rewrite the condition (2.6.27) as

$$g(x(t), \dot{x}(t), u(t)) = \dot{x}(t) + x(t) - u(t) = 0. \tag{2.6.36}$$

Now, we form an augmented functional as

$$J = \int_0^1 \left[x^2(t) + u^2(t) + \lambda(t)\{\dot{x}(t) + x(t) - u(t)\} \right] dt$$
$$= \int_0^1 \mathcal{L}(x(t), \dot{x}(t), u(t), \lambda(t)) dt \tag{2.6.37}$$

where, $\lambda(t)$ is the Lagrange multiplier, and

$$\mathcal{L}(x(t), \dot{x}(t), u(t), \lambda(t)) = x^2(t) + u^2(t)$$
$$+ \lambda(t)\{\dot{x}(t) + x(t) - u(t)\} \tag{2.6.38}$$

is the Lagrangian. Now, we apply the Euler-Lagrange equation to the previous Lagrangian to get

$$\left(\frac{\partial \mathcal{L}}{\partial x}\right)_* - \frac{d}{dt}\left(\frac{\partial \mathcal{L}}{\partial \dot{x}}\right)_* = 0 \longrightarrow 2x^*(t) + \lambda^*(t) - \dot{\lambda}^*(t) = 0 \quad (2.6.39)$$

$$\left(\frac{\partial \mathcal{L}}{\partial u}\right)_* - \frac{d}{dt}\left(\frac{\partial \mathcal{L}}{\partial \dot{u}}\right)_* = 0 \longrightarrow 2u^*(t) - \lambda^*(t) = 0 \quad (2.6.40)$$

$$\left(\frac{\partial \mathcal{L}}{\partial \lambda}\right)_* - \frac{d}{dt}\left(\frac{\partial \mathcal{L}}{\partial \dot{\lambda}}\right)_* = 0 \longrightarrow \dot{x}^*(t) + x^*(t) - u^*(t) = 0 \quad (2.6.41)$$

and solve for optimal $x^*(t)$, $u^*(t)$, and $\lambda^*(t)$. We get first from (2.6.40) and (2.6.41)

$$\lambda^*(t) = 2u^*(t) = 2(\dot{x}^*(t) + x^*(t)). \quad (2.6.42)$$

Using the equation (2.6.42) in (2.6.39)

$$2x^*(t) + 2(\dot{x}^*(t) + x^*(t)) - 2(\ddot{x}^*(t) + \dot{x}^*(t)) = 0. \quad (2.6.43)$$

Solving the previous equation, we get

$$\ddot{x}^*(t) - 2x^*(t) = 0 \longrightarrow x^*(t) = C_1 e^{-\sqrt{2}t} + C_2 e^{\sqrt{2}t}. \quad (2.6.44)$$

Once we know $x^*(t)$, we get $\lambda^*(t)$ and hence $u^*(t)$ from (2.6.42) as

$$u^*(t) = \dot{x}^*(t) + x^*(t)$$
$$= C_1(1 - \sqrt{2})e^{-\sqrt{2}t} + C_2(1 + \sqrt{2})e^{\sqrt{2}t}. \quad (2.6.45)$$

Thus, we get the same results as in direct method. The constants C_1 and C_2, evaluated using the boundary conditions (2.6.26) are the same as given in (2.6.34).

   The solution for the set of differential equations (2.6.32) with the boundary conditions (2.6.26) for Example 2.11 using Symbolic Toolbox of the MATLAB©, Version 6, is shown below.

```
***********************************************************
x=dsolve('D2x-2*x=0','x(0)=1,x(1)=0')

x =

-(exp(2^(1/2))^2+1)/(exp(2^(1/2))^2-1)*sinh(2^(1/2)*t)+
cosh(2^(1/2)*t)
```

$$-\frac{(\exp(2^{1/2})^2 + 1)\,\sinh(2^{1/2}\,t)}{\exp(2^{1/2})^2 - 1} + \cosh(2^{1/2}\,t)$$

```
u =

-(exp(2^(1/2))^2+1)/(exp(2^(1/2))^2-1)*cosh(2^(1/2)*t)*2^(1/2)+
sinh(2^(1/2)*t)*2^(1/2)-(exp(2^(1/2))^2+1)/(exp(2^(1/2))^2-
1)*sinh(2^(1/2)*t)+cosh(2^(1/2)*t)
```

$$-\frac{(\exp(2^{1/2})^2 + 1)\,\cosh(2^{1/2}\,t)\,2^{1/2}}{\exp(2^{1/2})^2 - 1} + \sinh(2^{1/2}\,t)\,2^{1/2}$$

$$-\frac{(\exp(2^{1/2})^2 + 1)\,\sinh(2^{1/2}\,t)}{\exp(2^{1/2})^2 - 1} + \cosh(2^{1/2}\,t)$$

```
***********************************************************
```

It is easy to see that the previous solution for optimal $x^*(t)$ is the same as given in (2.6.33) and (2.6.34).

Let us note once again that the Lagrange multiplier $\lambda(t)$ helped us to treat the augmented functional (2.6.38) as if it contained *independent* variables $x(t)$ and $u(t)$, although they are *dependent* as per the plant equation (2.6.36).

2.7 Variational Approach to Optimal Control Systems

In this section, we approach the optimal control system by variational techniques, and in the process introduce the Hamiltonian function, which was used by Pontryagin and his associates to develop the famous *Minimum Principle* [109].

2.7.1 Terminal Cost Problem

Here we consider the optimal control system where the performance index is of general form containing a final (terminal) cost function in addition to the integral cost function. Such an optimal control problem is called the *Bolza* problem. Consider the plant as

$$\dot{\mathbf{x}}(t) = \mathbf{f}(\mathbf{x}(t), \mathbf{u}(t), t), \qquad (2.7.1)$$

the performance index as

$$J(\mathbf{u}(t)) = S(\mathbf{x}(t_f), t_f) + \int_{t_0}^{t_f} V(\mathbf{x}(t), \mathbf{u}(t), t) dt \qquad (2.7.2)$$

and given boundary conditions as

$$\mathbf{x}(t_0) = \mathbf{x}_0; \qquad \mathbf{x}(t_f) \text{ is free and } t_f \text{ is free} \qquad (2.7.3)$$

where, $\mathbf{x}(t)$ and $\mathbf{u}(t)$ are n- and r- dimensional state and control vectors respectively. This *problem of Bolza* is the one with the most general form of the performance index.

> *The Lagrange problem was first discussed in 1762, Mayer considered his problem in 1878, and the problem of Bolza was formulated in 1913.*

Before we begin illustrating the Pontryagin procedure for this problem, let us note that

$$\int_{t_0}^{t_f} \frac{dS(\mathbf{x}(t), t)}{dt} dt = S(\mathbf{x}(t), t)|_{t_0}^{t_f} = S(\mathbf{x}(t_f), t_f) - S(\mathbf{x}(t_0), t_0). \quad (2.7.4)$$

Using the equation (2.7.4) in the original performance index (2.7.2), we get

$$J_2(\mathbf{u}(t)) = \int_{t_0}^{t_f} \left[V(\mathbf{x}(t), \mathbf{u}(t), t) + \frac{dS}{dt} \right] dt$$

$$= \int_{t_0}^{t_f} V(\mathbf{x}(t), \mathbf{u}(t), t) dt + S(\mathbf{x}(t_f), t_f) - S(\mathbf{x}(t_0), t_0). \quad (2.7.5)$$

Since $S(\mathbf{x}(t_0), t_0)$ is a fixed quantity, the optimization of the original performance index J in (2.7.2) is equivalent to that of the performance index J_2 in (2.7.5). However, the *optimal cost* given by (2.7.2) is different from the optimal cost (2.7.5). Here, we are interested in finding the optimal control only. Once the optimal control is determined, the optimal cost is found using the original performance index J in (2.7.2) and not J_2 in (2.7.5). Also note that

$$\frac{d[S(\mathbf{x}(t), t)]}{dt} = \left(\frac{\partial S}{\partial \mathbf{x}}\right)' \dot{\mathbf{x}}(t) + \frac{\partial S}{\partial t}. \tag{2.7.6}$$

We now illustrate the procedure in the following steps. Also, we first introduce the Lagrangian and then, a little later, introduce the Hamiltonian. Let us first list the various steps and then describe the same in detail.

- **Step 1:** *Assumption of Optimal Conditions*

- **Step 2:** *Variations of Control and State Vectors*

- **Step 3:** *Lagrange Multiplier*

- **Step 4:** *Lagrangian*

- **Step 5:** *First Variation*

- **Step 6:** *Condition for Extrema*

- **Step 7:** *Hamiltonian*

- **Step 1:** *Assumptions of Optimal Conditions:* We assume optimum values $\mathbf{x}^*(t)$ and $\mathbf{u}^*(t)$ for state and control, respectively. Then

$$J(\mathbf{u}^*(t)) = \int_{t_0}^{t_f} \left[V(\mathbf{x}^*(t), \mathbf{u}^*(t), t) + \frac{dS(\mathbf{x}^*(t), t)}{dt} \right] dt$$
$$\dot{\mathbf{x}}^*(t) = \mathbf{f}(\mathbf{x}^*(t), \mathbf{u}^*(t), t). \tag{2.7.7}$$

- **Step 2:** *Variations of Controls and States:* We consider the variations (perturbations) in control and state vectors as (see Figure 2.7)

$$\mathbf{x}(t) = \mathbf{x}^*(t) + \delta\mathbf{x}(t); \quad \mathbf{u}(t) = \mathbf{u}^*(t) + \delta\mathbf{u}(t). \tag{2.7.8}$$

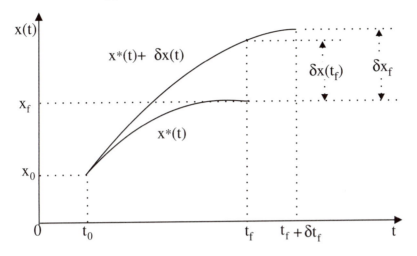

Figure 2.7   Free-Final Time and Free-Final State System

Then, the state equation (2.7.1) and the performance index (2.7.5) become

$$\dot{\mathbf{x}}^*(\mathbf{t}) + \delta\dot{\mathbf{x}}(\mathbf{t}) = \mathbf{f}(\mathbf{x}^*(\mathbf{t}) + \delta\mathbf{x}(\mathbf{t}), \mathbf{u}^*(\mathbf{t}) + \delta\mathbf{u}(\mathbf{t}), \mathbf{t})$$

$$J(\mathbf{u}(t)) = \int_{t_0}^{t_f + \delta t_f} \left[V(\mathbf{x}^*(t) + \delta\mathbf{x}(t), \mathbf{u}^*(t) + \delta\mathbf{u}(t), t) + \frac{dS}{dt} \right] dt$$

$$(2.7.9)$$

- **Step 3:** *Lagrange Multiplier:* Introducing the Lagrange multiplier vector $\boldsymbol{\lambda}(t)$ (also called costate vector) and using (2.7.6), we introduce the augmented performance index at the optimal condition as

$$J_a(\mathbf{u}^*(t)) = \int_{t_0}^{t_f} [V(\mathbf{x}^*(t), \mathbf{u}^*(t), t) + \left(\frac{\partial S}{\partial \mathbf{x}}\right)'_* \dot{\mathbf{x}}^*(t) + \left(\frac{\partial S}{\partial t}\right)_*$$
$$+ \boldsymbol{\lambda}'(t)\{\mathbf{f}(\mathbf{x}^*(t), \mathbf{u}^*(t), t) - \dot{\mathbf{x}}^*(t)\}]dt \qquad (2.7.10)$$

and at any other (perturbed) condition as

$$J_a(\mathbf{u}(t)) = \int_{t_0}^{t_f + \delta t_f} [V(\mathbf{x}^*(t) + \delta \mathbf{x}(t), \mathbf{u}^*(t) + \delta \mathbf{u}(t), t)$$

$$+ \left(\frac{\partial S}{\partial \mathbf{x}}\right)'_* [\dot{\mathbf{x}}^*(t) + \delta \dot{\mathbf{x}}(t)] + \left(\frac{\partial S}{\partial t}\right)_*$$

$$+ \boldsymbol{\lambda}'(t) [\mathbf{f}(\mathbf{x}^*(t) + \delta \mathbf{x}(t), \mathbf{u}^*(t) + \delta \mathbf{u}(t), t)$$

$$- \{\dot{\mathbf{x}}^*(t) + \delta \dot{\mathbf{x}}(t)\}]] dt. \tag{2.7.11}$$

- **Step 4:** *Lagrangian:* Let us define the Lagrangian function at optimal condition as

$$\mathcal{L} = \mathcal{L}(\mathbf{x}^*(t), \dot{\mathbf{x}}^*(t), \mathbf{u}^*(t), \boldsymbol{\lambda}(t), t)$$

$$= V(\mathbf{x}^*(t), \mathbf{u}^*(t), t) + \left(\frac{\partial S}{\partial \mathbf{x}}\right)'_* \dot{\mathbf{x}}^*(t) + \frac{\partial S}{\partial t}$$

$$+ \boldsymbol{\lambda}'(t) \{\mathbf{f}(\mathbf{x}^*(t), \mathbf{u}^*(t), t) - \dot{\mathbf{x}}^*(t)\} \tag{2.7.12}$$

and at any other condition as

$$\mathcal{L}^\delta = \mathcal{L}^\delta(\mathbf{x}^*(t) + \delta \mathbf{x}(t), \dot{\mathbf{x}}^*(t) + \delta \dot{\mathbf{x}}(t), \mathbf{u}^*(t) + \delta \mathbf{u}(t), \boldsymbol{\lambda}(t), t)$$

$$= V(\mathbf{x}^*(t) + \delta \mathbf{x}(t), \mathbf{u}^*(t) + \delta \mathbf{u}(t), t)$$

$$+ \left(\frac{\partial S}{\partial \mathbf{x}}\right)'_* [\dot{\mathbf{x}}^*(t) + \delta \dot{\mathbf{x}}(t)] + \left(\frac{\partial S}{\partial t}\right)_*$$

$$+ \boldsymbol{\lambda}'(t) [\mathbf{f}(\mathbf{x}^*(t) + \delta \mathbf{x}(t), \mathbf{u}^*(t) + \delta \mathbf{u}(t), t)$$

$$- \{\dot{\mathbf{x}}^*(t) + \delta \dot{\mathbf{x}}(t)\}] . \tag{2.7.13}$$

With these, the augmented performance index at the optimal and any other condition becomes

$$J_a(\mathbf{u}^*(t)) = \int_{t_0}^{t_f} \mathcal{L}(\mathbf{x}^*(t), \dot{\mathbf{x}}^*(t), \mathbf{u}^*(t), \boldsymbol{\lambda}(t), t) dt = \int_{t_0}^{t_f} \mathcal{L} dt$$

$$J_a(\mathbf{u}(t)) = \int_{t_0}^{t_f + \delta t_f} \mathcal{L}^\delta dt = \int_{t_0}^{t_f} \mathcal{L}^\delta dt + \int_{t_f}^{t_f + \delta t_f} \mathcal{L}^\delta dt. \tag{2.7.14}$$

Using mean-value theorem and Taylor series, and retaining the *linear* terms only, we have

$$\int_{t_f}^{t_f+\delta t_f} \mathcal{L}^\delta dt = \mathcal{L}^\delta \Big|_{t_f} \delta t_f$$

$$\approx \left\{ \mathcal{L} + \left(\frac{\partial \mathcal{L}}{\partial \mathbf{x}}\right)'_* \delta \mathbf{x}(t) + \left(\frac{\partial \mathcal{L}}{\partial \dot{\mathbf{x}}}\right)'_* \delta \dot{\mathbf{x}}(t) \right.$$

$$\left. + \left(\frac{\partial \mathcal{L}}{\partial \mathbf{u}}\right)'_* \delta \mathbf{u}(t) \right\} \Big|_{t_f} \delta t_f$$

$$\approx \mathcal{L}|_{t_f} \delta t_f. \qquad (2.7.15)$$

- **Step 5:** *First Variation:* Defining increment ΔJ, using Taylor series expansion, extracting the first variation δJ by retaining only the first order terms, we get the first variation as

$$\Delta J = J_a(\mathbf{u}(t)) - J_a(\mathbf{u}^*(t))$$

$$= \int_{t_0}^{t_f} (\mathcal{L}^\delta - \mathcal{L})dt + \mathcal{L}|_{t_f} \delta t_f$$

$$\delta J = \int_{t_0}^{t_f} \left\{ \left(\frac{\partial \mathcal{L}}{\partial \mathbf{x}}\right)'_* \delta \mathbf{x}(t) + \left(\frac{\partial \mathcal{L}}{\partial \dot{\mathbf{x}}}\right)'_* \delta \dot{\mathbf{x}}(t) + \left(\frac{\partial \mathcal{L}}{\partial \mathbf{u}}\right)'_* \delta \mathbf{u}(t) \right\} dt$$

$$+ \mathcal{L}|_{t_f} \delta t_f. \qquad (2.7.16)$$

Considering the $\delta \dot{\mathbf{x}}(t)$ term in the first variation (2.7.16) and integrating by parts (using $\int u dv = uv - \int v du$),

$$\int_{t_0}^{t_f} \left(\frac{\partial \mathcal{L}}{\partial \dot{\mathbf{x}}}\right)'_* \delta \dot{\mathbf{x}}(t) dt = \int_{t_0}^{t_f} \left(\frac{\partial \mathcal{L}}{\partial \dot{\mathbf{x}}}\right)'_* \frac{d}{dt} (\delta \mathbf{x}(t)) dt$$

$$= \left[\left(\frac{\partial \mathcal{L}}{\partial \dot{\mathbf{x}}}\right)'_* \delta \mathbf{x}(t) \right] \Big|_{t_0}^{t_f}$$

$$- \int_{t_0}^{t_f} \left[\frac{d}{dt} \left(\frac{\partial \mathcal{L}}{\partial \dot{\mathbf{x}}}\right)_* \right]' \delta \mathbf{x}(t) dt. \qquad (2.7.17)$$

Also note that since $\mathbf{x}(t_0)$ is specified, $\delta\mathbf{x}(t_0) = 0$. Thus, using (2.7.17) the first variation δJ in (2.7.16) becomes

$$\delta J = \int_{t_0}^{t_f} \left[\left(\frac{\partial \mathcal{L}}{\partial \mathbf{x}}\right)_* - \frac{d}{dt}\left(\frac{\partial \mathcal{L}}{\partial \dot{\mathbf{x}}}\right)_* \right]' \delta\mathbf{x}(t)dt$$

$$+ \int_{t_0}^{t_f} \left(\frac{\partial \mathcal{L}}{\partial \mathbf{u}}\right)_*' \delta\mathbf{u}(t)dt$$

$$+ \mathcal{L}|_{t_f}\, \delta t_f + \left[\left(\frac{\partial \mathcal{L}}{\partial \dot{\mathbf{x}}}\right)_*' \delta\mathbf{x}(t) \right]\Bigg|_{t_f} . \qquad (2.7.18)$$

- **Step 6:** *Condition for Extrema:* For extrema of the functional J, the *first variation* δJ should vanish according to the fundamental theorem (Theorem 2.1) of the CoV. Also, in a typical control system such as (2.7.1), we note that $\delta\mathbf{u}(t)$ is the *independent* control variation and $\delta\mathbf{x}(t)$ is the *dependent* state variation. First, we choose $\boldsymbol{\lambda}(t) = \boldsymbol{\lambda}^*(t)$ which is at our disposal and hence $\mathcal{L}^*$ such that the coefficient of the *dependent* variation $\delta\mathbf{x}(t)$ in (2.7.18) be zero. Then, we have the Euler-Lagrange equation

$$\left(\frac{\partial \mathcal{L}}{\partial \mathbf{x}}\right)_* - \frac{d}{dt}\left(\frac{\partial \mathcal{L}}{\partial \dot{\mathbf{x}}}\right)_* = 0 \qquad (2.7.19)$$

where the partials are evaluated at the optimal (*) condition. Next, since the independent control variation $\delta\mathbf{u}(t)$ is arbitrary, the coefficient of the control variation $\delta\mathbf{u}(t)$ in (2.7.18) should be set to zero. That is

$$\left(\frac{\partial \mathcal{L}}{\partial \mathbf{u}}\right)_* = 0. \qquad (2.7.20)$$

Finally, the first variation (2.7.18) reduces to

$$\mathcal{L}^*|_{t_f}\, \delta t_f + \left[\left(\frac{\partial \mathcal{L}}{\partial \dot{\mathbf{x}}}\right)_*' \delta\mathbf{x}(t) \right]\Bigg|_{t_f} = 0. \qquad (2.7.21)$$

Let us note that the condition (or plant) equation (2.7.1) can be written in terms of the Lagrangian (2.7.12) as

$$\left(\frac{\partial \mathcal{L}}{\partial \boldsymbol{\lambda}}\right)_* = 0. \qquad (2.7.22)$$

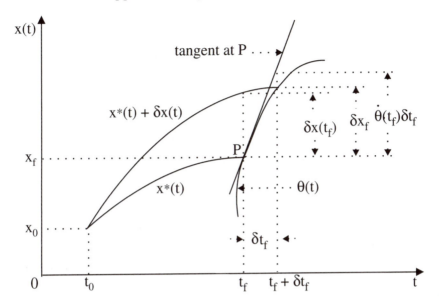

Figure 2.8   Final-Point Condition with a Moving Boundary $\theta(t)$

In order to convert the expression containing $\delta\mathbf{x}(t)$ in (2.7.21) into an expression containing $\delta\mathbf{x}_f$ (see Figure 2.8), we note that the slope of $\dot{\mathbf{x}}^*(t) + \delta\dot{\mathbf{x}}(t)$ at t_f is approximated as

$$\dot{\mathbf{x}}^*(t_f) + \delta\dot{\mathbf{x}}(t_f) \approx \frac{\delta\mathbf{x}_f - \delta\mathbf{x}(t_f)}{\delta t_f} \qquad (2.7.23)$$

which is rewritten as

$$\delta\mathbf{x}_f = \delta\mathbf{x}(t_f) + \{\dot{\mathbf{x}}^*(t) + \delta\dot{\mathbf{x}}(t)\}\,\delta t_f \qquad (2.7.24)$$

and retaining only the linear (in δ) terms in the relation (2.7.24), we have

$$\delta\mathbf{x}(t_f) = \delta\mathbf{x}_f - \dot{\mathbf{x}}^*(t_f)\delta t_f. \qquad (2.7.25)$$

Using (2.7.25) in the boundary condition (2.7.21), we have the general boundary condition in terms of the Lagrangian as

$$\left[\mathcal{L}^* - \left(\frac{\partial\mathcal{L}}{\partial\dot{\mathbf{x}}}\right)'_* \dot{\mathbf{x}}(t)\right]\Bigg|_{t_f} \delta t_f + \left(\frac{\partial\mathcal{L}}{\partial\dot{\mathbf{x}}}\right)'_*\Bigg|_{t_f} \delta\mathbf{x}_f = 0. \qquad (2.7.26)$$

- **Step 7:** *Hamiltonian:* We define the Hamiltonian $\mathcal{H}^*$ (also called the Pontryagin $\mathcal{H}$ function) at the optimal condition as

$$\boxed{\mathcal{H}^* = V(\mathbf{x}^*(t), \mathbf{u}^*(t), t) + \boldsymbol{\lambda}^{*'}(t)\mathbf{f}(\mathbf{x}^*(t), \mathbf{u}^*(t), t),} \qquad (2.7.27)$$

where,

$$\mathcal{H}^* = \mathcal{H}^*(\mathbf{x}^*(t), \mathbf{u}^*(t), \boldsymbol{\lambda}^*(t), t).$$

Then from (2.7.12) the Lagrangian $\mathcal{L}^*$ in terms of the Hamiltonian $\mathcal{H}^*$ becomes

$$
\begin{aligned}
\mathcal{L}^* &= \mathcal{L}^*(\mathbf{x}^*(t), \dot{\mathbf{x}}^*(t), \mathbf{u}^*(t), \boldsymbol{\lambda}^*(t), t) \\
&= \mathcal{H}^*(\mathbf{x}^*(t), \mathbf{u}^*(t), \boldsymbol{\lambda}^*(t), t) \\
&\quad + \left(\frac{\partial S}{\partial \mathbf{x}}\right)_*' \dot{\mathbf{x}}^*(t) + \left(\frac{\partial S}{\partial t}\right)_* - \boldsymbol{\lambda}^{*'}(t)\dot{\mathbf{x}}^*(t).
\end{aligned}
\tag{2.7.28}
$$

Using (2.7.28) in (2.7.20), (2.7.19), and (2.7.22) and noting that the terminal cost function $S = S(\mathbf{x}(t), t)$, we have the control, state and costate equations, respectively expressed in terms of the Hamiltonian. Thus, for the optimal control $\mathbf{u}^*(t)$, the relation (2.7.20) becomes

$$\left(\frac{\partial \mathcal{L}}{\partial \mathbf{u}}\right)_* = 0 \longrightarrow \boxed{\left(\frac{\partial \mathcal{H}}{\partial \mathbf{u}}\right)_* = 0} \tag{2.7.29}$$

for the optimal state $\mathbf{x}^*(t)$, the relation (2.7.19) becomes

$$
\left(\frac{\partial \mathcal{L}}{\partial \mathbf{x}}\right)_* - \frac{d}{dt}\left(\frac{\partial \mathcal{L}^*}{\partial \dot{\mathbf{x}}}\right)_* = 0 \longrightarrow
$$

$$
\left(\frac{\partial \mathcal{H}}{\partial \mathbf{x}}\right)_* + \left(\frac{\partial^2 S}{\partial \mathbf{x}^2}\right)_*' \dot{\mathbf{x}}^*(t) + \left(\frac{\partial^2 S}{\partial \mathbf{x}\partial t}\right)_* - \frac{d}{dt}\left\{\left(\frac{\partial S}{\partial \mathbf{x}}\right)_*' - \boldsymbol{\lambda}^*(t)\right\} = 0 \longrightarrow
$$

$$
\left(\frac{\partial \mathcal{H}}{\partial \mathbf{x}}\right)_* + \left(\frac{\partial^2 S}{\partial \mathbf{x}^2}\right)_*' \dot{\mathbf{x}}^*(t) + \left(\frac{\partial^2 S}{\partial \mathbf{x}\partial t}\right)_* - \left[\left(\frac{\partial^2 S}{\partial \mathbf{x}^2}\right)_*' \dot{\mathbf{x}}^*(t) + \left(\frac{\partial^2 S}{\partial \mathbf{x}\partial t}\right)_* - \dot{\boldsymbol{\lambda}}^*(t)\right] = 0
$$

leading to

$$\boxed{\left(\frac{\partial \mathcal{H}}{\partial \mathbf{x}}\right)_* = -\dot{\boldsymbol{\lambda}}^*(t)} \tag{2.7.30}$$

and for the costate $\boldsymbol{\lambda}^*(t)$,

$$\left(\frac{\partial \mathcal{L}}{\partial \boldsymbol{\lambda}}\right)_* = 0 \longrightarrow \boxed{\left(\frac{\partial \mathcal{H}}{\partial \boldsymbol{\lambda}}\right)_* = \dot{\mathbf{x}}^*(t).} \tag{2.7.31}$$

Looking at the similar structure of the relation (2.7.30) for the optimal costate $\boldsymbol{\lambda}^*(t)$ and (2.7.31) for the optimal state $\mathbf{x}^*(t)$ it is clear why $\lambda(t)$ is called the *costate* vector. Finally, using the relation (2.7.28), the boundary condition (2.7.26) at the optimal condition reduces to

$$\left[\mathcal{H}^* + \frac{\partial S}{\partial t}\right]_{t_f} \delta t_f + \left[\left(\frac{\partial S}{\partial \mathbf{x}}\right)_* - \lambda^*(t)\right]_{t_f}' \delta \mathbf{x}_f = 0.$$

(2.7.32)

This is the general boundary condition for free-end point system in terms of the Hamiltonian.

2.7.2   Different Types of Systems

We now obtain different cases depending on the statement of the problem regarding the final time t_f and the final state $\mathbf{x}(t_f)$ (see Figure 2.9).

- **Type (a):** *Fixed-Final Time and Fixed-Final State System:* Here, since t_f and $\mathbf{x}(t_f)$ are fixed or specified (Figure 2.9(a)), both δt_f and $\delta \mathbf{x}_f$ are zero in the general boundary condition (2.7.32), and there is no extra boundary condition to be used other than those given in the problem formulation.

- **Type (b):** *Free-Final Time and Fixed-Final State System:* Since t_f is free or not specified in advance, δt_f is arbitrary, and since $\mathbf{x}(t_f)$ is fixed or specified, $\delta \mathbf{x}_f$ is zero as shown in Figure 2.9(b). Then, the coefficient of the arbitrary δt_f in the general boundary condition (2.7.32) is zero resulting in

$$\left(\mathcal{H} + \frac{\partial S}{\partial t}\right)_{*t_f} = 0.$$

(2.7.33)

- **Type (c):** *Fixed-Final Time and Free-Final State System:* Here t_f is specified and $\mathbf{x}(t_f)$ is free (see Figure 2.9(c)). Then δt_f is zero and $\delta \mathbf{x}_f$ is arbitrary, which in turn means that the coefficient of $\delta \mathbf{x}_f$ in the general boundary condition (2.7.32) is zero. That is

$$\left(\frac{\partial S}{\partial \mathbf{x}} - \lambda^*(t)\right)_{*t_f} = 0 \longrightarrow \lambda^*(t_f) = \left(\frac{\partial S}{\partial \mathbf{x}}\right)_{*t_f}.$$

(2.7.34)

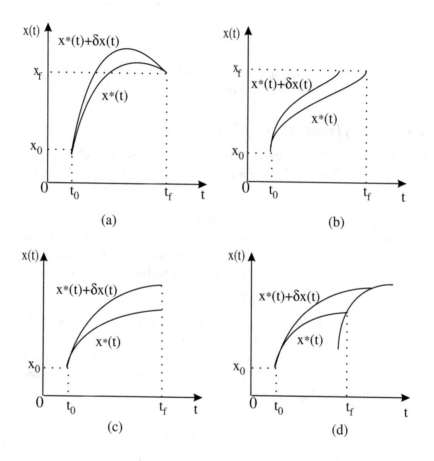

Figure 2.9 Different Types of Systems: (a) Fixed-Final Time and Fixed-Final State System, (b) Free-Final Time and Fixed-Final State System, (c) Fixed-Final Time and Free-Final State System, (d) Free-Final Time and Free-Final State System

- **Type (d):** *Free-Final Time and Dependent Free-Final State System:* If t_f and $\mathbf{x}(t_f)$ are *related* such that $\mathbf{x}(t_f)$ lies on a moving curve $\boldsymbol{\theta}(t)$ as shown in Figure 2.8, then

$$\mathbf{x}(t_f) = \boldsymbol{\theta}(t_f) \quad \text{and} \quad \delta\mathbf{x}_f \approx \dot{\boldsymbol{\theta}}(t_f)\delta t_f. \quad (2.7.35)$$

Using (2.7.35), the boundary condition (2.7.32) for the optimal condition becomes

$$\left[\left(\mathcal{H} + \frac{\partial S}{\partial t}\right)_* + \left(\frac{\partial S}{\partial \mathbf{x}} - \boldsymbol{\lambda}^*(t)\right)'_* \dot{\boldsymbol{\theta}}(t)\right]_{t_f} \delta t_f = 0. \quad (2.7.36)$$

Since t_f is free, δt_f is arbitrary and hence the coefficient of δt_f in (2.7.36) is zero. That is

$$\boxed{\left[\left(\mathcal{H} + \frac{\partial S}{\partial t}\right)_* + \left(\frac{\partial S}{\partial \mathbf{x}} - \boldsymbol{\lambda}^*(t)\right)'_* \dot{\boldsymbol{\theta}}(t)\right]_{t_f} = 0.} \quad (2.7.37)$$

- **Type (e):** *Free-Final Time and Independent Free-Final State:* If t_f and $\mathbf{x}(t_f)$ are *not related*, then δt_f and $\delta\mathbf{x}_f$ are unrelated, and the boundary condition (2.7.32) at the optimal condition becomes

$$\left(\mathcal{H} + \frac{\partial S}{\partial t}\right)_{*t_f} = 0 \quad (2.7.38)$$

$$\left(\frac{\partial S}{\partial \mathbf{x}} - \boldsymbol{\lambda}^*(t)\right)_{*t_f} = 0. \quad (2.7.39)$$

2.7.3   Sufficient Condition

In order to determine the nature of optimization, i.e., whether it is minimum or maximum, we need to consider the second variation and examine its sign. In other words, we have to find a *sufficient* condition for extremum. Using (2.7.14), (2.7.28) and (2.7.37), we have the second variation in (2.7.16) and using the relation (2.7.28), we get

$$\delta^2 J = \int_{t_0}^{t_f} \left[\frac{\partial^2 \mathcal{H}}{\partial \mathbf{x}^2} (\delta \mathbf{x}(t))^2 + \frac{\partial^2 \mathcal{H}}{\partial \mathbf{u}^2} (\delta \mathbf{u}(t))^2 + 2 \frac{\partial^2 \mathcal{H}}{\partial \mathbf{u} \partial \mathbf{x}} (\delta \mathbf{u}(t) \delta \mathbf{x}(t)) \right]_* dt$$

$$= \int_{t_0}^{t_f} \left[\delta \mathbf{x}'(t) \; \delta \mathbf{u}'(t) \right] \begin{bmatrix} \frac{\partial^2 \mathcal{H}}{\partial \mathbf{x}^2} & \frac{\partial^2 \mathcal{H}}{\partial \mathbf{x} \partial \mathbf{u}} \\ \frac{\partial^2 \mathcal{H}}{\partial \mathbf{x} \partial \mathbf{u}} & \frac{\partial^2 \mathcal{H}}{\partial \mathbf{u}^2} \end{bmatrix}_* \begin{bmatrix} \delta \mathbf{x}(t) \\ \delta \mathbf{u}(t) \end{bmatrix} dt$$

$$= \int_{t_0}^{t_f} \left[\delta \mathbf{x}'(t) \; \delta \mathbf{u}'(t) \right] \Pi \begin{bmatrix} \delta \mathbf{x}(t) \\ \delta \mathbf{u}(t) \end{bmatrix} dt. \qquad (2.7.40)$$

For the *minimum*, the second variation $\delta^2 J$ must be *positive*. This means that the matrix Π in (2.7.40)

$$\Pi = \begin{bmatrix} \frac{\partial^2 \mathcal{H}}{\partial \mathbf{x}^2} & \frac{\partial^2 \mathcal{H}}{\partial \mathbf{x} \partial \mathbf{u}} \\ \frac{\partial^2 \mathcal{H}}{\partial \mathbf{x} \partial \mathbf{u}} & \frac{\partial^2 \mathcal{H}}{\partial \mathbf{u}^2} \end{bmatrix}_* \qquad (2.7.41)$$

must be *positive definite*. But the important condition is that the second partial derivative of $\mathcal{H}^*$ w.r.t. $\mathbf{u}(t)$ must be positive. That is

$$\boxed{ \left(\frac{\partial^2 \mathcal{H}}{\partial \mathbf{u}^2} \right)_* > 0 } \qquad (2.7.42)$$

and for the *maximum*, the sign of (2.7.42) is reversed.

2.7.4 Summary of Pontryagin Procedure

Consider a free-final time and free-final state problem with general cost function (Bolza problem), where we want to minimize the performance index

$$J = S(\mathbf{x}(t_f), t_f) + \int_{t_0}^{t_f} V(\mathbf{x}(t), \mathbf{u}(t), t) dt \qquad (2.7.43)$$

for the plant described by

$$\dot{\mathbf{x}}(t) = \mathbf{f}(\mathbf{x}(t), \mathbf{u}(t), t) \qquad (2.7.44)$$

with the boundary conditions as

$$\mathbf{x}(t = t_0) = \mathbf{x}_0; \quad t = t_f \text{ is free and } \mathbf{x}(t_f) \text{ is free.} \qquad (2.7.45)$$

Table 2.1 Procedure Summary of Pontryagin Principle for Bolza Problem

A. Statement of the Problem
Given the plant as $\dot{\mathbf{x}}(t) = \mathbf{f}(\mathbf{x}(t), \mathbf{u}(t), t)$, the performance index as $J = S(\mathbf{x}(t_f), t_f) + \int_{t_0}^{t_f} V(\mathbf{x}(t), \mathbf{u}(t), t)dt$, and the boundary conditions as $\mathbf{x}(t_0) = \mathbf{x}_0$ and t_f and $\mathbf{x}(t_f) = \mathbf{x}_f$ are free, find the optimal control.

B. Solution of the Problem	
Step 1	Form the Pontryagin $\mathcal{H}$ function $\mathcal{H}(\mathbf{x}(t), \mathbf{u}(t), \boldsymbol{\lambda}(t), t) = V(\mathbf{x}(t), \mathbf{u}(t), t) + \boldsymbol{\lambda}'(t)\mathbf{f}(\mathbf{x}(t), \mathbf{u}(t), t)$.
Step 2	Minimize $\mathcal{H}$ w.r.t. $\mathbf{u}(t)$ $\left(\frac{\partial \mathcal{H}}{\partial \mathbf{u}}\right)_* = 0$ and obtain $\mathbf{u}^*(t) = \mathbf{h}(\mathbf{x}^*(t), \boldsymbol{\lambda}^*(t), t)$.
Step 3	Using the results of Step 2 in Step 1, find the optimal $\mathcal{H}^*$ $\mathcal{H}^*(\mathbf{x}^*(t), \mathbf{h}(\mathbf{x}^*(t), \boldsymbol{\lambda}^*(t), t), \boldsymbol{\lambda}^*(t), t) = \mathcal{H}^*(\mathbf{x}^*(t), \boldsymbol{\lambda}^*(t), t)$.
Step 4	Solve the set of 2n differential equations $\dot{\mathbf{x}}^*(t) = +\left(\frac{\partial \mathcal{H}}{\partial \boldsymbol{\lambda}}\right)_*$ and $\dot{\boldsymbol{\lambda}}^*(t) = -\left(\frac{\partial \mathcal{H}}{\partial \mathbf{x}}\right)_*$ with initial conditions $\mathbf{x}_0$ and the final conditions $\left[\mathcal{H}^* + \frac{\partial S}{\partial t}\right]_{t_f}\delta t_f + \left[\left(\frac{\partial S}{\partial \mathbf{x}}\right)_* - \boldsymbol{\lambda}^*(t)\right]'_{t_f}\delta \mathbf{x}_f = 0$.
Step 5	Substitute the solutions of $\mathbf{x}^*(t)$, $\boldsymbol{\lambda}^*(t)$ from Step 4 into the expression for the optimal control $\mathbf{u}^*(t)$ of Step 2.

C. Types of Systems
(a). Fixed-final time and fixed-final state system, Fig. 2.9(a)
(b). Free-final time and fixed-final state system, Fig. 2.9(b)
(c). Fixed-final time and free-final state system, Fig. 2.9(c)
(d). Free-final time and dependent free-final state system, Fig. 2.9(d).
(e). Free-final time and independent free-final state system

Type	Substitutions	Boundary Conditions
(a)	$\delta t_f = 0, \delta \mathbf{x}_f = 0$	$\mathbf{x}(t_0) = \mathbf{x}_0, \mathbf{x}(t_f) = \mathbf{x}_f$
(b)	$\delta t_f \neq 0, \delta \mathbf{x}_f = 0$	$\mathbf{x}(t_0) = \mathbf{x}_0, \mathbf{x}(t_f) = \mathbf{x}_f, \left[\mathcal{H}^* + \frac{\partial S}{\partial t}\right]_{t_f} = 0$
(c)	$\delta t_f = 0, \delta \mathbf{x}_f \neq 0$	$\mathbf{x}(t_0) = \mathbf{x}_0, \boldsymbol{\lambda}^*(t_f) = \left(\frac{\partial S}{\partial \mathbf{x}}\right)_{*t_f}$
(d)	$\delta \mathbf{x}_f = \dot{\boldsymbol{\theta}}(t_f)\delta t_f$	$\mathbf{x}(t_0) = \mathbf{x}_0, \mathbf{x}(t_f) = \boldsymbol{\theta}(t_f)$ $\left[\mathcal{H}^* + \frac{\partial S}{\partial t} + \left\{\left(\frac{\partial S}{\partial \mathbf{x}}\right)_* - \boldsymbol{\lambda}^*(t)\right\}' \dot{\boldsymbol{\theta}}(t)\right]_{t_f} = 0$
(e)	$\delta t_f \neq 0$ $\delta \mathbf{x}_f \neq 0$	$\delta \mathbf{x}(t_0) = \mathbf{x}_0$ $\left[\mathcal{H}^* + \frac{\partial S}{\partial t}\right]_{t_f} = 0, \left[\left(\frac{\partial S}{\partial \mathbf{x}}\right)_* - \boldsymbol{\lambda}^*(t)\right]_{t_f} = 0$

Here, $\mathbf{x}(t)$ and $\mathbf{u}(t)$ are $n-$ and $r-$ dimensional state and control vectors respectively. Let us note that $\mathbf{u}(t)$ is *unconstrained*. The entire procedure (called Pontryagin Principle) is now summarized in Table 2.1.

Note: From Table 2.1 we note that the only difference in the procedure between the *free-final point system without the final cost function* (Lagrange problem) and *free-final point system with final cost function* (Bolza problem) is in the application of the general boundary condition.

   To illustrate the Pontryagin method described previously, consider the following simple examples describing a second order system. Specifically, we selected a double integral plant whose analytical solutions for the optimal condition can be obtained and the same verified by using MATLAB©.

   First we consider the fixed-final time and fixed-final state problem (Figure 2.9(a), Table 2.1, Type (a)).

Example 2.12

Given a second order (double integrator) system as

$$\dot{x}_1(t) = x_2(t)$$
$$\dot{x}_2(t) = u(t) \qquad\qquad (2.7.46)$$

and the performance index as

$$J = \frac{1}{2}\int_{t_0}^{t_f} u^2(t)dt \qquad\qquad (2.7.47)$$

find the optimal control and optimal state, given the boundary (initial and final) conditions as

$$\mathbf{x}(0) = [1 \quad 2]'; \quad \mathbf{x}(2) = [1 \quad 0]'. \qquad\qquad (2.7.48)$$

Assume that the control and state are unconstrained.

Solution: We follow the step-by-step procedure given in Table 2.1. First, by comparing the present plant (2.7.46) and the PI (2.7.47) with the general formulation of the plant (2.7.1) and the PI (2.7.2), we identify

$$V(\mathbf{x}(t), \mathbf{u}(t), t) = V(u(t)) = \frac{1}{2}u^2(t)$$
$$\mathbf{f}(\mathbf{x}(t), \mathbf{u}(t), t) = [f_1, \ f_2]', \qquad\qquad (2.7.49)$$

where, $f_1 = x_2(t), \quad f_2 = u(t)$.

- **Step 1:** Form the Hamiltonian function as

$$\mathcal{H} = \mathcal{H}(x_1(t), x_2(t), u(t), \lambda_1(t), \lambda_2(t))$$
$$= V(u(t)) + \boldsymbol{\lambda}'(t)\mathbf{f}(\mathbf{x}(t), \mathbf{u}(t))$$
$$= \frac{1}{2}u^2(t) + \lambda_1(t)x_2(t) + \lambda_2(t)u(t). \tag{2.7.50}$$

- **Step 2:** Find $u^*(t)$ from

$$\frac{\partial \mathcal{H}}{\partial u} = 0 \longrightarrow u^*(t) + \lambda_2^*(t) = 0 \longrightarrow u^*(t) = -\lambda_2^*(t). \tag{2.7.51}$$

- **Step 3:** Using the results of Step 2 in Step 1, find the optimal $\mathcal{H}^*$ as

$$\mathcal{H}^*(x_1^*(t), x_2^*(t), \lambda_1^*(t), \lambda_2^*(t)) = \frac{1}{2}\lambda_2^{*2}(t) + \lambda_1^*(t)x_2^*(t) - \lambda_2^{*2}(t)$$
$$= \lambda_1^*(t)x_2^*(t) - \frac{1}{2}\lambda_2^{*2}(t). \tag{2.7.52}$$

- **Step 4:** Obtain the state and costate equations from

$$\dot{x}_1^*(t) = +\left(\frac{\partial \mathcal{H}}{\partial \lambda_1}\right)_* = x_2^*(t)$$
$$\dot{x}_2^*(t) = +\left(\frac{\partial \mathcal{H}}{\partial \lambda_2}\right)_* = -\lambda_2^*(t)$$
$$\dot{\lambda}_1^*(t) = -\left(\frac{\partial \mathcal{H}}{\partial x_1}\right)_* = 0$$
$$\dot{\lambda}_2^*(t) = -\left(\frac{\partial \mathcal{H}}{\partial x_2}\right)_* = -\lambda_1^*(t). \tag{2.7.53}$$

Solving the previous equations, we have the optimal state and costate as

$$x_1^*(t) = \frac{C_3}{6}t^3 - \frac{C_4}{2}t^2 + C_2 t + C_1$$
$$x_2^*(t) = \frac{C_3}{2}t^2 - C_4 t + C_2$$
$$\lambda_1^*(t) = C_3$$
$$\lambda_2^*(t) = -C_3 t + C_4. \tag{2.7.54}$$

- **Step 5:** Obtain the optimal control from

$$u^*(t) = -\lambda_2^*(t) = C_3 t - C_4 \tag{2.7.55}$$

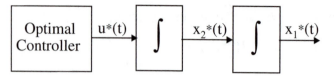

Figure 2.10   Optimal Controller for Example 2.12

where, C_1, C_2, C_3, and C_4 are constants evaluated using the given boundary conditions (2.7.48). These are found to be

$$C_1 = 1, \quad C_2 = 2, \quad C_3 = 3, \quad \text{and} \quad C_4 = 4. \quad (2.7.56)$$

Finally, we have the optimal states, costates and control as

$$x_1^*(t) = 0.5t^3 - 2t^2 + 2t + 1,$$
$$x_2^*(t) = 1.5t^2 - 4t + 2,$$
$$\lambda_1^*(t) = 3,$$
$$\lambda_2^*(t) = -3t + 4,$$
$$u^*(t) = 3t - 4. \quad (2.7.57)$$

The system with the optimal controller is shown in Figure 2.10.

The solution for the set of differential equations (2.7.53) with the boundary conditions (2.7.48) for Example 2.12 using Symbolic Toolbox of the MATLAB©, Version 6, is shown below.

```
************************************************************
%% Solution Using Symbolic Toolbox (STB) in
%% MATLAB Version 6.0
%%
S=dsolve('Dx1=x2,Dx2=-lambda2,Dlambda1=0,Dlambda2=-lambda1,...
x1(0)=1,x2(0)=2,x1(2)=1,x2(2)=0')
S.x1
S.x2
S.lambda1
S.lambda2

S =

     lambda1: [1x1 sym]
     lambda2: [1x1 sym]
          x1: [1x1 sym]
          x2: [1x1 sym]
S.x1

ans =
```

```
1+2*t-2*t^2+1/2*t^3

S.x2

ans =
2-4*t+3/2*t^2

S.lambda1

ans =
3
S.lambda2

ans =

4-3*t

Plot command is used for which we need to
%% convert the symbolic values to numerical values.
j=1;
for tp=0:.02:2
t=sym(tp);
x1p(j)=double(subs(S.x1));
%% subs substitutes S.x1 to x1p
x2p(j)=double(subs(S.x2));
%% double converts symbolic to numeric
up(j)=-double(subs(S.lambda2));
%% optimal control u = -lambda_2
t1(j)=tp;
j=j+1;
end
plot(t1,x1p,'k',t1,x2p,'k',t1,up,'k:')
xlabel('t')
gtext('x_1(t)')
gtext('x_2(t)')
gtext('u(t)')
**********************************************************
```

It is easy to see that the previous solutions for $x_1^*(t), x_2^*(t), \lambda_1^*(t),$ $\lambda_2^*(t),$ and $u^*(t) = -\lambda_2^*(t)$ obtained by using MATLAB© are the same as those given by the analytical solutions (2.7.57). The optimal control and state are plotted (using MATLAB©) in Figure 2.11.

Next, we consider the fixed-final time and free-final state case (Fig-

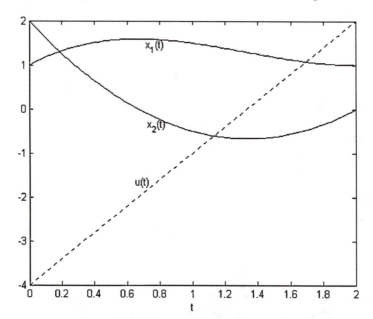

Figure 2.11   Optimal Control and States for Example 2.12

ure 2.9(b), Table 2.1, Type (c)) of the same system.

Example 2.13

Consider the same Example 2.12 with changed boundary conditions
as

$$\mathbf{x}(0) = [1 \ 2]'; \quad x_1(2) = 0; \quad x_2(2) \text{ is free.} \qquad (2.7.58)$$

Find the optimal control and optimal states.

Solution: Following the procedure illustrated in Table 2.1 (Type
(c)), we get the same optimal states, costates, and control as given
in (2.7.54) and (2.7.55) which are repeated here for convenience.

$$x_1^*(t) = \frac{C_3}{6}t^3 - \frac{C_4}{2}t^2 + C_2t + C_1,$$

$$x_2^*(t) = \frac{C_3}{2}t^2 - C_4t + C_2,$$

$$\lambda_1^*(t) = C_3,$$

$$\lambda_2^*(t) = -C_3t + C_4,$$

$$u^*(t) = -\lambda_2^*(t) = C_3t - C_4. \qquad (2.7.59)$$

The only difference is in solving for the constants C_1 to C_4. First
of all, note that the performance index (2.7.47) does not contain
the terminal cost function S. From the given boundary conditions

(2.7.58), we have t_f specified to be 2 and hence δt_f is zero in the general boundary condition (2.7.32).

Also, since $x_2(2)$ is free, δx_{2_f} is arbitrary and hence the corresponding final condition on the costate becomes

$$\lambda_2(t_f) = \left(\frac{\partial S}{\partial x_2}\right)_{*t_f} = 0 \qquad (2.7.60)$$

(since $S = 0$). Thus we have the four boundary conditions as

$$x_1(0) = 1; \quad x_2(0) = 2; \quad x_1(2) = 0; \quad \lambda_2(2) = 0. \quad (2.7.61)$$

With these boundary conditions substituted in (2.7.59), the constants are found to be

$$C_1 = 1; \quad C_2 = 2; \quad C_3 = 15/8; \quad C_4 = 15/4. \qquad (2.7.62)$$

Finally the optimal states, costates and control are given from (2.7.59) and (2.7.62) as

$$x_1^*(t) = \frac{5}{16}t^3 - \frac{15}{8}t^2 + 2t + 1,$$

$$x_2^*(t) = \frac{15}{16}t^2 - \frac{15}{4}t + 2,$$

$$\lambda_1^*(t) = \frac{15}{8},$$

$$\lambda_2^*(t) = -\frac{15}{8}t + \frac{15}{4},$$

$$u^*(t) = \frac{15}{8}t - \frac{15}{4}.$$

$$(2.7.63)$$

The solution for the set of differential equations (2.7.53) with the boundary conditions (2.7.58) for Example 2.13 using Symbolic Toolbox of the MATLAB©, Version 6, is shown below.

```
********************************************************************
%% Solution Using Symbolic Toolbox (STB) in
%% MATLAB Version 6.0
%%
S=dsolve('Dx1=x2,Dx2=-lambda2,Dlambda1=0,Dlambda2=-lambda1,
x1(0)=1,x2(0)=2,x1(2)=0,lambda2(2)=0')

S =
```

```
      lambda1: [1x1 sym]
      lambda2: [1x1 sym]
           x1: [1x1 sym]
           x2: [1x1 sym]
S.x1

ans =

5/16*t^3+2*t+1-15/8*t^2

S.x2

ans =

15/16*t^2+2-15/4*t

S.lambda1

ans =

15/8

S.lambda2

ans =

-15/8*t+15/4

%% Plot command is used for which we need to
%% convert the symbolic values to numerical values.
j=1;
for tp=0:.02:2
t=sym(tp);
x1p(j)=double(subs(S.x1));
%% subs substitutes S.x1 to x1p
x2p(j)=double(subs(S.x2));
%% double converts symbolic to numeric
up(j)=-double(subs(S.lambda2));
%% optimal control u = -lambda_2
t1(j)=tp;
j=j+1;
end
plot(t1,x1p,'k',t1,x2p,'k',t1,up,'k:')
xlabel('t')
gtext('x_1(t)')
```

```
gtext('x_2(t)')
gtext('u(t)')
```
****

It is easy to see that the previous solutions for $x_1^*(t), x_2^*(t), \lambda_1^*(t), \lambda_2^*(t)$, and $u^*(t) = -\lambda_2^*(t)$ obtained by using MATLAB© are the same as those given by (2.7.63) obtained analytically. The optimal control and states for Example 2.13 are plotted in Figure 2.12.

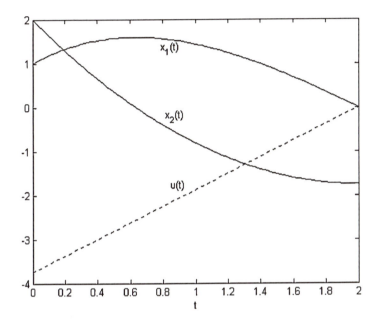

Figure 2.12   Optimal Control and States for Example 2.13

Next, we consider the free-final time and independent free-final state case (Figure 2.9(e), Table 2.1, Type (e)) of the same system.

Example 2.14

Consider the same Example 2.12 with changed boundary conditions as

$$\mathbf{x}(0) = [1 \ 2]'; \quad x_1(2) = 0; \quad x_1(t_f) = 3; \quad x_2(t_f) \text{ is free.} \quad (2.7.64)$$

Find the optimal control and optimal state.

Solution: Following the procedure illustrated in Table 2.1 (Type (e)), we get the same optimal control, states and costates as given in (2.7.54) and (2.7.55) which are repeated here for convenience.

$$x_1^*(t) = \frac{C_3}{6}t^3 - \frac{C_4}{2}t^2 + C_2t + C_1,$$

$$x_2^*(t) = \frac{C_3}{2}t^2 - C_4t + C_2,$$

$$\lambda_1^*(t) = C_3,$$

$$\lambda_2^*(t) = -C_3t + C_4,$$

$$u^*(t) = -\lambda_2^*(t) = C_3t - C_4. \qquad (2.7.65)$$

The only difference is in solving for the constants C_1 to C_4 and the unknown t_f. First of all, note that the performance index (2.7.47) does not contain the terminal cost function S, that is, $S = 0$. From the given boundary conditions (2.7.64), we have t_f unspecified and hence δt_f is free in the general boundary condition (2.7.32) leading to the specific final condition

$$\left(\mathcal{H} + \frac{\partial S}{\partial t}\right)_{t_f} = 0 \longrightarrow \lambda_1(t_f)x_2(t_f) - 0.5\lambda_2^2(t_f) = 0 \qquad (2.7.66)$$

Also, since $x_2(t_f)$ is free, δx_{2_f} is arbitrary and hence the general boundary condition (2.7.32) becomes

$$\lambda_2(t_f) = \left(\frac{\partial S}{\partial x_2}\right) = 0 \qquad (2.7.67)$$

where $\mathcal{H}$ is given by (2.7.52). Combining (2.7.64), (2.7.66) and (2.7.67), we have the following 5 boundary conditions for the 5 unknowns (4 constants of integration C_1 to C_4 and 1 unknown t_f) as

$$x_1(0) = 1; \quad x_2(0) = 2; \quad x_1(t_f) = 3;$$

$$\lambda_2(t_f) = 0; \quad \lambda_1(t_f)x_2(t_f) - 0.5\lambda_2^2(t_f) = 0. \qquad (2.7.68)$$

Using these boundary conditions along with (2.7.65), the constants are found to be

$$C_1 = 1; \quad C_2 = 2; \quad C_3 = 4/9; \quad C_4 = 4/3; \quad t_f = 3. \quad (2.7.69)$$

Finally, the optimal states, costates, and control are given from

(2.7.65) and (2.7.69) as

$$x_1^*(t) = \frac{4}{54}t^3 - \frac{2}{3}t^2 + 2t + 1,$$

$$x_2^*(t) = \frac{4}{18}t^2 - \frac{4}{3}t + 2,$$

$$\lambda_1^*(t) = \frac{4}{9},$$

$$\lambda_2^*(t) = -\frac{4}{9}t + \frac{4}{3},$$

$$u^*(t) = \frac{4}{9}t - \frac{4}{3}.$$

$$(2.7.70)$$

The solution for the set of differential equations (2.7.53) with the boundary conditions (2.7.68) for Example 2.14 using Symbolic Toolbox of the MATLAB©, Version 6 is shown below.

```
**********************************************************
%% Solution Using Symbolic Toolbox (STB) in
%% of MATLAB Version 6
%%
clear all
S=dsolve('Dx1=x2,Dx2=-lam2,Dlam1=0,Dlam2=-lam1,x1(0)=1,
    x2(0)=2,x1(tf)=3,lam2(tf)=0')
t='tf';
eq1=subs(S.x1)-'x1tf';
eq2=subs(S.x2)-'x2tf';
eq3=S.lam1-'lam1tf';
eq4=subs(S.lam2)-'lam2tf';
eq5='lam1tf*x2tf-0.5*lam2tf^2';
S2=solve(eq1,eq2,eq3,eq4,eq5,'tf,x1tf,x2tf,lam1tf,
    lam2tf','lam1tf<>0')
%% lam1tf<>0 means lam1tf is not equal to 0;
%% This is a condition derived from eq5.
%% Otherwise, without this condition in the above
%% SOLVE routine, we get two values for tf (1 and 3 in this case)
%%
tf=S2.tf
x1tf=S2.x1tf;
x2tf=S2.x2tf;
clear t
x1=subs(S.x1)
x2=subs(S.x2)
lam1=subs(S.lam1)
```

```
lam2=subs(S.lam2)
%% Convert the symbolic values to
%% numerical values as shown below.
j=1;
tf=double(subs(S2.tf))
%% coverts tf from symbolic to numerical
for tp=0:0.05:tf
t=sym(tp);
%% coverts tp from numerical to symbolic
x1p(j)=double(subs(S.x1));
%% subs substitutes S.x1 to x1p
x2p(j)=double(subs(S.x2));
%% double converts symbolic to numeric
up(j)=-double(subs(S.lam2));
%% optimal control u = -lambda_2
t1(j)=tp;
j=j+1;
end
plot(t1,x1p,'k',t1,x2p,'k',t1,up,'k:')
xlabel('t')
gtext('x_1(t)')
gtext('x_2(t)')
gtext('u(t)')
**********************************************************
```

The optimal control and states for Example 2.14 are plotted in Figure 2.13.

Finally, we consider the fixed-final time and free-final state system with a terminal cost (Figure 2.9 (b), Table 2.1, Type (b)).

Example 2.15

We consider the same Example 2.12 with changed performance index

$$J = \frac{1}{2}[x_1(2) - 4]^2 + \frac{1}{2}[x_2(2) - 2]^2 + \frac{1}{2}\int_0^2 u^2 dt \qquad (2.7.71)$$

and boundary conditions as

$$\mathbf{x}(0) = [1 \ \ 2]; \quad \mathbf{x}(2) = \text{is free}. \qquad (2.7.72)$$

Following the procedure illustrated in Table 2.1 (Type (b)), we get the same optimal control, states and costates as given in (2.7.54) and (2.7.55), which are reproduced here for ready reference. Thus

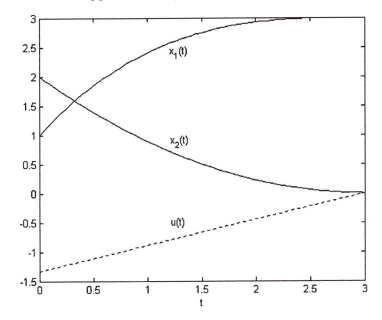

Figure 2.13   Optimal Control and States for Example 2.14

we have

$$x_1^*(t) = \frac{C_3}{6}t^3 - \frac{C_4}{2}t^2 + C_2t + C_1,$$

$$x_2^*(t) = \frac{C_3}{2}t^2 - C_4t + C_2,$$

$$\lambda_1^*(t) = C_3,$$

$$\lambda_2^*(t) = -C_3t + C_4,$$

$$u^*(t) = -\lambda_2(t) = C_3t - C_4. \qquad (2.7.73)$$

The only difference is in solving for the constants C_1 to C_4 using the given and obtained boundary conditions. Since t_f is specified as 2, δt_f is zero and since $\mathbf{x}(2)$ unspecified, $\delta\mathbf{x}_f$ is free in the boundary condition (2.7.32), which now reduces to

$$\lambda^*(t_f) = \left(\frac{\partial S}{\partial \mathbf{x}}\right)_{*t_f} \qquad (2.7.74)$$

where,

$$S(\mathbf{x}(t_f)) = \frac{1}{2}[x_1(2) - 4]^2 + \frac{1}{2}[x_2(2) - 2]^2. \qquad (2.7.75)$$

Thus, (2.7.74) becomes

$$\lambda_1^*(t_f) = \left(\frac{\partial S}{\partial x_1}\right)_{t_f} \longrightarrow \lambda_1^*(2) = x_1(2) - 4$$

$$\lambda_2^*(t_f) = \left(\frac{\partial S}{\partial x_2}\right)_{t_f} \longrightarrow \lambda_2^*(2) = x_2(2) - 2. \qquad (2.7.76)$$

Now, we have two initial conditions from (2.7.72) and two final conditions from (2.7.76) to solve for the four constants as

$$C_1 = 1, \quad C_2 = 2, \quad C_3 = \frac{3}{7}, \quad C_4 = \frac{4}{7}. \qquad (2.7.77)$$

Finally, we have the optimal states, costates and control given as

$$x_1^*(t) = \frac{1}{14}t^3 - \frac{2}{7}t^2 + 2t + 1,$$

$$x_2^*(t) = \frac{3}{14}t^2 - \frac{4}{7}t + 2,$$

$$\lambda_1^*(t) = \frac{3}{7},$$

$$\lambda_2^*(t) = -\frac{3}{7}t + \frac{4}{7},$$

$$u^*(t) = \frac{3}{7}t - \frac{4}{7}. \qquad (2.7.78)$$

The previous results can also obtained using Symbolic Math Toolbox of the MATLAB©, Version 6, as shown below.

```
****************************************************************
%% Solution Using Symbolic Math Toolbox (STB) in
%%   MATLAB Version 6
%%
S=dsolve('Dx1=x2,Dx2=-lambda2,Dlambda1=0,Dlambda2=-lambda1,
x1(0)=1,x2(0)=2,lambda1(2)=x12-4,lambda2(2)=x22-2')
t='2';
S2=solve(subs(S.x1)-'x12',subs(S.x2)-'x22','x12,x22');
%% solves for x1(t=2) and x2(t=2)
x12=S2.x12;
x22=S2.x22;
clear t

S =
```

```
        lambda1:  [1x1 sym]
        lambda2:  [1x1 sym]
            x1:  [1x1 sym]
            x2:  [1x1 sym]

x1=subs(S.x1)

x1 =

1-2/7*t^2+1/14*t^3+2*t


x2=subs(S.x2)

x2 =

-4/7*t+3/14*t^2+2

lambda1=subs(S.lambda1)

lambda1 =

3/7


lambda2=subs(S.lambda2)

lambda2 =

4/7-3/7*t

%% Plot command is used for which we need to
%% convert the symbolic values to numerical values.
j=1;
for tp=0:.02:2
t=sym(tp);
x1p(j)=double(subs(S.x1));
%% subs substitutes S.x1 to x1p
x2p(j)=double(subs(S.x2));
%% double converts symbolic to numeric
up(j)=-double(subs(S.lambda2));
%% optimal control u = -lambda_2
t1(j)=tp;
j=j+1;
end
```

```
plot(t1,x1p,'k',t1,x2p,'k',t1,up,'k:')
xlabel('t')
gtext('x_1(t)')
gtext('x_2(t)')
gtext('u(t)')
```
****

It is easy to see that the previous solutions for $x_1^*(t), x_2^*(t), \lambda_1^*(t), \lambda_2^*(t)$, and $u^*(t) = -\lambda_2^*(t)$ obtained by using MATLAB© are the same as those given by (2.7.78) obtained analytically.

The optimal control and states for Example 2.15 are plotted in Figure 2.14.

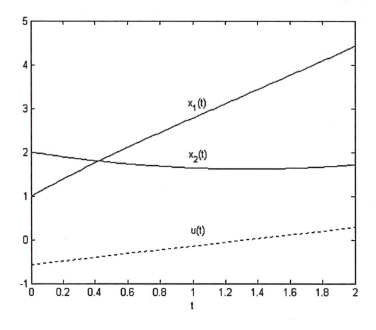

Figure 2.14   Optimal Control and States for Example 2.15

2.8   *Summary of Variational Approach*

In this section, we summarize the development of the topics covered so far in obtaining optimal conditions using the calculus of variations. The development is carried out in different stages as follows. Also shown is the systematic link between various stages of development.

2.8.1   Stage I: Optimization of a Functional

Consider the optimal of

$$J = \int_{t_0}^{t_f} V(\mathbf{x}(t), \dot{\mathbf{x}}(t), t)dt \qquad (2.8.1)$$

with the given boundary conditions

$$\mathbf{x}(t_0) \text{ fixed and } \mathbf{x}(t_f) \text{ free.} \qquad (2.8.2)$$

Then, the optimal function $\mathbf{x}^*(t)$ should satisfy the Euler-Lagrange equation

$$\boxed{\left(\frac{\partial V}{\partial \mathbf{x}}\right)_* - \frac{d}{dt}\left(\frac{\partial V}{\partial \dot{\mathbf{x}}}\right)_* = 0.} \qquad (2.8.3)$$

The general boundary condition to be satisfied at the free-final point is given by [79]

$$\boxed{\left[V - \dot{\mathbf{x}}'(t)\left(\frac{\partial V}{\partial \dot{\mathbf{x}}}\right)\right]_{*t_f} \delta t_f + \left(\frac{\partial V}{\partial \dot{\mathbf{x}}}\right)'_{*t_f} \delta \mathbf{x}_f = 0.} \qquad (2.8.4)$$

This boundary condition is to be modified depending on the nature of the given t_f and $\mathbf{x}(t_f)$. Although the previous general boundary condition is not derived in this book, it can be easily seen to be similar to the general boundary condition (2.7.26) in terms of the Lagrangian which embeds a performance index and a dynamical plant into a single augmented performance index with integrand $\mathcal{L}$.

The sufficient condition for optimum is the Legendre condition given by

$$\boxed{\left(\frac{\partial^2 V}{\partial \dot{\mathbf{x}}^2}\right)_* > 0} \text{ for minimum} \qquad (2.8.5)$$

and

$$\boxed{\left(\frac{\partial^2 V}{\partial \dot{\mathbf{x}}^2}\right)_* < 0} \text{ for maximum.} \qquad (2.8.6)$$

2.8.2 Stage II: Optimization of a Functional with Condition

Consider the optimization of a functional

$$J = \int_{t_0}^{t_f} V(\mathbf{x}(t), \dot{\mathbf{x}}(t), t) dt \tag{2.8.7}$$

with given boundary values as

$$\mathbf{x}(t_0) \text{ fixed and } \mathbf{x}(t_f) \text{ free}, \tag{2.8.8}$$

and the condition relation

$$\mathbf{g}(\mathbf{x}(t), \dot{\mathbf{x}}(t), t) = 0. \tag{2.8.9}$$

Here, the condition (2.8.9) is absorbed by forming the augmented functional

$$J_a = \int_{t_0}^{t_f} \mathcal{L}(\mathbf{x}(t), \dot{\mathbf{x}}(t), \boldsymbol{\lambda}(t), t) dt \tag{2.8.10}$$

where, $\boldsymbol{\lambda}(t)$ is the Lagrange multiplier (also called the costate function), and $\mathcal{L}$ is the Lagrangian given by

$$\boxed{\mathcal{L}(\mathbf{x}(t), \dot{\mathbf{x}}(t), \boldsymbol{\lambda}(t), t) = V(\mathbf{x}(t), \dot{\mathbf{x}}(t), t) + \boldsymbol{\lambda}'(t)\mathbf{g}(\mathbf{x}(t), \dot{\mathbf{x}}(t), t).}$$
$$\tag{2.8.11}$$

Now, we just *use the results of the previous Stage I* for the augmented functional (2.8.10) except its integrand is $\mathcal{L}$ instead of V. For optimal condition, we have the Euler-Lagrange equation (2.8.3) for the augmented functional (2.8.10) given in terms of $\mathbf{x}(t)$ and $\boldsymbol{\lambda}(t)$ as

$$\boxed{\left(\frac{\partial \mathcal{L}}{\partial \mathbf{x}}\right)_* - \frac{d}{dt}\left(\frac{\partial \mathcal{L}}{\partial \dot{\mathbf{x}}}\right)_* = 0} \quad state \text{ equation and} \tag{2.8.12}$$

$$\boxed{\left(\frac{\partial \mathcal{L}}{\partial \boldsymbol{\lambda}}\right)_* - \frac{d}{dt}\left(\frac{\partial \mathcal{L}}{\partial \dot{\boldsymbol{\lambda}}}\right)_* = 0} \quad costate \text{ equation.} \tag{2.8.13}$$

Let us note from (2.8.11) that the Lagrangian $\mathcal{L}$ is independent of $\dot{\boldsymbol{\lambda}}^*(t)$ and that the Euler-Lagrange equation (2.8.13) for the costate $\boldsymbol{\lambda}(t)$ is nothing but the constraint relation (2.8.9). The general boundary condition (2.8.4) to be satisfied at the free-final point (in terms of $\mathcal{L}$) is given by

$$\boxed{\left[\mathcal{L} - \dot{\mathbf{x}}'(t)\left(\frac{\partial \mathcal{L}}{\partial \dot{\mathbf{x}}}\right)\right]_{*t_f} \delta t_f + \left(\frac{\partial \mathcal{L}}{\partial \dot{\mathbf{x}}}\right)'_{*t_f} \delta \mathbf{x}_f = 0.} \qquad (2.8.14)$$

This boundary condition is to be modified depending on the nature of the given t_f and $\mathbf{x}(t_f)$.

2.8.3 Stage III: Optimal Control System with Lagrangian Formalism

Here, we consider the standard control system with a plant described by [56]

$$\dot{\mathbf{x}}(t) = \mathbf{f}(\mathbf{x}(t), \mathbf{u}(t), t), \qquad (2.8.15)$$

the given boundary conditions as

$$\mathbf{x}(t_0) \text{ is fixed and } \mathbf{x}(t_f) \text{ is free}, \qquad (2.8.16)$$

and the performance index as

$$J(\mathbf{u}(t)) = \int_{t_0}^{t_f} V(\mathbf{x}(t), \mathbf{u}(t), t) dt. \qquad (2.8.17)$$

Now, we rewrite the plant equation (2.8.15) as the condition relation (2.8.9) as

$$\mathbf{g}(\mathbf{x}(t), \dot{\mathbf{x}}(t), \mathbf{u}(t), t) = \mathbf{f}(\mathbf{x}(t), \mathbf{u}(t), t) - \dot{\mathbf{x}}(t) = 0. \qquad (2.8.18)$$

Then we form the augmented functional out of the performance index (2.8.17) and the condition relation (2.8.18) as

$$J_a(\mathbf{u}(t)) = \int_{t_0}^{t_f} \mathcal{L}(\mathbf{x}(t), \dot{\mathbf{x}}(t), \mathbf{u}(t), \boldsymbol{\lambda}(t), t) dt \qquad (2.8.19)$$

where, the Lagrangian $\mathcal{L}$ is given as

$$\mathcal{L} = \mathcal{L}(\mathbf{x}(t), \dot{\mathbf{x}}(t), \mathbf{u}(t), \boldsymbol{\lambda}(t), t)$$
$$= V(\mathbf{x}(t), \mathbf{u}(t), t) + \boldsymbol{\lambda}'(t)\{\mathbf{f}(\mathbf{x}(t), \mathbf{u}(t), t) - \dot{\mathbf{x}}(t)\}. \qquad (2.8.20)$$

Now we *just use the previous results of Stage II.* For optimal condition, we have the set of Euler-Lagrange equations (2.8.12) and (2.8.13) for the augmented functional (2.8.19) given in terms of $\mathbf{x}(t)$, $\boldsymbol{\lambda}(t)$, and $\mathbf{u}(t)$ as

$$\boxed{\left(\frac{\partial \mathcal{L}}{\partial \mathbf{x}}\right)_* - \frac{d}{dt}\left(\frac{\partial \mathcal{L}}{\partial \dot{\mathbf{x}}}\right)_* = 0} \quad \text{\textit{state} equation,} \qquad (2.8.21)$$

$$\boxed{\left(\frac{\partial \mathcal{L}}{\partial \boldsymbol{\lambda}}\right)_* - \frac{d}{dt}\left(\frac{\partial \mathcal{L}}{\partial \dot{\boldsymbol{\lambda}}}\right)_* = 0} \quad \text{\textit{costate} equation, and} \qquad (2.8.22)$$

$$\boxed{\left(\frac{\partial \mathcal{L}}{\partial \mathbf{u}}\right)_* - \frac{d}{dt}\left(\frac{\partial \mathcal{L}}{\partial \dot{\mathbf{u}}}\right)_* = 0} \quad \text{\textit{control} equation.} \qquad (2.8.23)$$

Note from (2.8.20) that the Lagrangian $\mathcal{L}$ is independent of $\dot{\boldsymbol{\lambda}}^*(t)$ and $\dot{\mathbf{u}}^*(t)$ and that the Euler-Lagrange equation (2.8.22) is the same as the constraint relation (2.8.18). The general boundary condition (2.8.14) to be satisfied at the free-final point becomes

$$\boxed{\left[\mathcal{L} - \dot{\mathbf{x}}'(t)\left(\frac{\partial \mathcal{L}}{\partial \dot{\mathbf{x}}}\right)\right]_{*t_f} \delta t_f + \left(\frac{\partial \mathcal{L}}{\partial \dot{\mathbf{x}}}\right)'_{*t_f} \delta \mathbf{x}_f = 0.} \qquad (2.8.24)$$

This boundary condition is to be modified depending on the nature of the given t_f and $\mathbf{x}(t_f)$.

2.8.4   Stage IV: Optimal Control System with Hamiltonian Formalism: Pontryagin Principle

Here, we just transform the previous Lagrangian formalism to Hamiltonian formalism by defining the Hamiltonian as [57]

$$\mathcal{H}(\mathbf{x}(t), \mathbf{u}(t), \boldsymbol{\lambda}(t), t) = V(\mathbf{x}(t), \mathbf{u}(t), t) + \boldsymbol{\lambda}'(t)\mathbf{f}(\mathbf{x}(t), \mathbf{u}(t), t) \quad (2.8.25)$$

which in terms of the Lagrangian (2.8.20) becomes

$$\mathcal{L}(\mathbf{x}(t), \dot{\mathbf{x}}(t), \mathbf{u}(t), \boldsymbol{\lambda}(t), t) = \mathcal{H}(\mathbf{x}(t), \mathbf{u}(t), \boldsymbol{\lambda}(t), t) - \boldsymbol{\lambda}'(t)\dot{\mathbf{x}}(t). \quad (2.8.26)$$

Now using (2.8.26), the set of Euler-Lagrange equations (2.8.21) to (2.8.23) which are in terms of the Lagrangian, are rewritten in terms of the Hamiltonian as

$$\left(\frac{\partial \mathcal{H}}{\partial \mathbf{x}}\right)_* - \frac{d}{dt}(-\boldsymbol{\lambda}^*) = 0 \tag{2.8.27}$$

$$\left(\frac{\partial \mathcal{H}}{\partial \boldsymbol{\lambda}}\right)_* - \dot{\mathbf{x}}^*(t) - \frac{d}{dt}(0) = 0 \tag{2.8.28}$$

$$\left(\frac{\partial \mathcal{H}}{\partial \mathbf{u}}\right)_* - \frac{d}{dt}(0) = 0 \tag{2.8.29}$$

which in turn are rewritten as

$$\boxed{\dot{\mathbf{x}}^*(t) = +\left(\frac{\partial \mathcal{H}}{\partial \boldsymbol{\lambda}}\right)_*} \quad \text{state equation,} \tag{2.8.30}$$

$$\boxed{\dot{\boldsymbol{\lambda}}^*(t) = -\left(\frac{\partial \mathcal{H}}{\partial \mathbf{x}}\right)_*} \quad \text{costate equation, and} \tag{2.8.31}$$

$$\boxed{0 = +\left(\frac{\partial \mathcal{H}}{\partial \mathbf{u}}\right)_*} \quad \text{control equation.} \tag{2.8.32}$$

Similarly using (2.8.26), the boundary condition (2.8.24) is rewritten in terms of the Hamiltonian as

$$[\mathcal{H} - \boldsymbol{\lambda}'(t)\dot{\mathbf{x}}(t) - \dot{\mathbf{x}}'(t)(-\boldsymbol{\lambda}(t))]\big|_{*t_f} \, \delta t_f + [-\boldsymbol{\lambda}'(t)]\big|_{*t_f} \, \delta \mathbf{x}_f = 0 \tag{2.8.33}$$

which becomes

$$\boxed{\mathcal{H}\big|_{*t_f} \, \delta t_f - \boldsymbol{\lambda}^{*\prime}(t_f)\delta \mathbf{x}_f = 0.} \tag{2.8.34}$$

The sufficient condition is

$$\boxed{\left(\frac{\partial^2 \mathcal{H}}{\partial \mathbf{u}^2}\right)_* > 0} \quad \text{for minimum and} \tag{2.8.35}$$

$$\boxed{\left(\frac{\partial^2 \mathcal{H}}{\partial \mathbf{u}^2}\right)_* < 0} \quad \text{for maximum.} \tag{2.8.36}$$

The state, costate, and control equations (2.8.30) to (2.8.32) are solved along with the given initial condition (2.8.16) and the final condition (2.8.34) leading us to a two-point boundary value problem (TPBVP).

Free-Final Point System with Final Cost Function

This problem is an extension of the problem in Stage IV, with the addition of final cost function. We summarize the result risking the repetition of some of the equations. Let the plant be described as

$$\dot{\mathbf{x}}(t) = \mathbf{f}(\mathbf{x}(t), \mathbf{u}(t), t) \tag{2.8.37}$$

and the performance index be

$$J(\mathbf{u}(t)) = S(\mathbf{x}(t_f), t_f) + \int_{t_0}^{t_f} V(\mathbf{x}(t), \mathbf{u}(t), t)dt \tag{2.8.38}$$

along with the boundary conditions

$$\mathbf{x}(t_0) \text{ is fixed and } \mathbf{x}(t_f) \text{ is free.} \tag{2.8.39}$$

Now, if we rewrite the performance index (2.8.38) to absorb the final cost function S within the integrand, then the results of Stage III can be used to get the optimal conditions. Thus we rewrite (2.8.38) as

$$J_1(\mathbf{u}(t)) = \int_{t_0}^{t_f} \left[V(\mathbf{x}(t), \mathbf{u}(t), t) + \left(\frac{\partial S}{\partial \mathbf{x}}\right)' \dot{\mathbf{x}}(t) + \frac{\partial S}{\partial t} \right] dt. \tag{2.8.40}$$

Now we *repeat the results of Stage III* except for the modification of the final condition equation (2.8.34). Thus the state, costate and control equations are

$$\boxed{\dot{\mathbf{x}}^*(t) = + \left(\frac{\partial \mathcal{H}}{\partial \lambda}\right)_*} \quad \text{state equation} \tag{2.8.41}$$

$$\boxed{\dot{\lambda}^*(t) = - \left(\frac{\partial \mathcal{H}}{\partial \mathbf{x}}\right)_*} \quad \text{costate equation} \tag{2.8.42}$$

$$\boxed{0 = + \left(\frac{\partial \mathcal{H}}{\partial \mathbf{u}}\right)_*} \quad \text{control equation} \tag{2.8.43}$$

and the final boundary condition is

$$\boxed{\left[\mathcal{H} + \frac{\partial S}{\partial t}\right]_{*t_f} \delta t_f + \left[\left(\frac{\partial S}{\partial \mathbf{x}}\right) - \lambda(t)\right]'_{*t_f} \delta \mathbf{x}_f = 0.} \tag{2.8.44}$$

The sufficient condition for optimum is

$$\left(\frac{\partial^2 \mathcal{H}}{\partial \mathbf{u}^2}\right)_* > 0 \quad \text{for minimum and} \tag{2.8.45}$$

$$\left(\frac{\partial^2 \mathcal{H}}{\partial \mathbf{u}^2}\right)_* < 0 \quad \text{for maximum.} \tag{2.8.46}$$

The state, costate, and control equations (2.8.41) to (2.8.43) are solved along with the given initial condition (2.8.39) and the final condition (2.8.44), thus this formulation leads us to a TPBVP.

2.8.5   Salient Features

We now discuss the various features of the methodology used so far in obtaining the optimal conditions through the use of the calculus of variations [6, 79, 120, 108]. Also, we need to consider the problems discussed above under the various stages of development. So we refer to the appropriate relations of, say Stage III or Stage IV during our discussion.

1. *Significance of Lagrange Multiplier:* The Lagrange multiplier $\lambda(t)$ is also called the costate (or adjoint) function.

   (a) The Lagrange multiplier $\lambda(t)$ is introduced to "take care of" the constraint relation imposed by the plant equation (2.8.15).
   (b) The costate variable $\lambda(t)$ enables us to use the Euler-Lagrange equation for each of the variables $\mathbf{x}(t)$ and $\mathbf{u}(t)$ separately as if they were *independent* of each other although they are *dependent* of each other as per the plant equation.

2. *Lagrangian and Hamiltonian:* We defined the Lagrangian and Hamiltonian as

$$\begin{aligned}
\mathcal{L} &= \mathcal{L}(\mathbf{x}(t), \dot{\mathbf{x}}(t), \lambda(t), \mathbf{u}(t), t) \\
&= V(\mathbf{x}(t), \mathbf{u}(t), t) \\
&\quad + \lambda'(t)\{\mathbf{f}(\mathbf{x}(t), \mathbf{u}(t), t) - \dot{\mathbf{x}}(t)\} \\
\mathcal{H} &= \mathcal{H}(\mathbf{x}(t), \mathbf{u}(t), \lambda(t), t) \\
&= V(\mathbf{x}(t), \mathbf{u}(t), t) \\
&\quad + \lambda'(t)\mathbf{f}(\mathbf{x}(t), \mathbf{u}(t), t).
\end{aligned}$$

(2.8.47)

(2.8.48)

In defining the Lagrangian and Hamiltonian we use extensively the *vector* notation, still it should be noted that these $\mathcal{L}$ and $\mathcal{H}$ are *scalar* functions only.

3. *Optimization of Hamiltonian*

   (a) The control equation (2.8.32) indicates the optimization of the Hamiltonian w.r.t. the control $\mathbf{u}(t)$. That is, the optimization of the original performance index (2.8.17), which is a *functional* subject to the plant equation (2.8.15), is equivalent to the optimization of the Hamiltonian *function* w.r.t. $\mathbf{u}(t)$. Thus, we "reduced" our original *functional* optimization problem to an ordinary *function* optimization problem.

   (b) We note that we assumed *unconstrained or unbounded* control $\mathbf{u}(t)$ and obtained the control relation $\partial \mathcal{H}/\partial \mathbf{u} = 0$.

   (c) If $\mathbf{u}(t)$ is *constrained or bounded* as being a member of the set $\mathbf{U}$, i.e., $\mathbf{u}(t) \in \mathbf{U}$, we can no longer take $\partial \mathcal{H}/\partial \mathbf{u} = 0$, since $\mathbf{u}(t)$, so calculated, may lie outside the range of the permissible region $\mathbf{U}$. In practice, the control $\mathbf{u}(t)$ is always *limited* by such things as saturation of amplifiers, speed of a motor, or thrust of a rocket. The constrained optimal control systems are discussed in Chapter 7.

   (d) Regardless of any constraints on $\mathbf{u}(t)$, Pontryagin had shown that $\mathbf{u}(t)$ must still be chosen to minimize the Hamiltonian. A rigorous proof of the fact that $\mathbf{u}(t)$ must be chosen to optimize $\mathcal{H}$ function is Pontryagin's most notable contribution to optimal control theory. For this reason, the approach is often called *Pontryagin Principle*. So in the case of *constrained* control, it is shown that

$$\boxed{\min_{\mathbf{u} \in \mathbf{U}} \mathcal{H}(\mathbf{x}^*(t), \boldsymbol{\lambda}^*(t), \mathbf{u}(t), t) = \mathcal{H}(\mathbf{x}^*(t), \boldsymbol{\lambda}^*(t), \mathbf{u}^*(t), t)}$$

(2.8.49)

or equivalently

$$\boxed{\mathcal{H}(\mathbf{x}^*(t), \boldsymbol{\lambda}^*(t), \mathbf{u}^*(t), t) \leq \mathcal{H}(\mathbf{x}^*(t), \boldsymbol{\lambda}^*(t), \mathbf{u}(t), t).}$$

(2.8.50)

4. *Pontryagin Maximum Principle:* Originally, Pontryagin used a slightly different performance index which is *maximized* rather

than *minimized* and hence it is called *Pontryagin Maximum Principle*. For this reason, the Hamiltonian is also sometimes called Pontryagin $\mathcal{H}$-function. Let us note that *minimization* of the performance index J is equivalent to the *maximization* of $-J$. Then, if we define the Hamiltonian as

$$\mathcal{H}(\mathbf{x}(t), \mathbf{u}(t), \boldsymbol{\lambda}(t), t) = -V(\mathbf{x}(t), \mathbf{u}(t), t) + \hat{\boldsymbol{\lambda}}'(t)\mathbf{f}(\mathbf{x}(t), \mathbf{u}(t), t)$$

$$(2.8.51)$$

we have *Maximum Principle*. Further discussion on Pontryagin Principle is given in Chapter 6.

5. *Hamiltonian at the Optimal Condition:* At the optimal condition the Hamiltonian can be written as

$$\mathcal{H}^* = \mathcal{H}^*(\mathbf{x}^*(t), \mathbf{u}^*(t), \boldsymbol{\lambda}^*(t), t)$$
$$\frac{d\mathcal{H}^*}{dt} = \frac{d\mathcal{H}^*}{dt}$$
$$= \left(\frac{\partial \mathcal{H}}{\partial \mathbf{x}}\right)'_* \dot{\mathbf{x}}^*(t) + \left(\frac{\partial \mathcal{H}}{\partial \boldsymbol{\lambda}}\right)'_* \dot{\boldsymbol{\lambda}}^*(t) + \left(\frac{\partial \mathcal{H}}{\partial \mathbf{u}}\right)'_* \dot{\mathbf{u}}^*(t)$$
$$+ \left(\frac{\partial \mathcal{H}}{\partial t}\right)_*.$$

$$(2.8.52)$$

Using the state, costate and control equations (2.8.30) to (2.8.32) in the previous equation, we get

$$\left(\frac{d\mathcal{H}}{dt}\right)_* = \left(\frac{\partial \mathcal{H}}{\partial t}\right)_*.$$

$$(2.8.53)$$

We observe that along the optimal trajectory, the *total* derivative of $\mathcal{H}$ w.r.t. time is the same as the *partial* derivative of $\mathcal{H}$ w.r.t. time. If $\mathcal{H}$ does not depend on t explicitly, then

$$\boxed{\left.\frac{d\mathcal{H}}{dt}\right|_* = 0}$$

$$(2.8.54)$$

and $\mathcal{H}$ is constant w.r.t. the time t along the optimal trajectory.

6. *Two-Point Boundary Value Problem (TPBVP):* As seen earlier, the optimal control problem of a dynamical system leads to a TPBVP.

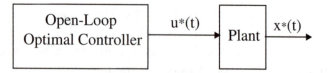

Figure 2.15   Open-Loop Optimal Control

(a) The state and costate equations (2.8.30) and (2.8.32) are solved using the initial and final conditions. In general, these are nonlinear, time varying and we may have to resort to numerical methods for their solutions.

(b) We note that the state and costate equations are the same for any kind of boundary conditions.

(c) For the optimal control system, although obtaining the state and costate equations is very easy, the computation of their solutions is quite tedious. This is the unfortunate feature of optimal control theory. It is the *price* one must pay for demanding the *best* performance from a system. One has to weigh the optimization of the system against the computational burden.

7. *Open-Loop Optimal Control:* In solving the TPBVP arising due to the state and costate equations, and then substituting in the control equation, we get only the open-loop optimal control as shown in Figure 2.15. Here, one has to construct or realize an open-loop optimal controller (OLOC) and in many cases it is very tedious. Also, changes in plant parameters are not taken into account by the OLOC. This prompts us to think in terms of a closed-loop optimal control (CLOC), i.e., to obtain optimal control $\mathbf{u}^*(t)$ in terms of the state $\mathbf{x}^*(t)$ as shown in Figure 2.16. This CLOC will have many advantages such as sensitive to plant parameter variations and simplified construction of the controller. The closed-loop optimal control systems are discussed in Chapter 7.

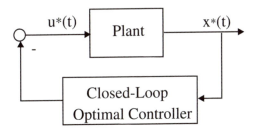

Figure 2.16   Closed-Loop Optimal Control

2.9 Problems

1. Make reasonable assumptions wherever necessary.

2. Use MATLAB© wherever possible to solve the problems and plot all the optimal controls and states for all problems. Provide the relevant MATLAB© m files.

Problem 2.1 Find the extremal of the following functional

$$J = \int_0^2 [2x^2(t) + \dot{x}^2(t)]dt$$

with the initial condition as $x(0) = 0$ and the final condition as $x(2) = 5$.

Problem 2.2 Find the extremal of the functional

$$J = \int_{-2}^0 \left[12tx(t) + \dot{x}^2(t) \right] dt$$

to satisfy the boundary conditions $x(-2) = 3$, and $x(0) = 0$.

Problem 2.3 Find the extremal for the following functional

$$J = \int_1^2 \frac{\dot{x}^2(t)}{2t^3} dt$$

with $x(1) = 1$ and $x(2) = 10$.

Problem 2.4 Consider the extremization of a functional which is dependent on derivatives higher than the first derivative $\dot{x}(t)$ such as

$$J(x(t), t) = \int_{t_0}^{t_f} V(x(t), \dot{x}(t), \ddot{x}(t), t)dt.$$

with fixed-end point conditions. Show that the corresponding Euler-Lagrange equation is given by

$$\frac{\partial V}{\partial x} - \frac{d}{dt}\left(\frac{\partial V}{\partial \dot{x}}\right) + \frac{d^2}{dt^2}\left(\frac{\partial V}{\partial \ddot{x}}\right) = 0.$$

Similarly, show that, in general, for extremization of

$$J = \int_{t_0}^{t_f} V\left(x(t), \dot{x}(t), \ddot{x}(t), \ldots, x^{(r)}(t), t\right) dt$$

with fixed-end point conditions, the Euler-Lagrange equation becomes

$$\sum_{i=0}^{r} (-1)^i \frac{d^i}{dt^i} \left(\frac{\partial V}{\partial x^{(i)}}\right) = 0.$$

Problem 2.5 A first order system is given by

$$\dot{x}(t) = ax(t) + bu(t)$$

and the performance index is

$$J = \frac{1}{2} \int_0^{t_f} (qx^2(t) + ru^2(t))dt$$

where, $x(t_0) = x_0$ and $x(t_f)$ is free and t_f being fixed. Show that the optimal state $x^*(t)$ is given by

$$x^*(t) = x_0 \frac{\sinh\beta(t_f - t)}{\sinh\beta t_f}, \qquad \beta = \sqrt{a^2 + b^2 q/r}.$$

Problem 2.6 A mechanical system is described by

$$\ddot{x}(t) = u(t)$$

find the optimal control and the states by minimizing

$$J = \frac{1}{2} \int_0^5 u^2(t)dt$$

such that the boundary conditions are

$$x(t = 0) = 2; \quad x(t = 5) = 0; \quad \dot{x}(t = 0) = 2; \quad \dot{x}(t = 5) = 0.$$

Problem 2.7 For the first order system

$$\frac{dx}{dt} = -x(t) + u(t)$$

find the optimal control $u^*(t)$ to minimize

$$J = \int_0^{t_f} [x^2(t) + u^2(t)]dt$$

where, t_f is unspecified, and $x(0) = 5$ and $x(t_f) = 0$. Also find t_f.

Problem 2.8 Find the optimal control $u^*(t)$ of the plant

$$\dot{x}_1(t) = x_2(t); \quad x_1(0) = 3, \quad x_1(2) = 0$$
$$\dot{x}_2(t) = -2x_1(t) + 5u(t); \quad x_2(0) = 5, \quad x_2(2) = 0$$

which minimizes the performance index

$$J = \frac{1}{2} \int_0^2 \left[x_1^2(t) + u^2(t) \right] dt.$$

Problem 2.9 A second order plant is described by

$$\dot{x}_1(t) = x_2(t)$$
$$\dot{x}_2(t) = -2x_1(t) - 3x_2(t) + 5u(t)$$

and the cost function is

$$J = \int_0^\infty [x_1^2(t) + u^2(t)]dt.$$

Find the optimal control, when $x_1(0) = 3$ and $x_2(0) = 2$.

Problem 2.10 For a second order system

$$\dot{x}_1(t) = x_2(t)$$
$$\dot{x}_2(t) = -2x_1(t) + 3u(t)$$

with performance index

$$J = 0.5x_1^2(\pi/2) + \int_0^{\pi/2} 0.5u^2(t)dt$$

and boundary conditions $\mathbf{x}(0) = [0 \ 1]'$ and $\mathbf{x}(t_f)$ is free, find the optimal control.

Problem 2.11 Find the optimal control for the plant

$$\dot{x}_1(t) = x_2(t)$$
$$\dot{x}_2(t) = -2x_1(t) + 3u(t)$$

with performance criterion

$$J = \frac{1}{2}F_{11}\left[x_1(t_f) - 4\right]^2 + \frac{1}{2}F_{22}\left[x_2(t_f) - 2\right]^2$$
$$+ \frac{1}{2}\int_0^{t_f} \left[x_1^2(t) + 2x_2^2(t) + 4u^2(t)\right] dt$$

and initial conditions as $\mathbf{x}(0) = [1 \ 2]'$. The additional conditions are given below.

1. Fixed-final conditions $F_{11} = 0, F_{22} = 0, t_f = 2, \mathbf{x}(2) = [4 \ 6]'$.

2. Free-final time conditions $F_{11} = 3, F_{22} = 5, \mathbf{x}(t_f) = [4 \ 6]'$ and t_f is free.

3. Free-final state conditions, $F_{11} = F_{22} = 0, x_1(2)$ is free and $x_2(2) = 6$.

4. Free-final time and free-final state conditions, $F_{11} = 3, F_{22} = 5$ and the final state to have $x_1(t_f) = 4$ and $x_2(t_f)$ to lie on $\theta(t) = -5t + 15$.

Problem 2.12 For the D.C. motor speed control system described in Problem 1.1, find the open-loop optimal control to keep the speed constant at a particular value and the system to respond for any disturbances from the regulated value.

Problem 2.13 For the liquid-level control system described in Problem 1.2, find the open-loop optimal control to keep the liquid level constant at a reference value and the system to act only if there is a change in the liquid level.

Problem 2.14 For the inverted pendulum control system described in Problem 1.3, find the open-loop, optimal control to keep the pendulum in a vertical position.

Problem 2.15 For the mechanical control system described in Problem 1.4, find the open-loop, optimal control to keep the system at equilibrium condition and act only if there is a disturbance.

Problem 2.16 For the automobile suspension control system described in Problem 1.5, find the open-loop, optimal control to provide minimum control energy and passenger comfort.

Problem 2.17 For the chemical control system described in Problem 1.6, find the open-loop, optimal control to keep the system at equilibrium condition and act only if there is a disturbance.

@@@@@@@@@@@@@@@@

Chapter 3

Linear Quadratic Optimal Control Systems I

In this chapter, we present the closed-loop *optimal control* of *linear* plants or systems with *quadratic* performance index or measure. This leads to the *linear quadratic regulator* (LQR) system dealing with state regulation, output regulation, and tracking. Broadly speaking, we are interested in the design of optimal linear systems with quadratic performance indices. It is suggested that the student reviews the material in Appendices A and B given at the end of the book. This chapter is inspired by [6, 3, 89][1].

3.1 Problem Formulation

We discuss the plant and the quadratic performance index with particular reference to physical significance. This helps us to obtain some elegant mathematical conditions on the choice of various matrices in the quadratic cost functional. Thus, we will be dealing with an optimization problem from the engineering perspective.

Consider a linear, time-varying (LTV) system

$$\dot{\mathbf{x}}(t) = \mathbf{A}(t)\mathbf{x}(t) + \mathbf{B}(t)\mathbf{u}(t) \tag{3.1.1}$$

$$\mathbf{y}(t) = \mathbf{C}(t)\mathbf{x}(t) \tag{3.1.2}$$

---

[1]The permissions given by John Wiley for F. L. Lewis, *Optimal Control*, John Wiley & Sons, Inc., New York, NY, 1986, and McGraw-Hill for M. Athans and P. L. Falb, *Optimal Control: An Introduction to the Theory and Its Applications*, McGraw-Hill Book Company, New York, NY, 1966, are hereby acknowledged.

with a cost functional (CF) or performance index (PI)

$$J(\mathbf{u}(t)) = J(\mathbf{x}(t_0), \mathbf{u}(t), t_0)$$
$$= \frac{1}{2} [\mathbf{z}(t_f) - \mathbf{y}(t_f)]' \, \mathbf{F}(t_f) \, [\mathbf{z}(t_f) - \mathbf{y}(t_f)]$$
$$+ \frac{1}{2} \int_{t_0}^{t_f} \left[[\mathbf{z}(t) - \mathbf{y}(t)]' \, \mathbf{Q}(t) \, [\mathbf{z}(t) - \mathbf{y}(t)] + \mathbf{u}'(t)\mathbf{R}(t)\mathbf{u}(t) \right] dt$$

$$(3.1.3)$$

where, $\mathbf{x}(t)$ is nth *state* vector, $\mathbf{y}(t)$ is mth *output* vector, $\mathbf{z}(t)$ is mth reference or *desired* output vector (or nth desired state vector, if the state $\mathbf{x}(t)$ is available), $\mathbf{u}(t)$ is rth *control* vector, and $\mathbf{e}(t) = \mathbf{z}(t) - \mathbf{y}(t)$ (or $\mathbf{e}(t) = \mathbf{z}(t) - \mathbf{x}(t)$, if the state $\mathbf{x}(t)$ is directly available) is the mth *error* vector. $\mathbf{A}(t)$ is $n \times n$ *state* matrix, $\mathbf{B}(t)$ is $n \times r$ *control* matrix, and $\mathbf{C}(t)$ is $m \times n$ *output* matrix. We assume that the control $\mathbf{u}(t)$ is *unconstrained*, $0 < m \leq r \leq n$, and all the states and/or outputs are completely measurable. The preceding cost functional (3.1.3) contains quadratic terms in error $\mathbf{e}(t)$ and control $\mathbf{u}(t)$ and hence called the *quadratic* cost functional[2]. We also make certain assumptions to be described below on the various matrices in the quadratic cost functional (3.1.3). Under these assumptions, we will find that the optimal control is closed-loop in nature, that is, the optimal control $\mathbf{u}(t)$ is a function of the state $\mathbf{x}(t)$ or the output $\mathbf{y}(t)$. Also, depending on the final time t_f being finite (infinite), the system is called finite- (infinite-) time horizon system. Further, we have the following categories of systems.

1. If our objective is to keep the *state* $\mathbf{x}(t)$ near zero (i.e., $\mathbf{z}(t) = 0$ and $\mathbf{C} = I$), then we call it *state regulator* system. In other words, the objective is to obtain a control $\mathbf{u}(t)$ which takes the plant described by (3.1.1) and (3.1.2) from a nonzero state to zero state. This situation may arise when a plant is subjected to unwanted disturbances that perturb the state (for example, sudden load changes in an electrical voltage regulator system, sudden wind gust in a radar antenna positional control system).

2. If our interest is to keep the *output* $\mathbf{y}(t)$ near zero (i.e., $\mathbf{z}(t) = 0$), then it is termed the *output regulator* system.

---

[2]See Appendix A for more details on Quadratic Forms and Definiteness and other related topics.

3. If we try to keep the *output or state* near a *desired* state or output, then we are dealing with a *tracking* system. We see that in both state and output regulator systems, the desired or reference state is zero and in tracking system the error is to be made zero. For example, consider again the antenna control system to track an aircraft.

Let us consider the various matrices in the cost functional (3.1.3) and their implications.

1. *The Error Weighted Matrix* $\mathbf{Q}(t)$: In order to keep the error $\mathbf{e}(t)$ small and error squared non-negative, the integral of the expression $\frac{1}{2}\mathbf{e}'(t)\mathbf{Q}(t)\mathbf{e}(t)$ should be nonnegative and small. Thus, the matrix $\mathbf{Q}(t)$ must be *positive semidefinite*. Due to the quadratic nature of the weightage, we have to pay more attention to large errors than small errors.

2. *The Control Weighted Matrix* $\mathbf{R}(t)$: The quadratic nature of the control cost expression $\frac{1}{2}\mathbf{u}'(t)\mathbf{R}(t)\mathbf{u}(t)$ indicates that one has to pay higher cost for larger control effort. Since the cost of the control has to be a positive quantity, the matrix $\mathbf{R}(t)$ should be *positive definite*.

3. *The Control Signal* $\mathbf{u}(t)$: The assumption that there are *no constraints* on the control $\mathbf{u}(t)$ is very important in obtaining the closed loop optimal configuration.

   Combining all the previous assumptions, we would like on one hand, to keep the error small, but on the other hand, we must not pay higher cost to large controls.

4. *The Terminal Cost Weighted Matrix* $\mathbf{F}(t_f)$: The main purpose of this term is to ensure that the error $\mathbf{e}(t)$ at the final time t_f is as small as possible. To guarantee this, the corresponding matrix $\mathbf{F}(t_f)$ should be *positive semidefinite*.

   Further, without loss of generality, we assume that the weighted matrices $\mathbf{Q}(t), \mathbf{R}(t)$, and $\mathbf{F}(t)$ are *symmetric*. The *quadratic* cost functional described previously has some attractive features:

   (a) It provides an elegant procedure for the design of *closed-loop* optimal controller.

(b) It results in the optimal feed-back control that is *linear* in state function.

That is why we often say that the "quadratic performance index fits like a glove" [6].

5. *Infinite Final Time*: When the final time t_f is infinity, the terminal cost term involving $\mathbf{F}(t_f)$ does not provide any realistic sense since we are always interested in the solutions over finite time. Hence, $\mathbf{F}(t_f)$ must be zero.

3.2   Finite-Time Linear Quadratic Regulator

Now we proceed with the linear quadratic regulator (LQR) system, that is, to keep the state near zero during the interval of interest. For the sake of completeness we shall repeat the plant and performance index equations described in the earlier section. Consider a linear, time-varying plant described by

$$\dot{\mathbf{x}}(t) = \mathbf{A}(t)\mathbf{x}(t) + \mathbf{B}(t)\mathbf{u}(t) \tag{3.2.1}$$

with a cost functional

$$
\begin{aligned}
J(\mathbf{u}) &= J(\mathbf{x}(t_0), \mathbf{u}(t), t_0) \\
&= \frac{1}{2}\mathbf{x}'(t_f)\mathbf{F}(t_f)\mathbf{x}(t_f) \\
&\quad + \frac{1}{2}\int_{t_0}^{t_f} \left[\mathbf{x}'(t)\mathbf{Q}(t)\mathbf{x}(t) + \mathbf{u}'(t)\mathbf{R}(t)\mathbf{u}(t)\right] dt \\
&= \frac{1}{2}\mathbf{x}'(t_f)\mathbf{F}(t_f)\mathbf{x}(t_f) \\
&\quad + \frac{1}{2}\int_{t_0}^{t_f} \left[\mathbf{x}'(t) \ \mathbf{u}'(t)\right] \begin{bmatrix} \mathbf{Q}(t) & \mathbf{0} \\ \mathbf{0} & \mathbf{R}(t) \end{bmatrix} \begin{bmatrix} \mathbf{x}(t) \\ \mathbf{u}(t) \end{bmatrix} dt \quad (3.2.2)
\end{aligned}
$$

where, the various vectors and matrices are defined in the last section. Let us note that here, the reference or desired state $\mathbf{z}(t) = 0$ and hence the error $\mathbf{e}(t) = 0 - \mathbf{x}(t)$ itself is the state, thereby implying a state regulator system. We summarize again various assumptions as follows.

1. The control $\mathbf{u}(t)$ is *unconstrained*. However, in many physical situations, there are limitations on the control and state and the case of *unconstrained* control is discussed in a later chapter.

2. The initial condition $\mathbf{x}(t = t_0) = \mathbf{x}_0$ is given. The terminal time t_f is *specified*, and the final state $\mathbf{x}(t_f)$ is *not specified*.

3. The terminal cost matrix $\mathbf{F}(t_f)$ and the error weighted matrix $\mathbf{Q}(t)$ are *nxn* positive *semidefinite* matrices, respectively; and the control weighted matrix $\mathbf{R}(t)$ is an *rxr* positive *definite* matrix.

4. Finally, the fraction $\frac{1}{2}$ in the cost functional (3.2.2) is associated mainly to cancel a 2 that would have otherwise been carried on throughout the result, as seen later.

We follow the Pontryagin procedure described in Chapter 2 (Table 2.1) to obtain optimal solution and then propose the closed-loop configuration. First, let us list the various steps under which we present the method.

- **Step 1:** *Hamiltonian*

- **Step 2:** *Optimal Control*

- **Step 3:** *State and Costate System*

- **Step 4:** *Closed-Loop Optimal Control*

- **Step 5:** *Matrix Differential Riccati Equation*

Now let us discuss the preceding steps in detail.

- **Step 1:** *Hamiltonian*: Using the definition of the Hamiltonian given by (2.7.27) in Chapter 2 along with the performance index (3.2.2), formulate the Hamiltonian as

$$\mathcal{H}(\mathbf{x}(t), \mathbf{u}(t), \boldsymbol{\lambda}(t)) = \frac{1}{2}\mathbf{x}'(t)\mathbf{Q}(t)\mathbf{x}(t) + \frac{1}{2}\mathbf{u}'(t)\mathbf{R}(t)\mathbf{u}(t)$$
$$+ \boldsymbol{\lambda}'(t)\left[\mathbf{A}(t)\mathbf{x}(t) + \mathbf{B}(t)\mathbf{u}(t)\right] \quad (3.2.3)$$

where, $\boldsymbol{\lambda}(t)$ is the costate vector of *n*th order.

- **Step 2:** *Optimal Control*: Obtain the optimal control $\mathbf{u}^*(t)$ using the control relation (2.7.29) as

$$\frac{\partial \mathcal{H}}{\partial \mathbf{u}} = 0 \longrightarrow \mathbf{R}(t)\mathbf{u}^*(t) + \mathbf{B}'(t)\boldsymbol{\lambda}^*(t) = 0 \quad (3.2.4)$$

leading to

$$\mathbf{u}^*(t) = -\mathbf{R}^{-1}(t)\mathbf{B}'(t)\boldsymbol{\lambda}^*(t) \quad (3.2.5)$$

where, we used

$$\frac{\partial}{\partial \mathbf{u}} \left\{ \frac{1}{2} \mathbf{u}'(t) \mathbf{R}(t) \mathbf{u}(t) \right\} = \mathbf{R}(t) \mathbf{u}(t) \quad \text{and}$$

$$\frac{\partial}{\partial \mathbf{u}} \left\{ \boldsymbol{\lambda}'(t) \mathbf{B}(t) \mathbf{u}(t) \right\} = \mathbf{B}'(t) \boldsymbol{\lambda}(t).$$

Similar expressions are used throughout the rest of the book. Further details on such relations are found in Appendix A. We immediately notice from (3.2.5) the need for $\mathbf{R}(t)$ to be positive *definite* and not positive *semidefinite* so that the inverse $\mathbf{R}^{-1}(t)$ exists.

- **Step 3:** *State and Costate System*: Obtain the state and costate equations according to (2.7.30) and (2.7.31) as

$$\dot{\mathbf{x}}^*(t) = + \left(\frac{\partial \mathcal{H}}{\partial \boldsymbol{\lambda}} \right)_* \longrightarrow \dot{\mathbf{x}}^*(t) = \mathbf{A}(t)\mathbf{x}^*(t) + \mathbf{B}(t)\mathbf{u}^*(t) \qquad (3.2.6)$$

$$\dot{\boldsymbol{\lambda}}^*(t) = - \left(\frac{\partial \mathcal{H}}{\partial \mathbf{x}} \right)_* \longrightarrow \dot{\boldsymbol{\lambda}}^*(t) = -\mathbf{Q}(t)\mathbf{x}^*(t) - \mathbf{A}'(t)\boldsymbol{\lambda}^*(t). \ (3.2.7)$$

Substitute the control relation (3.2.5) in the state equation (3.2.6) to obtain the (state and costate) canonical system (also called Hamiltonian system) of equations

$$\begin{bmatrix} \dot{\mathbf{x}}^*(t) \\ \dot{\boldsymbol{\lambda}}^*(t) \end{bmatrix} = \begin{bmatrix} \mathbf{A}(t) & -\mathbf{E}(t) \\ -\mathbf{Q}(t) & -\mathbf{A}'(t) \end{bmatrix} \begin{bmatrix} \mathbf{x}^*(t) \\ \boldsymbol{\lambda}^*(t) \end{bmatrix} \qquad (3.2.8)$$

where $\mathbf{E}(t) = \mathbf{B}(t)\mathbf{R}^{-1}(t)\mathbf{B}'(t)$. The general boundary condition given by the relation (2.7.32) is reproduced here as

$$\left[\mathcal{H}^* + \frac{\partial S}{\partial t} \right]_{t_f} \delta t_f + \left[\left(\frac{\partial S}{\partial \mathbf{x}} \right)_* - \boldsymbol{\lambda}^*(t) \right]'_{t_f} \delta \mathbf{x}_f = 0 \quad (3.2.9)$$

where, S equals the entire terminal cost term in the cost functional (3.2.2). Here, for our present system t_f is *specified* which makes δt_f equal to zero in (3.2.9), and $\mathbf{x}(t_f)$ is *not specified* which makes $\delta \mathbf{x}_f$ arbitrary in (3.2.9). Hence, the coefficient of $\delta \mathbf{x}_f$ in (3.2.9) becomes zero, that is,

$$\boldsymbol{\lambda}^*(t_f) = \left(\frac{\partial S}{\partial \mathbf{x}(t_f)} \right)_*$$

$$= \frac{\partial \left[\frac{1}{2} \mathbf{x}'(t_f) \mathbf{F}(t_f) \mathbf{x}(t_f) \right]}{\partial \mathbf{x}(t_f)} = \mathbf{F}(t_f)\mathbf{x}^*(t_f). \quad (3.2.10)$$

This final condition on the costate $\boldsymbol{\lambda}^*(t_f)$ together with the initial condition on the state $\mathbf{x}_0$ and the canonical system of equations (3.2.8) form a two-point, boundary value problem (TPBVP). The state-space representation of the set of relations for the state and costate system (3.2.8) and the control (3.2.5) is shown in Figure 3.1.

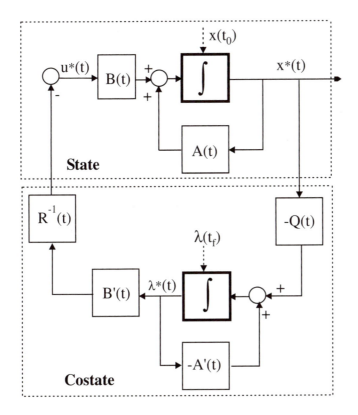

Figure 3.1   State and Costate System

- **Step 4:** *Closed-Loop Optimal Control:* The state space representation shown in Figure 3.1 prompts us to think that we can obtain the optimal control $\mathbf{u}^*(t)$ as a function (negative feedback) of the optimal state $\mathbf{x}^*(t)$. Now to formulate a closed-loop optimal control, that is, to obtain the optimal control $\mathbf{u}^*(t)$ which is a function of the costate $\boldsymbol{\lambda}^*(t)$ as seen from (3.2.5), as a function of the state $\mathbf{x}^*(t)$, let us examine the final condition on $\boldsymbol{\lambda}^*(t)$ given by (3.2.10). This in fact relates the costate in terms of the state at the final time t_f. Similarly, we may like to connect the costate with the state for the complete interval of time $[t_0, t_f]$. Thus, let us assume a transformation [113, 102]

$$\boldsymbol{\lambda}^*(t) = \mathbf{P}(t)\mathbf{x}^*(t) \qquad (3.2.11)$$

where, $\mathbf{P}(t)$ is yet to be determined. Then, we can easily see that with (3.2.11), the optimal control (3.2.5) becomes

$$\mathbf{u}^*(t) = -\mathbf{R}^{-1}(t)\mathbf{B}'(t)\mathbf{P}(t)\mathbf{x}^*(t) \qquad (3.2.12)$$

which is now a *negative feedback* of the state $\mathbf{x}^*(t)$. Note that this negative feedback resulted from our "theoretical development" or "mathematics" of optimal control procedure and not introduced intentionally [6].

Differentiating (3.2.11) w.r.t. time t, we get

$$\dot{\boldsymbol{\lambda}}^*(t) = \dot{\mathbf{P}}(t)\mathbf{x}^*(t) + \mathbf{P}(t)\dot{\mathbf{x}}^*(t). \qquad (3.2.13)$$

Using the transformation (3.2.11) in the control, state and costate system of equations (3.2.5), (3.2.6) and (3.2.7), respectively, we get

$$\dot{\mathbf{x}}^*(t) = \mathbf{A}(t)\mathbf{x}^*(t) - \mathbf{B}(t)\mathbf{R}^{-1}(t)\mathbf{B}'(t)\mathbf{P}(t)\mathbf{x}^*(t), \qquad (3.2.14)$$
$$\dot{\boldsymbol{\lambda}}^*(t) = -\mathbf{Q}(t)\mathbf{x}^*(t) - \mathbf{A}'(t)\mathbf{P}(t)\mathbf{x}^*(t). \qquad (3.2.15)$$

Now, substituting state and costate relations (3.2.14) and (3.2.15) in (3.2.13), we have

$$-\mathbf{Q}(t)\mathbf{x}^*(t) - \mathbf{A}'(t)\mathbf{P}(t)\mathbf{x}^*(t) = \dot{\mathbf{P}}(t)\mathbf{x}^*(t) +$$
$$\mathbf{P}(t)\left[\mathbf{A}(t)\mathbf{x}^*(t) - \mathbf{B}(t)\mathbf{R}^{-1}(t)\mathbf{B}'(t)\mathbf{P}(t)\mathbf{x}^*(t)\right] \longrightarrow$$
$$\left[\dot{\mathbf{P}}(t) + \mathbf{P}(t)\mathbf{A}(t) + \mathbf{A}'(t)\mathbf{P}(t) + \mathbf{Q}(t) -\right.$$
$$\left.\mathbf{P}(t)\mathbf{B}(t)\mathbf{R}^{-1}(t)\mathbf{B}'(t)\mathbf{P}(t)\right]\mathbf{x}^*(t) = 0 \qquad (3.2.16)$$

Essentially, we eliminated the costate function $\boldsymbol{\lambda}^*(t)$ from the control (3.2.5), the state (3.2.6) and the costate (3.2.7) equations by introducing the transformation (3.2.11).

- **Step 5:** *Matrix Differential Riccati Equation:* Now the relation (3.2.16) should be satisfied for all $t \in [t_0, t_f]$ and for any choice of the initial state $\mathbf{x}^*(t_0)$. Also, $\mathbf{P}(t)$ is not dependent on the initial state. It follows that the equation (3.2.16) should hold good for any value of $\mathbf{x}^*(t)$. This clearly means that the function $\mathbf{P}(t)$ should satisfy the matrix differential equation

$$\dot{\mathbf{P}}(t) + \mathbf{P}(t)\mathbf{A}(t) + \mathbf{A}'(t)\mathbf{P}(t) + \mathbf{Q}(t) -$$
$$\mathbf{P}(t)\mathbf{B}(t)\mathbf{R}^{-1}(t)\mathbf{B}'(t)\mathbf{P}(t) = 0. \qquad (3.2.17)$$

This is the matrix differential equation of the *Riccati type*, and often called the matrix *differential Riccati equation* (DRE). Also, the transformation (3.2.11) is called the *Riccati transformation*, $\mathbf{P}(t)$ is called the Riccati coefficient matrix or simply Riccati matrix or Riccati coefficient, and (3.2.12) is the *optimal control (feedback) law*. The matrix DRE (3.2.17) can also be written in a compact form as

$$\dot{\mathbf{P}}(t) = -\mathbf{P}(t)\mathbf{A}(t) - \mathbf{A}'(t)\mathbf{P}(t) - \mathbf{Q}(t) + \mathbf{P}(t)\mathbf{E}(t)\mathbf{P}(t) \quad (3.2.18)$$

where $\mathbf{E}(t) = \mathbf{B}(t)\mathbf{R}^{-1}(t)\mathbf{B}'(t)$.

Comparing the boundary condition(3.2.10) and the Riccati transformation (3.2.11), we have the final condition on $\mathbf{P}(t)$ as

$$\boldsymbol{\lambda}^*(t_f) = \mathbf{P}(t_f)\mathbf{x}^*(t_f) = \mathbf{F}(t_f)\mathbf{x}^*(t_f) \longrightarrow$$
$$\boxed{\mathbf{P}(t_f) = \mathbf{F}(t_f).} \qquad (3.2.19)$$

Thus, the matrix DRE (3.2.17) or (3.2.18) is to be solved *backward in time* using the final condition (3.2.19) to obtain the solution $\mathbf{P}(t)$ for the entire interval $[t_0, t_f]$.

3.2.1 Symmetric Property of the Riccati Coefficient Matrix

Here, we first show an important property of the Riccati matrix $\mathbf{P}(t)$. The fact that the nxn matrix $\mathbf{P}(t)$ is *symmetric* for all $t \in [t_0, t_f]$, i.e., $\mathbf{P}(t) = \mathbf{P}'(t)$ can be easily shown as follows. First of all, let us note that from the formulation of the problem itself, the matrices $\mathbf{F}(t_f)$, $\mathbf{Q}(t)$, and $\mathbf{R}(t)$ are symmetric and therefore, the matrix $\mathbf{B}(t)\mathbf{R}^{-1}(t)\mathbf{B}'(t)$ is also symmetric. Now transposing both sides of the matrix DRE (3.2.18), we notice that *both* $\mathbf{P}(t)$ and $\mathbf{P}'(t)$ are solutions of the *same* differential equation and that both satisfy the same final condition (3.2.19).

3.2.2　Optimal Control

Is the optimal control $\mathbf{u}^*(t)$ a minimum? This can be answered by considering the *second partials* of the Hamiltonian (3.2.3). Let us recall from Chapter 2 that this is done by examining the *second variation* of the cost functional. Thus, the condition (2.7.41) (reproduced here for convenience) for examining the nature of optimal control is that the matrix

$$\Pi = \begin{bmatrix} \dfrac{\partial^2 \mathcal{H}}{\partial x^2} & \dfrac{\partial^2 \mathcal{H}}{\partial x \partial u} \\[3mm] \dfrac{\partial^2 \mathcal{H}}{\partial u \partial x} & \dfrac{\partial^2 \mathcal{H}}{\partial u^2} \end{bmatrix}_* \tag{3.2.20}$$

must be *positive definite (negative definite)* for *minimum (maximum)*. In most of the cases this reduces to the condition that

$$\left(\frac{\partial^2 \mathcal{H}}{\partial u^2} \right)_* \tag{3.2.21}$$

must be positive definite (negative definite) for minimum (maximum). Now using the Hamiltonian (3.2.3) and calculating the various partials,

$$\left(\frac{\partial^2 \mathcal{H}}{\partial x^2} \right)_* = \mathbf{Q}(t), \quad \left(\frac{\partial^2 \mathcal{H}}{\partial x \partial u} \right)_* = 0,$$

$$\left(\frac{\partial^2 \mathcal{H}}{\partial u \partial x} \right)_* = 0, \quad \left(\frac{\partial^2 \mathcal{H}}{\partial u^2} \right)_* = \mathbf{R}(t). \tag{3.2.22}$$

Substituting the previous partials in the condition (3.2.20), we have

$$\Pi = \begin{bmatrix} \mathbf{Q}(t) & 0 \\ 0 & \mathbf{R}(t) \end{bmatrix}. \tag{3.2.23}$$

Since $\mathbf{R}(t)$ is positive definite, and $\mathbf{Q}(t)$ is positive semidefinite, it follows that the preceding matrix (3.2.23) is only positive semidefinite. However, the condition that the second partial of $\mathcal{H}$ w.r.t. $\mathbf{u}^*(t)$, which is $\mathbf{R}(t)$, is positive definite, is enough to guarantee that the control $\mathbf{u}^*(t)$ is *minimum*.

3.2.3　Optimal Performance Index

Here, we show how to obtain an expression for the optimal value of the performance index.

THEOREM 3.1

The optimal value of the PI (3.2.2) is given by

$$J^*(\mathbf{x}^*(t), t) = \frac{1}{2}\mathbf{x}^{*\prime}(t)\mathbf{P}(t)\mathbf{x}^*(t).$$ (3.2.24)

Proof: *First let us note that*

$$\int_{t_0}^{t_f} \frac{d}{dt}\left(\mathbf{x}^{*\prime}(t)\mathbf{P}(t)\mathbf{x}^*(t)\right) dt = -\frac{1}{2}\mathbf{x}^{*\prime}(t_0)\mathbf{P}(t_0)\mathbf{x}^*(t_0)$$

$$+\frac{1}{2}\mathbf{x}^{*\prime}(t_f)\mathbf{P}(t_f)\mathbf{x}^*(t_f).$$ (3.2.25)

Substituting for $\frac{1}{2}\mathbf{x}^{*\prime}(t_f)\mathbf{P}(t_f)\mathbf{x}^*(t_f)$ *from (3.2.25) into the PI (3.2.2), and noting that* $\mathbf{P}(t_f) = \mathbf{F}(t_f)$ *from (3.2.19) we get*

$$J^*(\mathbf{x}^*(t_0), t_0) = \frac{1}{2}\mathbf{x}^{*\prime}(t_0)\mathbf{P}(t_0)\mathbf{x}^*(t_0)$$

$$+\frac{1}{2}\int_{t_0}^{t_f} \left[\mathbf{x}^{*\prime}(t)\mathbf{Q}(t)\mathbf{x}^*(t) + \mathbf{u}^{*\prime}(t)\mathbf{R}(t)\mathbf{u}^*(t)\right.$$

$$\left.+ \frac{d}{dt}\left(\mathbf{x}^{*\prime}(t)\mathbf{P}(t)\mathbf{x}^*(t)\right)\right] dt$$

$$= \frac{1}{2}\mathbf{x}^{*\prime}(t_0)\mathbf{P}(t_0)\mathbf{x}(t_0)$$

$$+\frac{1}{2}\int_{t_0}^{t_f} \left[\mathbf{x}^{*\prime}(t)\mathbf{Q}(t)\mathbf{x}^*(t) + \mathbf{u}^{*\prime}(t)\mathbf{R}(t)\mathbf{u}^*(t)\right.$$

$$+ \dot{\mathbf{x}}^{*\prime}(t)\mathbf{P}(t)\mathbf{x}^*(t) + \mathbf{x}^{*\prime}(t)\dot{\mathbf{P}}(t)\mathbf{x}^*(t)$$

$$\left.+ \mathbf{x}^{*\prime}(t)\mathbf{P}(t)\dot{\mathbf{x}}^*(t)\right] dt.$$ (3.2.26)

Now, using the state equation (3.2.14) for $\dot{\mathbf{x}}^*(t)$, *we get*

$$J^*(\mathbf{x}^*(t_0), t_0) = \frac{1}{2}\mathbf{x}^{*\prime}(t_0)\mathbf{P}(t_0)\mathbf{x}^*(t_0)$$

$$+\frac{1}{2}\int_{t_0}^{t_f} \mathbf{x}^{*\prime}(t)\left[\mathbf{Q}(t) + \mathbf{A}'(t)\mathbf{P}(t) + \mathbf{P}(t)\mathbf{A}(t)\right.$$

$$\left.- \mathbf{P}(t)\mathbf{B}(t)\mathbf{R}^{-1}(t)\mathbf{B}'(t)\mathbf{P}(t) + \dot{\mathbf{P}}(t)\right]\mathbf{x}^*(t)dt.$$ (3.2.27)

Finally, using the matrix DRE (3.2.18) in the previous relations, the integral part becomes zero. Thus,

$$J^*(\mathbf{x}(t_0), t_0) = \frac{1}{2}\mathbf{x}^{*\prime}(t_0)\mathbf{P}(t_0)\mathbf{x}^*(t_0).$$ (3.2.28)

Now, the previous relation is also valid for any $\mathbf{x}^*(t)$. *Thus,*

$$J^*(\mathbf{x}^*(t), t) = \frac{1}{2}\mathbf{x}^{*\prime}(t)\mathbf{P}(t)\mathbf{x}^*(t). \tag{3.2.29}$$

In terms of the final time t_f, *the previous optimal cost becomes*

$$J^*(\mathbf{x}(t_f), t_f) = \frac{1}{2}\mathbf{x}^{*\prime}(t_f)\mathbf{P}(t_f)\mathbf{x}^*(t_f). \tag{3.2.30}$$

Since we are normally given the initial state $\mathbf{x}(t_0)$ *and the Riccati coefficient* $\mathbf{P}(t)$ *is solved for all time* t, *it is more convenient to use the relation (3.2.28).*

3.2.4 Finite-Time Linear Quadratic Regulator: Time-Varying Case: Summary

Given a linear, time-varying plant

$$\dot{\mathbf{x}}(t) = \mathbf{A}(t)\mathbf{x}(t) + \mathbf{B}(t)\mathbf{u}(t) \tag{3.2.31}$$

and a quadratic performance index

$$J = \frac{1}{2}\mathbf{x}'(t_f)\mathbf{F}(t_f)\mathbf{x}(t_f)$$
$$+ \frac{1}{2}\int_{t_0}^{t_f} \left[\mathbf{x}'(t)\mathbf{Q}(t)\mathbf{x}(t) + \mathbf{u}'(t)\mathbf{R}(t)\mathbf{u}(t)\right] dt \tag{3.2.32}$$

where, $\mathbf{u}(t)$ is not constrained, t_f is specified, and $\mathbf{x}(t_f)$ is not specified, $\mathbf{F}(t_f)$ and $\mathbf{Q}(t)$ are $n \times n$ symmetric, positive semidefinite matrices, and $\mathbf{R}(t)$ is $r \times r$ symmetric, positive definite matrix, the optimal control is given by

$$\mathbf{u}^*(t) = -\mathbf{R}^{-1}(t)\mathbf{B}'(t)\mathbf{P}(t)\mathbf{x}^*(t) = -\mathbf{K}(t)\mathbf{x}^*(t) \tag{3.2.33}$$

where $\mathbf{K}(t) = \mathbf{R}^{-1}(t)\mathbf{B}'(t)\mathbf{P}(t)$ is called *Kalman gain* and $\mathbf{P}(t)$, the $n \times n$ symmetric, *positive definite* matrix (for all $t \in [t_0, t_f]$), is the solution of the matrix differential Riccati equation (DRE)

$$\dot{\mathbf{P}}(t) = -\mathbf{P}(t)\mathbf{A}(t) - \mathbf{A}'(t)\mathbf{P}(t) - \mathbf{Q}(t) + \mathbf{P}(t)\mathbf{B}(t)\mathbf{R}^{-1}(t)\mathbf{B}'(t)\mathbf{P}(t) \tag{3.2.34}$$

satisfying the final condition

$$\mathbf{P}(t = t_f) = \mathbf{F}(t_f) \tag{3.2.35}$$

Table 3.1 Procedure Summary of Finite-Time Linear Quadratic Regulator System: Time-Varying Case

A. Statement of the Problem
Given the plant as $\dot{\mathbf{x}}(t) = \mathbf{A}(t)\mathbf{x}(t) + \mathbf{B}(t)\mathbf{u}(t)$, the performance index as $J = \frac{1}{2}\mathbf{x}'(t_f)\mathbf{F}(t_f)\mathbf{x}(t_f) + \frac{1}{2}\int_{t_0}^{t_f}\left[\mathbf{x}'(t)\mathbf{Q}(t)\mathbf{x}(t) + \mathbf{u}'(t)\mathbf{R}(t)\mathbf{u}(t)\right]dt$, and the boundary conditions as $\mathbf{x}(t_0) = \mathbf{x}_0, \quad t_f$ is fixed, and $\mathbf{x}(t_f)$ is free, find the optimal control, state and performance index.

	B. Solution of the Problem
Step 1	Solve the matrix differential Riccati equation $\dot{\mathbf{P}}(t) = -\mathbf{P}(t)\mathbf{A}(t) - \mathbf{A}'(t)\mathbf{P}(t) - \mathbf{Q}(t) + \mathbf{P}(t)\mathbf{B}(t)\mathbf{R}^{-1}(t)\mathbf{B}'(t)\mathbf{P}(t)$ with final condition $\mathbf{P}(t = t_f) = \mathbf{F}(t_f)$.
Step 2	Solve the optimal state $\mathbf{x}^*(t)$ from $\dot{\mathbf{x}}^*(t) = \left[\mathbf{A}(t) - \mathbf{B}(t)\mathbf{R}^{-1}(t)\mathbf{B}'(t)\mathbf{P}(t)\right]\mathbf{x}^*(t)$ with initial condition $\mathbf{x}(t_0) = \mathbf{x}_0$.
Step 3	Obtain the optimal control $\mathbf{u}^*(t)$ as $\mathbf{u}^*(t) = -\mathbf{K}(t)\mathbf{x}^*(t)$ where, $\mathbf{K}(t) = \mathbf{R}^{-1}(t)\mathbf{B}'(t)\mathbf{P}(t)$.
Step 4	Obtain the optimal performance index from $J^* = \frac{1}{2}\mathbf{x}^{*\prime}(t)\mathbf{P}(t)\mathbf{x}^*(t)$.

the optimal state is the solution of

$$\boxed{\dot{\mathbf{x}}^*(t) = \left[\mathbf{A}(t) - \mathbf{B}(t)\mathbf{R}^{-1}(t)\mathbf{B}'(t)\mathbf{P}(t)\right]\mathbf{x}^*(t)} \qquad (3.2.36)$$

and the optimal cost is

$$\boxed{J^* = \frac{1}{2}\mathbf{x}^{*\prime}(t)\mathbf{P}(t)\mathbf{x}^*(t).} \qquad (3.2.37)$$

The optimal control $\mathbf{u}^*(t)$, given by (3.2.33), is *linear* in the optimal state $\mathbf{x}^*(t)$. The entire procedure is now summarized in Table 3.1.
Note: It is simple to see that one can absorb the $\frac{1}{2}$ that is associated with J by redefining a performance measure as

$$J_2 = 2J = \mathbf{x}'(t_f)\mathbf{F}(t_f)\mathbf{x}(t_f)$$
$$+ \int_{t_0}^{t_f}\left[\mathbf{x}'(t)\mathbf{Q}(t)\mathbf{x}(t) + \mathbf{u}'(t)\mathbf{R}(t)\mathbf{u}(t)\right]dt, \qquad (3.2.38)$$

get the corresponding matrix differential Riccati equation for J_2 as

$$
\frac{\dot{\mathbf{P}}_2(t)}{2} = -\frac{\mathbf{P}_2(t)}{2}\mathbf{A}(t) - \mathbf{A}'(t)\frac{\mathbf{P}_2(t)}{2} - \mathbf{Q}(t)
$$
$$
+ \frac{\mathbf{P}_2(t)}{2}\mathbf{B}(t)\mathbf{R}^{-1}(t)\mathbf{B}'(t)\frac{\mathbf{P}_2(t)}{2} \tag{3.2.39}
$$

with final condition

$$
\frac{\mathbf{P}_2(t = t_f)}{2} = \mathbf{F}(t_f). \tag{3.2.40}
$$

Comparing the previous DRE for J_2 with the corresponding DRE (3.2.34) for J, we can easily see that $\mathbf{P}_2(t) = 2\mathbf{P}(t)$ and hence the optimal control becomes

$$
\mathbf{u}^*(t) = -\mathbf{R}^{-1}(t)\mathbf{B}'(t)\frac{\mathbf{P}_2(t)}{2}\mathbf{x}^*(t) = -\frac{\mathbf{K}_2(t)}{2}\mathbf{x}^*(t)
$$
$$
= -\mathbf{R}^{-1}(t)\mathbf{B}'(t)\mathbf{P}(t)\mathbf{x}^*(t) = -\mathbf{K}(t)\mathbf{x}^*(t). \tag{3.2.41}
$$

Thus, using J_2 without the $\frac{1}{2}$ in the performance index, we get the same optimal control (3.2.41) for the original plant (3.2.31), but the only difference being that the Riccati coefficient matrix $\mathbf{P}_2(t)$ is *twice* that of $\mathbf{P}(t)$ and J_2 is *twice* that of J(for example, see [3, 42]).

However, we will retain the $\frac{1}{2}$ in J throughout the book due to the obvious simplifications in obtaining the optimal control, state and costate equations (3.2.4), (3.2.6) and (3.2.7), respectively. Precisely, the factor $\frac{1}{2}$ in the PI (3.2.2) and hence in the Hamiltonian (3.2.3) gets eliminated while taking partial derivatives of the Hamiltonian w.r.t. the control, state and costate functions.

3.2.5   Salient Features

We next discuss the various salient features of the state regulator system and the matrix differential Riccati equation.

1. *Riccati Coefficient:* The Riccati coefficient matrix $\mathbf{P}(t)$ is a time-varying matrix which depends upon the system matrices $\mathbf{A}(t)$ and $\mathbf{B}(t)$, the performance index (design) matrices $\mathbf{Q}(t)$, $\mathbf{R}(t)$ and $\mathbf{F}(t_f)$, and the terminal time t_f, but $\mathbf{P}(t)$ does not depend upon the initial state $\mathbf{x}(t_0)$ of the system.

2. $\mathbf{P}(t)$ is *symmetric* and hence it follows that the nxn order matrix DRE (3.2.18) represents a system of $n(n+1)/2$ *first order, nonlinear, time-varying, ordinary differential* equations.

3. *Optimal Control:* From (3.2.21), we see that the optimal control $\mathbf{u}^*(t)$ is minimum (maximum) if the control weighted matrix $\mathbf{R}(t)$ is *positive definite (negative definite)*.

4. *Optimal State:* Using the optimal control (3.2.12) in the state equation (3.2.1), we have

$$\boxed{\dot{\mathbf{x}}^*(t) = \left[\mathbf{A}(t) - \mathbf{B}(t)\mathbf{R}^{-1}(t)\mathbf{B}'(t)\mathbf{P}(t)\right]\mathbf{x}^*(t) = \mathbf{G}(t)\mathbf{x}^*(t)}$$

(3.2.42)

where

$$\mathbf{G}(t) = \mathbf{A}(t) - \mathbf{B}(t)\mathbf{R}^{-1}(t)\mathbf{B}'(t)\mathbf{P}(t). \qquad (3.2.43)$$

The solution of this state differential equation along with the initial condition $\mathbf{x}(t_0)$ gives the optimal state $\mathbf{x}^*(t)$. Let us note that there is no condition on the closed-loop matrix $\mathbf{G}(t)$ regarding *stability* as long as we are considering the *finite* final time (t_f) system.

5. *Optimal Cost:* It is shown in (3.2.29) that the minimum cost J^* is given by

$$J^* = \frac{1}{2}\mathbf{x}^{*\prime}(t)\mathbf{P}(t)\mathbf{x}^*(t) \quad \text{for all} \quad t \in [t_0, t_f] \quad (3.2.44)$$

where, $\mathbf{P}(t)$ is the solution of the matrix DRE (3.2.18), and $\mathbf{x}^*(t)$ is the solution of the closed-loop optimal system (3.2.42).

6. *Definiteness of the Matrix* $\mathbf{P}(t)$: Since $\mathbf{F}(t_f)$ is positive semidefinite, and $\mathbf{P}(t_f) = \mathbf{F}(t_f)$, we can easily say that $\mathbf{P}(t_f)$ is *positive semidefinite*. We can argue that $\mathbf{P}(t)$ is *positive definite* for all $t \in [t_0, t_f)$. Suppose that $\mathbf{P}(t)$ is not positive definite for some $t = t_s < t_f$, then there exists the corresponding state $\mathbf{x}^*(t_s)$ such that the cost function $\frac{1}{2}\mathbf{x}^{*\prime}(t_s)\mathbf{P}(t_s)\mathbf{x}^*(t_s) \leq 0$, which clearly violates that fact that *minimum cost has to be a positive quantity*. Hence, $\mathbf{P}(t)$ is *positive definite* for all $t \in [t_0, t_f)$. Since we already know that $\mathbf{P}(t)$ is symmetric, we now have that $\mathbf{P}(t)$ is *positive definite, symmetric* matrix.

7. *Computation of Matrix DRE:* Under some conditions we can get analytical solution for the nonlinear matrix DRE as shown later. But in general, we may try to solve the matrix DRE (3.2.18) by integrating *backwards* from its known final condition (3.2.19).

8. *Independence of the Riccati Coefficient Matrix* $\mathbf{P}(t)$: The matrix $\mathbf{P}(t)$ is independent of the optimal state $\mathbf{x}^*(t)$, so that once the system and the cost are specified, that is, once we are given the system/plant matrices $\mathbf{A}(t)$ and $\mathbf{B}(t)$, and the performance index matrices $\mathbf{F}(t_f)$, $\mathbf{Q}(t)$, and $\mathbf{R}(t)$, we can independently compute the matrix $\mathbf{P}(t)$ before the optimal system operates in the forward direction from its initial condition. Typically, we compute (off-line) the matrix $\mathbf{P}(t)$ *backward* in the interval $t \in [t_f, t_0]$ and store them separately, and feed these stored values when the system is operating in the *forward* direction in the interval $t \in [t_0, t_f]$.

9. *Implementation of the Optimal Control:* The block diagram implementing the closed-loop optimal controller (CLOC) is shown in Figure 3.2. The figure shows clearly that the CLOC gets its values of $\mathbf{P}(t)$ externally, after solving the matrix DRE backward in time from $t = t_f$ to $t = t_0$ and hence there is no way that we can implement the closed-loop optimal control configuration *on-line*.

It is to be noted that the optimal control $\mathbf{u}^*(t)$ can be solved and implemented in *open-loop* configuration by using the Pontryagin procedure given in Chapter 2. In that case, the open-loop optimal controller (OLOC) is quite cumbersome compared to the equivalent closed-loop optimal controller as will be illustrated later in this chapter.

10. *Linear Optimal Control:* The optimal feedback control $\mathbf{u}^*(t)$ given by (3.2.12) is written as

$$\boxed{\mathbf{u}^*(t) = -\mathbf{K}(t)\mathbf{x}^*(t)} \qquad (3.2.45)$$

where, the *Kalman gain* $\mathbf{K}(t) = \mathbf{R}^{-1}(t)\mathbf{B}'(t)\mathbf{P}(t)$. Or alternatively, we can write

$$\mathbf{u}^*(t) = -\mathbf{K}'_a(t)\mathbf{x}^*(t) \qquad (3.2.46)$$

where, $\mathbf{K}_a(t) = \mathbf{P}(t)\mathbf{B}(t)\mathbf{R}^{-1}(t)$. The previous optimal control is *linear* in state $\mathbf{x}^*(t)$. This is one of the nice features of the optimal control of linear systems with quadratic cost functionals. Also, note that the negative feedback in the optimal control relation (3.2.46) emerged from the *theory* of optimal control and was not introduced intentionally in our development.

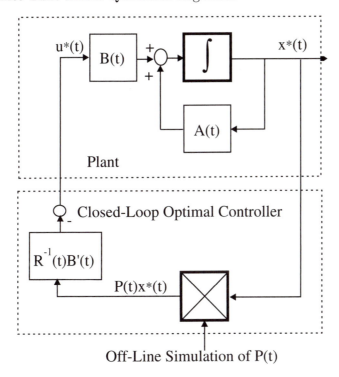

Figure 3.2 Closed-Loop Optimal Control Implementation

11. *Controllability:* Do we need the controllability condition on the system for implementing the optimal feedback control? No, as long as we are dealing with a finite time (t_f) system, because the contribution of those uncontrollable states (which are also unstable) to the cost function is still a *finite* quantity only. However, if we consider an *infinite* time interval, we certainly need the controllability condition, as we will see in the next section.

A historical note is very appropriate on the Riccati equation [22, 132].

The matrix Riccati equation has its origin in the scalar version of the equation

$$\dot{x}(t) = ax^2(t) + bx(t) + c \qquad (3.2.47)$$

with time varying coefficients, proposed by Jacopo Franceso Riccati around 1715. Riccati (1676-1754) gave the methods of solutions to the Riccati equation. However, the original paper by Riccati was not published immediately because he

*had the "suspicion" that the work was already known to
people such as the Bernoullis.*

*The importance of the Riccati equation, which has been
studied in the last two centuries by an extensive number
of scientists and engineers, need not be overstressed. The
matrix Riccati equation, which is a generalization in matrix
form of the original scalar equation, plays a very important
role in a range of control and systems theory areas such as
linear quadratic optimal control, stability, stochastic filtering and control, synthesis of passive networks, differential
games and more recently, H_∞-control and robust stabiliza-
tion and control.*

*Did Riccati ever imagine that his equation, proposed more
than a quarter millennium ago, would play such an important and ubiquitous role in modern control engineering and
other related fields?*

3.2.6 LQR System for General Performance Index

In this subsection, we address the state regulator system with a more
general performance index than given by (3.2.2). Consider a linear,
time-varying plant described by

$$\dot{\mathbf{x}}(t) = \mathbf{A}(t)\mathbf{x}(t) + \mathbf{B}(t)\mathbf{u}(t), \qquad (3.2.48)$$

with a cost functional

$$
\begin{aligned}
J(\mathbf{u}) &= \frac{1}{2}\mathbf{x}'(t_f)\mathbf{F}(t_f)\mathbf{x}(t_f) \\
&\quad + \frac{1}{2}\int_{t_0}^{t_f} \left[\mathbf{x}'(t)\mathbf{Q}(t)\mathbf{x}(t) + 2\mathbf{x}'(t)\mathbf{S}\mathbf{u}(t) + \mathbf{u}'(t)\mathbf{R}(t)\mathbf{u}(t)\right] dt \\
&= \frac{1}{2}\mathbf{x}'(t_f)\mathbf{F}(t_f)\mathbf{x}(t_f) \\
&\quad + \frac{1}{2}\int_{t_0}^{t_f} \begin{bmatrix}\mathbf{x}'(t) & \mathbf{u}'(t)\end{bmatrix} \begin{bmatrix}\mathbf{Q}(t) & \mathbf{S}(t) \\ \mathbf{S}'(t) & \mathbf{R}(t)\end{bmatrix} \begin{bmatrix}\mathbf{x}(t) \\ \mathbf{u}(t)\end{bmatrix} dt, \qquad (3.2.49)
\end{aligned}
$$

where, the various vectors and matrices are defined in earlier sections
and the nxr matrix $\mathbf{S}(t)$ is only a positive definite matrix.

Using the identical procedure as for the LQR system, we get the matrix differential Riccati equation as

$$\dot{\mathbf{P}}(t) = -\mathbf{P}(t)\mathbf{A}(t) - \mathbf{A}'(t)\mathbf{P}(t) - \mathbf{Q}(t)$$
$$+ [\mathbf{P}(t)\mathbf{B}(t) + \mathbf{S}(t)]\,\mathbf{R}^{-1}(t)[\mathbf{B}'(t)\mathbf{P}(t) + \mathbf{S}'(t)] \qquad (3.2.50)$$

with the final condition on $\mathbf{P}(t)$ as

$$\mathbf{P}(t_f) = \mathbf{F}(t_f). \qquad (3.2.51)$$

The optimal control is then given by

$$\mathbf{u}(t) = -\mathbf{R}^{-1}(t)\mathbf{B}'(t)\,[\mathbf{S}'(t) + \mathbf{P}(t)]\,\mathbf{x}(t). \qquad (3.2.52)$$

Obviously, when $\mathbf{S}(t)$ is made zero in the previous analysis, we get the previous results shown in Table 3.1.

3.3 Analytical Solution to the Matrix Differential Riccati Equation

In this section, we explore an analytical solution for the matric differential Riccati equation (DRE). This material is based on [138, 89]. Let us rewrite the Hamiltonian system (3.2.8) of the state and costate equations for the time-invariant case as (omitting * for the sake of simplicity)

$$\begin{bmatrix} \dot{\mathbf{x}}(t) \\ \dot{\boldsymbol{\lambda}}(t) \end{bmatrix} = \begin{bmatrix} \mathbf{A} & -\mathbf{E} \\ -\mathbf{Q} & -\mathbf{A}' \end{bmatrix} \begin{bmatrix} \mathbf{x}(t) \\ \boldsymbol{\lambda}(t) \end{bmatrix} \qquad (3.3.1)$$

where, $\mathbf{E} = \mathbf{B}\mathbf{R}^{-1}\mathbf{B}'$. Let

$$\Delta = \begin{bmatrix} \mathbf{A} & -\mathbf{E} \\ -\mathbf{Q} & -\mathbf{A}' \end{bmatrix}. \qquad (3.3.2)$$

Let us also recall that by the transformation $\boldsymbol{\lambda}(t) = \mathbf{P}(t)\mathbf{x}(t)$, we get the differential matrix Riccati equation (3.2.18), rewritten for (time-invariant matrices $\mathbf{A}, \mathbf{B}, \mathbf{Q}$ and $\mathbf{R}$) as

$$\dot{\mathbf{P}}(t) = -\mathbf{P}(t)\mathbf{A} - \mathbf{A}'\mathbf{P}(t) - \mathbf{Q} + \mathbf{P}(t)\mathbf{B}\mathbf{R}^{-1}\mathbf{B}'\mathbf{P}(t), \qquad (3.3.3)$$

with the final condition

$$\mathbf{P}(t_f) = \mathbf{F}(t_f). \qquad (3.3.4)$$

The solution $\mathbf{P}(t)$ can be obtained *analytically* (in contrast to numerical integration) in terms of the eigenvalues and eigenvectors of the Hamiltonian matrix Δ. In order to find analytical solution to the differential Riccati equation (3.3.3), it is necessary to show that if μ is an eigenvalue of the Hamiltonian matrix Δ in (3.3.2), then it implies that $-\mu$ is also the eigenvalue of Δ [89, 3]. For this, let us define

$$\Gamma = \begin{bmatrix} \mathbf{0} & \mathbf{I} \\ -\mathbf{I} & \mathbf{0} \end{bmatrix} \tag{3.3.5}$$

so that $\Gamma^{-1} = -\Gamma$. Then by a simple pre- and post-multiplication with Γ we get

$$\Delta = \Gamma\Delta'\Gamma = -\Gamma\Delta'\Gamma^{-1}. \tag{3.3.6}$$

Now, if μ is an eigenvalue of Δ with corresponding eigenvector $\mathbf{v}$,

$$\Delta\mathbf{v} = \mu\mathbf{v} \tag{3.3.7}$$

then

$$\Gamma\Delta'\Gamma\mathbf{v} = \mu\mathbf{v}, \qquad \Delta'\Gamma\mathbf{v} = -\mu\Gamma\mathbf{v} \tag{3.3.8}$$

where, we used $\Gamma^{-1} = -\Gamma$. Rearranging

$$(\Gamma\mathbf{v})'\Delta = -\mu(\Gamma\mathbf{v})' \tag{3.3.9}$$

Next, rearranging the eigenvalues of Δ as

$$\mathbf{D} = \begin{bmatrix} -\mathbf{M} & \mathbf{0} \\ \mathbf{0} & \mathbf{M} \end{bmatrix} \tag{3.3.10}$$

where, $\mathbf{M}(-\mathbf{M})$ is a diagonal matrix with right-half-plane (left-half plane) eigenvalues. Let $\mathbf{W}$, the modal matrix of eigenvectors corresponding to $\mathbf{D}$, be defined as

$$\mathbf{W} = \begin{bmatrix} \mathbf{W}_{11} & \mathbf{W}_{12} \\ \mathbf{W}_{21} & \mathbf{W}_{22} \end{bmatrix}, \tag{3.3.11}$$

where, $[\mathbf{W}_{11} \ \ \mathbf{W}_{21}]'$ are the n eigenvectors of the left-half-plane (stable) eigenvalues of Δ. Also,

$$\mathbf{W}^{-1}\Delta\mathbf{W} = \mathbf{D}. \tag{3.3.12}$$

Let us now define a state transformation

$$\begin{bmatrix} \mathbf{x}(t) \\ \boldsymbol{\lambda}(t) \end{bmatrix} = \mathbf{W} \begin{bmatrix} \mathbf{w}(t) \\ \mathbf{z}(t) \end{bmatrix} = \begin{bmatrix} \mathbf{W}_{11} & \mathbf{W}_{12} \\ \mathbf{W}_{21} & \mathbf{W}_{22} \end{bmatrix} \begin{bmatrix} \mathbf{w}(t) \\ \mathbf{z}(t) \end{bmatrix}. \tag{3.3.13}$$

Then, using (3.3.12) and (3.3.13), the Hamiltonian system (3.3.1) becomes

$$\begin{bmatrix} \dot{\mathbf{w}}(t) \\ \dot{\mathbf{z}}(t) \end{bmatrix} = \mathbf{W}^{-1} \begin{bmatrix} \dot{\mathbf{x}}(t) \\ \dot{\boldsymbol{\lambda}}(t) \end{bmatrix} = \mathbf{W}^{-1} \Delta \begin{bmatrix} \mathbf{x}(t) \\ \boldsymbol{\lambda}(t) \end{bmatrix} = \mathbf{W}^{-1} \Delta \mathbf{W} \begin{bmatrix} \mathbf{w}(t) \\ \mathbf{z}(t) \end{bmatrix}$$

$$= \mathbf{D} \begin{bmatrix} \mathbf{w}(t) \\ \mathbf{z}(t) \end{bmatrix}. \tag{3.3.14}$$

Solving (3.3.14) in terms of the known final conditions, we have

$$\begin{bmatrix} \mathbf{w}(t) \\ \mathbf{z}(t) \end{bmatrix} = \begin{bmatrix} e^{-\mathbf{M}(t-t_f)} & \mathbf{0} \\ \mathbf{0} & e^{\mathbf{M}(t-t_f)} \end{bmatrix} \begin{bmatrix} \mathbf{w}(t_f) \\ \mathbf{z}(t_f) \end{bmatrix}. \tag{3.3.15}$$

Rewriting (3.3.15)

$$\begin{bmatrix} \mathbf{w}(t_f) \\ \mathbf{z}(t) \end{bmatrix} = \begin{bmatrix} e^{\mathbf{M}(t-t_f)} & \mathbf{0} \\ \mathbf{0} & e^{\mathbf{M}(t-t_f)} \end{bmatrix} \begin{bmatrix} \mathbf{w}(t) \\ \mathbf{z}(t_f) \end{bmatrix}. \tag{3.3.16}$$

Next, from (3.3.13) and using the final condition (3.3.4)

$$\boldsymbol{\lambda}(t_f) = \mathbf{W}_{21} \mathbf{w}(t_f) + \mathbf{W}_{22} \mathbf{z}(t_f)$$
$$= \mathbf{F} \mathbf{x}(t_f)$$
$$= \mathbf{F} \left[\mathbf{W}_{11} \mathbf{w}(t_f) + \mathbf{W}_{12} \mathbf{z}(t_f) \right]. \tag{3.3.17}$$

Solving the previous relation for $\mathbf{z}(t_f)$ in terms of $\mathbf{w}(t_f)$

$$\mathbf{z}(t_f) = \mathbf{T}(t_f) \mathbf{w}(t_f), \text{ where}$$
$$\mathbf{T}(t_f) = - \left[\mathbf{W}_{22} - \mathbf{F} \mathbf{W}_{12} \right]^{-1} \left[\mathbf{W}_{21} - \mathbf{F} \mathbf{W}_{11} \right]. \tag{3.3.18}$$

Again, from (3.3.16)

$$\mathbf{z}(t) = e^{-\mathbf{M}(t_f - t)} \mathbf{z}(t_f)$$
$$= e^{-\mathbf{M}(t_f - t)} \mathbf{T}(t_f) \mathbf{w}(t_f)$$
$$= e^{-\mathbf{M}(t_f - t)} \mathbf{T}(t_f) e^{-\mathbf{M}(t_f - t)} \mathbf{w}(t). \tag{3.3.19}$$

Rewriting the previous relation as

$$\mathbf{z}(t) = \mathbf{T}(t) \mathbf{w}(t), \text{ where,}$$
$$\mathbf{T}(t) = e^{-\mathbf{M}(t_f - t)} \mathbf{T}(t_f) e^{-\mathbf{M}(t_f - t)}. \tag{3.3.20}$$

Finally, to relate $\mathbf{P}(t)$ in (3.3.3) to the relation (3.3.20) for $\mathbf{T}(t)$, let us use (3.3.13) to write

$$
\begin{aligned}
\boldsymbol{\lambda}(t) &= \mathbf{W}_{21}\mathbf{w}(t) + \mathbf{W}_{22}\mathbf{z}(t) \\
&= \mathbf{P}(t)\mathbf{x}(t) \\
&= \mathbf{P}(t)\left[\mathbf{W}_{11}\mathbf{w}(t) + \mathbf{W}_{12}\mathbf{z}(t)\right]
\end{aligned}
\tag{3.3.21}
$$

and by (3.3.20), the previous relation can be written as

$$
\left[\mathbf{W}_{21} + \mathbf{W}_{22}\mathbf{T}(t)\right]\mathbf{w}(t) = \mathbf{P}(t)\left[\mathbf{W}_{11} + \mathbf{W}_{12}\mathbf{T}(t)\right]\mathbf{w}(t). \tag{3.3.22}
$$

Since the previous relation should hold good for all $\mathbf{x}(t_0)$ and hence for all states $\mathbf{w}(t)$, it implies that the analytical expression to the solution of $\mathbf{P}(t)$ is given by

$$
\boxed{\mathbf{P}(t) = \left[\mathbf{W}_{21} + \mathbf{W}_{22}\mathbf{T}(t)\right]\left[\mathbf{W}_{11} + \mathbf{W}_{12}\mathbf{T}(t)\right]^{-1}.} \tag{3.3.23}
$$

3.3.1 MATLAB© Implementation of Analytical Solution to Matrix DRE

The solution of the matrix DRE (3.2.34) is not readily available with MATLAB and hence a MATLAB-based program was developed for solving the matrix DRE based on the analytical solution of matrix DRE [138] described earlier. The MATLAB solution is illustrated by the following example.

Example 3.1

Let us illustrate the previous procedure with a simple second order example. Given a double integral system

$$
\begin{aligned}
\dot{x}_1(t) &= x_2(t), \quad x_1(0) = 2 \\
\dot{x}_2(t) &= -2x_1(t) + x_2(t) + u(t), \quad x_2(0) = -3,
\end{aligned}
\tag{3.3.24}
$$

and the performance index (PI)

$$
\begin{aligned}
J = &\frac{1}{2}\left[x_1^2(5) + x_1(5)x_2(5) + 2x_2^2(5)\right] \\
&+ \frac{1}{2}\int_0^5 \left[2x_1^2(t) + 6x_1(t)x_2(t) + 5x_2^2(t) + 0.25u^2(t)\right]dt,
\end{aligned}
\tag{3.3.25}
$$

obtain the feedback control law.

Solution: Comparing the present plant (3.3.24) and the PI (3.3.25) of the problem with the corresponding general formulations of the plant (3.2.31) and the PI (3.2.32), respectively, let us first identify the various quantities as

$$\mathbf{A}(t) = \begin{bmatrix} 0 & 1 \\ -2 & 1 \end{bmatrix}; \quad \mathbf{B}(t) = \begin{bmatrix} 0 \\ 1 \end{bmatrix}; \quad \mathbf{F}(t_f) = \begin{bmatrix} 1 & 0.5 \\ 0.5 & 2 \end{bmatrix}$$

$$\mathbf{Q}(t) = \begin{bmatrix} 2 & 3 \\ 3 & 5 \end{bmatrix}; \quad \mathbf{R}(t) = r(t) = \frac{1}{4}; \quad t_0 = 0; \quad t_f = 5.$$

It is easy to check that the system (3.3.24) is unstable. Let $\mathbf{P}(t)$ be the 2x2 symmetric matrix

$$\mathbf{P}(t) = \begin{bmatrix} p_{11}(t) & p_{12}(t) \\ p_{12}(t) & p_{22}(t) \end{bmatrix}. \tag{3.3.26}$$

Then, the optimal control (3.2.33) is given by

$$u^*(t) = -4 \begin{bmatrix} 0 & 1 \end{bmatrix} \begin{bmatrix} p_{11}(t) & p_{12}(t) \\ p_{12}(t) & p_{22}(t) \end{bmatrix} \begin{bmatrix} x_1^*(t) \\ x_2^*(t) \end{bmatrix}$$

$$= -4[p_{12}(t)x_1^*(t) + p_{22}(t)x_2^*(t)] \tag{3.3.27}$$

where, $\mathbf{P}(t)$, the 2x2 symmetric, positive definite matrix, is the solution of the matrix DRE (3.2.34)

$$\begin{bmatrix} \dot{p}_{11}(t) & \dot{p}_{12}(t) \\ \dot{p}_{12}(t) & \dot{p}_{22}(t) \end{bmatrix} = - \begin{bmatrix} p_{11}(t) & p_{12}(t) \\ p_{12}(t) & p_{22}(t) \end{bmatrix} \begin{bmatrix} 0 & 1 \\ -2 & 1 \end{bmatrix}$$

$$- \begin{bmatrix} 0 & -2 \\ 1 & 1 \end{bmatrix} \begin{bmatrix} p_{11}(t) & p_{12}(t) \\ p_{12}(t) & p_{22}(t) \end{bmatrix}$$

$$+ \begin{bmatrix} p_{11}(t) & p_{12}(t) \\ p_{12}(t) & p_{22}(t) \end{bmatrix} \begin{bmatrix} 0 \\ 1 \end{bmatrix} 4 \begin{bmatrix} 0 & 1 \end{bmatrix} \begin{bmatrix} p_{11}(t) & p_{12}(t) \\ p_{12}(t) & p_{22}(t) \end{bmatrix}$$

$$- \begin{bmatrix} 2 & 3 \\ 3 & 5 \end{bmatrix} \tag{3.3.28}$$

satisfying the final condition (3.2.35)

$$\begin{bmatrix} p_{11}(5) & p_{12}(5) \\ p_{12}(5) & p_{22}(5) \end{bmatrix} = \begin{bmatrix} 1 & 0.5 \\ 0.5 & 2 \end{bmatrix}. \tag{3.3.29}$$

Simplifying the matrix DRE (3.3.28), we get

$$\dot{p}_{11}(t) = 4p_{12}^2(t) + 4p_{12}(t) - 2,$$
$$p_{11}(5) = 1,$$
$$\dot{p}_{12}(t) = -p_{11}(t) - p_{12}(t) + 2p_{22}(t) + 4p_{12}(t)p_{22}(t) - 3,$$
$$p_{12}(5) = 0.5$$
$$\dot{p}_{22}(t) = -2p_{12}(t) - 2p_{22}(t) + 4p_{22}^2(t) - 5,$$
$$p_{22}(5) = 2.$$

$$(3.3.30)$$

Solving the previous set of nonlinear, differential equations *backward* in time with the given final conditions, one can obtain the numerical solutions for the Riccati coefficient matrix $\mathbf{P}(t)$. However, here the solutions are obtained using the analytical solution as given earlier in this section. The solutions for the Riccati coefficients are plotted in Figure 3.3. Using these Riccati coefficients, the closed-loop optimal control system is shown in Figure 3.4. Using the optimal control $u^*(t)$ given by (3.3.27), the plant equations (3.3.24) are solved *forward* in time to obtain the optimal states $x_1^*(t)$ and $x_2^*(t)$ as shown in Figure 3.5 for the initial conditions $[2 - 3]'$. Finally, the optimal control $u^*(t)$ is shown in Figure 3.6. The previous results are obtained using Control System Toolbox of the MATLAB©, Version 6 as shown below.

The following MATLAB© *m* file for Example 3.1 requires two additional MATLAB© files *lqrnss.m* which itself requires *lqrnssf.m* given in Appendix C.

```
****************************************************
%% Solution using Control System Toolbox of
%% the MATLAB. Version 6
%% The following file example.m requires
%% two other files lqrnss.m and lqrnssf.m
%% which are given in Appendix
clear all
A=[0.,1.;-2.,1.];
B=[0.;1.];
Q=[2.,3.;3.,5.];
F=[1.,0.5;0.5,2.];
R=[.25];
tspan=[0 5];
x0=[2.,-3.];
[x,u,K]=lqrnss(A,B,F,Q,R,x0,tspan);
****************************************************
```

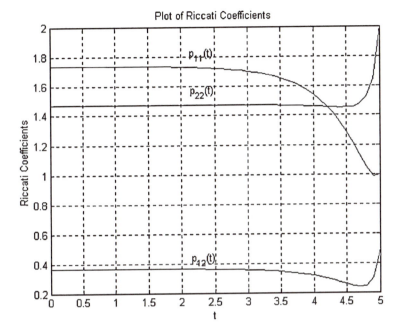

Figure 3.3   Riccati Coefficients for Example 3.1

3.4   *Infinite-Time LQR System I*

In this section, let us make the terminal (final) time t_f to be infinite in the previous linear, *time-varying*, quadratic regulator system. Then, this is called the *infinite-time* (or infinite horizon) linear quadratic regulator system [6, 3].

Consider a linear, time-varying plant

$$\dot{\mathbf{x}}(t) = \mathbf{A}(t)\mathbf{x}(t) + \mathbf{B}(t)\mathbf{u}(t), \qquad (3.4.1)$$

and a quadratic performance index

$$J = \frac{1}{2}\int_{t_0}^{\infty} \left[\mathbf{x}'(t)\mathbf{Q}(t)\mathbf{x}(t) + \mathbf{u}'(t)\mathbf{R}(t)\mathbf{u}(t)\right] dt, \qquad (3.4.2)$$

where, $\mathbf{u}(t)$ is not constrained. Also, $\mathbf{Q}(t)$ is $n \times n$ symmetric, positive *semidefinite* matrix, and $\mathbf{R}(t)$ is an $r \times r$ symmetric, positive *definite* matrix. Note, it makes no engineering sense to have a terminal cost term with terminal time being infinite.

This problem cannot always be solved without some special conditions. For example, if any one of the states is uncontrollable and/or unstable, the corresponding performance measure J will become infinite

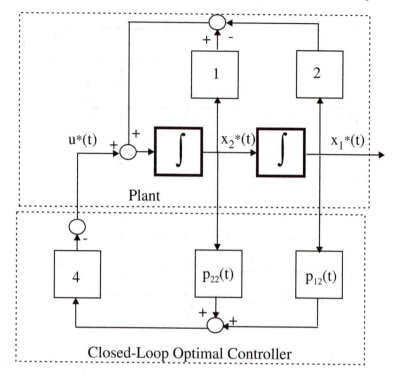

Figure 3.4   Closed-Loop Optimal Control System for Example 3.1

and makes no physical sense. On the other hand, with the finite-time system, the performance measure is always finite. Thus, we need to impose the condition that the system (3.4.1) is *completely controllable*.

Using results similar to the previous case of finite final time t_f (see Table 3.1), the optimal control for the infinite-horizon linear regulator system is obtained as

$$\mathbf{u}^*(t) = -\mathbf{R}^{-1}(t)\mathbf{B}'(t)\hat{\mathbf{P}}(t)\mathbf{x}^*(t), \tag{3.4.3}$$

where,

$$\hat{\mathbf{P}}(t) = \lim_{t_f \to \infty} \{\mathbf{P}(t)\}, \tag{3.4.4}$$

the $n{\times}n$ symmetric, positive definite matrix (for all $t \in [t_0,\, t_f]$) is the solution of the matrix differential Riccati equation (DRE)

$$\dot{\hat{\mathbf{P}}}(t) = -\hat{\mathbf{P}}(t)\mathbf{A}(t) - \mathbf{A}'(t)\hat{\mathbf{P}}(t) - \mathbf{Q}(t) + \hat{\mathbf{P}}(t)\mathbf{B}(t)\mathbf{R}^{-1}(t)\mathbf{B}'(t)\hat{\mathbf{P}}(t),$$

$$\tag{3.4.5}$$

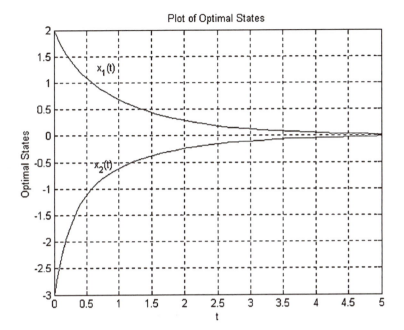

Figure 3.5   Optimal States for Example 3.1

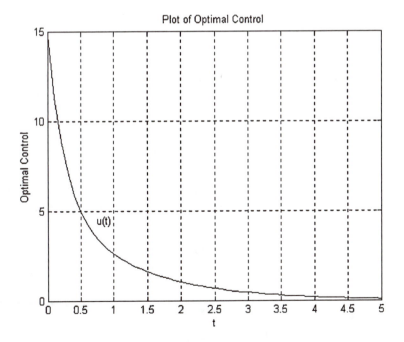

Figure 3.6   Optimal Control for Example 3.1

satisfying the final condition

$$\lim_{t_f \to \infty} \hat{\mathbf{P}}(t_f) = 0. \tag{3.4.6}$$

The optimal cost is given by

$$J^* = \frac{1}{2}\mathbf{x}^{*\prime}(t)\hat{\mathbf{P}}(t)\mathbf{x}^*(t). \tag{3.4.7}$$

The proofs for the previous results are found in optimal control text specializing in quadratic methods [3]. Example 3.1 can be easily solved for $t_f \to \infty$ and $\mathbf{F} = 0$.

3.4.1　Infinite-Time Linear Quadratic Regulator: Time-Varying Case: Summary

Consider a linear, time-varying plant

$$\dot{\mathbf{x}}(t) = \mathbf{A}(t)\mathbf{x}(t) + \mathbf{B}(t)\mathbf{u}(t), \tag{3.4.8}$$

and a quadratic performance index

$$J = \frac{1}{2}\int_{t_0}^{\infty} [\mathbf{x}'(t)\mathbf{Q}(t)\mathbf{x}(t) + \mathbf{u}'(t)\mathbf{R}(t)\mathbf{u}(t)]\,dt, \tag{3.4.9}$$

where, $\mathbf{u}(t)$ is *not constrained* and $\mathbf{x}(t_f), t_f \to \infty$ is not specified. Also, $\mathbf{Q}(t)$ is *nxn* symmetric, positive semidefinite matrix, and $\mathbf{R}(t)$ is *rxr* symmetric, positive definite matrix. Then, the optimal control is given by

$$\boxed{\mathbf{u}^*(t) = -\mathbf{R}^{-1}(t)\mathbf{B}'(t)\hat{\mathbf{P}}(t)\mathbf{x}^*(t)} \tag{3.4.10}$$

where, $\hat{\mathbf{P}}(t)$, the *nxn* symmetric, positive definite matrix (for all $t \in [t_0, t_f]$, is the solution of the matrix differential Riccati equation (DRE)

$$\boxed{\dot{\hat{\mathbf{P}}}(t) = -\hat{\mathbf{P}}(t)\mathbf{A}(t) - \mathbf{A}'(t)\hat{\mathbf{P}}(t) - \mathbf{Q}(t) + \hat{\mathbf{P}}(t)\mathbf{B}(t)\mathbf{R}^{-1}(t)\mathbf{B}'(t)\hat{\mathbf{P}}(t)}$$
$$\tag{3.4.11}$$

satisfying the final condition

$$\hat{\mathbf{P}}(t = t_f \to \infty) = 0. \tag{3.4.12}$$

Table 3.2   Procedure Summary of Infinite-Time Linear Quadratic Regulator System: Time-Varying Case

A. Statement of the Problem	
Given the plant as $\dot{\mathbf{x}}(t) = \mathbf{A}(t)\mathbf{x}(t) + \mathbf{B}(t)\mathbf{u}(t)$, the performance index as $J = \frac{1}{2}\int_{t_0}^{\infty}\left[\mathbf{x}'(t)\mathbf{Q}(t)\mathbf{x}(t) + \mathbf{u}'(t)\mathbf{R}(t)\mathbf{u}(t)\right]dt$, and the boundary conditions as $\mathbf{x}(t_0) = \mathbf{x}_0;\quad \mathbf{x}(\infty)$ is free, find the optimal control, state and performance index.	
B. Solution of the Problem	
Step 1	Solve the matrix differential Riccati equation (DRE) $\dot{\hat{\mathbf{P}}}(t) = -\hat{\mathbf{P}}(t)\mathbf{A}(t) - \mathbf{A}'(t)\hat{\mathbf{P}}(t) - \mathbf{Q}(t) + \hat{\mathbf{P}}(t)\mathbf{B}(t)\mathbf{R}^{-1}(t)\mathbf{B}'(t)\hat{\mathbf{P}}(t)$ with final condition $\hat{\mathbf{P}}(t = t_f) = 0$.
Step 2	Solve the optimal state $\mathbf{x}^*(t)$ from $\mathbf{x}^*(t) = \left[\mathbf{A}(t) - \mathbf{B}(t)\mathbf{R}^{-1}(t)\mathbf{B}'(t)\hat{\mathbf{P}}(t)\right]\mathbf{x}^*(t)$ with initial condition $\mathbf{x}(t_0) = \mathbf{x}_0$.
Step 3	Obtain the optimal control $\mathbf{u}^*(t)$ from $\mathbf{u}^*(t) = -\mathbf{R}^{-1}(t)\mathbf{B}'(t)\hat{\mathbf{P}}(t)\mathbf{x}^*(t)$.
Step 4	Obtain the optimal performance index from $J^* = \frac{1}{2}\mathbf{x}^{*\prime}(t)\hat{\mathbf{P}}(t)\mathbf{x}^*(t)$.

The optimal state is the solution of

$$\boxed{\dot{\mathbf{x}}^*(t) = \left[\mathbf{A}(t) - \mathbf{B}(t)\mathbf{R}^{-1}(t)\mathbf{B}'(t)\hat{\mathbf{P}}(t)\right]\mathbf{x}^*(t)} \qquad (3.4.13)$$

and the optimal cost is

$$\boxed{J^* = \frac{1}{2}\mathbf{x}^{*\prime}(t)\hat{\mathbf{P}}(t)\mathbf{x}^*(t).} \qquad (3.4.14)$$

The optimal control $\mathbf{u}^*(t)$ given by (3.4.10) is *linear* in the optimal state $\mathbf{x}^*(t)$. The entire procedure is now summarized in Table 3.2.

3.5   *Infinite-Time LQR System II*

In this section, we examine the state regulator system with infinite time interval for a *linear time-invariant* (LTI) system. Let us consider the plant as

$$\dot{\mathbf{x}}(t) = \mathbf{A}\mathbf{x}(t) + \mathbf{B}\mathbf{u}(t) \tag{3.5.1}$$

and the cost functional as

$$J = \frac{1}{2} \int_0^\infty \left[\mathbf{x}'(t)\mathbf{Q}\mathbf{x}(t) + \mathbf{u}'(t)\mathbf{R}\mathbf{u}(t) \right] dt \tag{3.5.2}$$

where, $\mathbf{x}(t)$ is nth order state vector; $\mathbf{u}(t)$ is rth order control vector; $\mathbf{A}$ is nxn-order state matrix; $\mathbf{B}$ is rxr-order control matrix; $\mathbf{Q}$ is nxn-order, symmetric, positive *semidefinite* matrix; $\mathbf{R}$ is rxr-order, symmetric, positive *definite* matrix. First of all, let us discuss some of the implications of the time-invariance and the infinite final-time.

1. The infinite time interval case is considered for the following reasons:

   (a) We wish to make sure that the state-regulator stays near zero state after the initial transient.

   (b) We want to include any special case of large final time.

2. With infinite final-time interval, to include the *final cost function* does not make any practical sense. Hence, the final cost term involving $\mathbf{F}(t_f)$ does not exist in the cost functional (3.5.2).

3. With infinite final-time interval, the system (3.5.1) has to be completely *controllable*. Let us recall that this controllability condition of the plant (3.5.1) requires that the controllability matrix (see Appendix B)

$$\begin{bmatrix} \mathbf{B} & \mathbf{A}\mathbf{B} \cdots \mathbf{A}^{n-1}\mathbf{B} \end{bmatrix} \tag{3.5.3}$$

must be *nonsingular* or contain n *linearly independent* column vectors. The controllability requirement guarantees that the optimal cost is *finite*. On the other hand, if the system is not controllable and some or all of those uncontrollable states are unstable, then the cost functional would be *infinite* since the control interval is infinite. In such situations, we cannot distinguish optimal control from the other controls. Alternatively, we can assume that the system (3.5.1) is completely stabilizable.

As before in the case of finite final-time interval, we can proceed and obtain the closed-loop optimal control and the associated Riccati equation. Still $\mathbf{P}(t)$ must be the solution of the matrix differential Riccati equation (3.2.34) with boundary condition $\mathbf{P}(t_f) = 0$. It was shown that the assumptions of [70]

1. controllability and

2. $F(t_f) = 0$

imply that

$$\lim_{t_f \to \infty} \{\mathbf{P}(t_f)\} = \bar{\mathbf{P}} \qquad (3.5.4)$$

where, $\bar{\mathbf{P}}$ is the *nxn positive definite*, symmetric, *constant* matrix. If $\bar{\mathbf{P}}$ is constant, then $\bar{\mathbf{P}}$ is the solution of the nonlinear, matrix, *algebraic Riccati equation* (ARE),

$$\frac{d\bar{\mathbf{P}}}{dt} = 0 = -\bar{\mathbf{P}}\mathbf{A} - \mathbf{A}'\bar{\mathbf{P}} + \bar{\mathbf{P}}\mathbf{B}\mathbf{R}^{-1}\mathbf{B}'\bar{\mathbf{P}} - \mathbf{Q}. \qquad (3.5.5)$$

Alternatively, we can write (3.5.5) as

$$\bar{\mathbf{P}}\mathbf{A} + \mathbf{A}'\bar{\mathbf{P}} + \mathbf{Q} - \bar{\mathbf{P}}\mathbf{B}\mathbf{R}^{-1}\mathbf{B}'\bar{\mathbf{P}} = 0. \qquad (3.5.6)$$

Note, for a *time-varying* system with finite-time interval, we have the *differential* Riccati equation (3.2.34), whereas for a linear *time-invariant* system with *infinite-time* horizon, we have the *algebraic* Riccati equation (3.5.6).

A historical note on R.E. Kalman is appropriate (from SIAM News, 6/94 - article about R.E. Kalman).

Rudolph E. Kalman is best known for the linear filtering technique that he and Richard Bucy [31] developed in 1960–1961 to strip unwanted noise out of a stream of data [71, 74, 76]. The Kalman filter, which is based on the use of state-space techniques and recursive algorithms, revolutionized the field of estimation. The Kalman filter is widely used in navigational and guidance systems, radar tracking, sonar ranging, and satellite orbit determination (for the Ranger, Apollo, and Mariner missions, for instance), as well as in fields as diverse as seismic data processing, nuclear power plant instrumentation, and econometrics. Among Kalman's many outstanding contributions were the formulation and study of most fundamental state-space notions [72, 73, 77] including controllability, observability, minimality, realizability from input and output data, matrix Riccati equations, linear-quadratic control [70, 75, 75], and the separation principle that are today ubiquitous in control. While some of these concepts were also encountered in other contexts, such as optimal control theory, it was Kalman who recognized the central role that they play in systems analysis.

Born in Hungary, Kalman received BS and MS degrees from the Massachusetts Institute of Technology (MIT) and a DSci in engineering from Columbia University in 1957. In the early years of his career he held research positions at International Business Machines (IBM) and at the Research Institute for Advanced Studies (RIAS) in Baltimore. From 1962 to 1971, he was at Stanford University. In 1971, he became a graduate research professor and director of the Center for Mathematical System Theory at the University of Florida, Gainesville, USA, and later retired with emeritus status. Kalman's contributions to control theory and to applied mathematics and engineering in general have been widely recognized with several honors and awards.

3.5.1 *Meaningful Interpretation of Riccati Coefficient*

Consider the matrix *differential* Riccati equation (3.2.34) with final condition $\mathbf{P}(t_f) = 0$. Now consider a simple time transformation $\tau = t_f - t$. Then, in τ scale we can think of the final time t_f as the "starting time," $\mathbf{P}(t_f)$ as the "initial condition," and $\bar{\mathbf{P}}$ as the "steady-state solution" of the matrix DRE. As the time $t_f \to \infty$, the "transient solution" is pushed to near t_f which is at infinity. Then for most of the practical time interval the matrix $\mathbf{P}(t)$ becomes a steady state, i.e., a constant matrix $\bar{\mathbf{P}}$, as shown in Figure 3.7 [6]. Then the optimal control is given by

$$\mathbf{u}^*(t) = -\mathbf{R}^{-1}\mathbf{B}'\bar{\mathbf{P}}\mathbf{x}^*(t) = -\bar{\mathbf{K}}\mathbf{x}^*(t), \qquad (3.5.7)$$

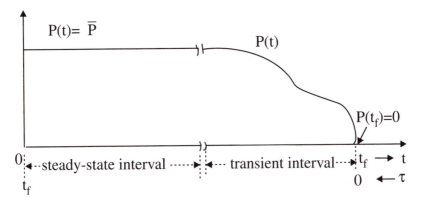

Figure 3.7   Interpretation of the Constant Matrix $\bar{\mathbf{P}}$

where, $\bar{\mathbf{K}} = \mathbf{R}^{-1}\mathbf{B}'\bar{\mathbf{P}}$ is called the *Kalman gain*. Alternatively, we can write

$$\mathbf{u}^*(t) = -\bar{\mathbf{K}}'_a\mathbf{x}^*(t) \tag{3.5.8}$$

where, $\bar{\mathbf{K}}_a = \bar{\mathbf{P}}\mathbf{B}\mathbf{R}^{-1}$. The optimal state is the solution of the system obtained by using the control (3.5.8) in the plant (3.5.1)

$$\dot{\mathbf{x}}^*(t) = \left[\mathbf{A} - \mathbf{B}\mathbf{R}^{-1}\mathbf{B}'\bar{\mathbf{P}}\right]\mathbf{x}^*(t) = \mathbf{G}\mathbf{x}^*(t), \tag{3.5.9}$$

where, the matrix $\mathbf{G} = \mathbf{A} - \mathbf{B}\mathbf{R}^{-1}\mathbf{B}'\bar{\mathbf{P}}$ must have *stable* eigenvalues so that the closed-loop optimal system (3.5.9) is *stable*. This is required since any unstable states with infinite time interval would lead to an *infinite* cost functional J^*. Let us note that we have no constraint on the stability of the *original* system (3.5.1). This means that although the *original* system may be *unstable*, the *optimal* system must be definitely *stable*.

Finally, the minimum cost (3.2.29) is given by

$$J^* = \frac{1}{2}\mathbf{x}^{*\prime}(t)\bar{\mathbf{P}}\mathbf{x}^*(t). \tag{3.5.10}$$

3.5.2   *Analytical Solution of the Algebraic Riccati Equation*

The next step is to find the analytical expression to the steady-state (limiting) solution of the differential Riccati equation (3.3.3). Thus, we are interested in finding the analytical solution to the algebraic Riccati equation (3.5.5). Obviously, one can let the terminal time t_f tend to

∞ in the solution (3.3.23) for $\mathbf{P}(t_f)$. As $t_f \to \infty$, $e^{-\mathbf{M}(t_f-t)}$ goes to zero, which in turn makes $\mathbf{T}(t)$ tend to zero. Thus, under the usual conditions of $(\mathbf{A}, \mathbf{B})$ being stabilizable and $(\mathbf{A}, \sqrt{Q})$ being reachable, and $\mathbf{T} = 0$, we have from (3.3.23)

$$\lim_{t_f \to \infty} \mathbf{P}(t, t_f) = \bar{\mathbf{P}} = \mathbf{W}_{21}\mathbf{W}_{11}^{-1}. \qquad (3.5.11)$$

Thus, the solution to the ARE is constructed by using the stable eigenvectors of the Hamiltonian matrix. For further treatment on this topic, consult [3] and the references therein.

3.5.3   Infinite-Interval Regulator System: Time-Invariant Case: Summary

For a controllable, linear, time-invariant plant

$$\dot{\mathbf{x}}(t) = \mathbf{A}\mathbf{x}(t) + \mathbf{B}\mathbf{u}(t), \qquad (3.5.12)$$

and the infinite interval cost functional

$$J = \frac{1}{2} \int_0^\infty \left[\mathbf{x}'(t)\mathbf{Q}\mathbf{x}(t) + \mathbf{u}'(t)\mathbf{R}\mathbf{u}(t)\right] dt, \qquad (3.5.13)$$

the optimal control is given by

$$\boxed{\mathbf{u}^*(t) = -\mathbf{R}^{-1}\mathbf{B}'\bar{\mathbf{P}}\mathbf{x}^*(t)} \qquad (3.5.14)$$

where, $\bar{\mathbf{P}}$, the $n \times n$ constant, *positive definite*, symmetric matrix, is the solution of the nonlinear, matrix *algebraic Riccati equation* (ARE)

$$\boxed{-\bar{\mathbf{P}}\mathbf{A} - \mathbf{A}'\bar{\mathbf{P}} + \bar{\mathbf{P}}\mathbf{B}\mathbf{R}^{-1}\mathbf{B}'\bar{\mathbf{P}} - \mathbf{Q} = 0} \qquad (3.5.15)$$

the optimal trajectory is the solution of

$$\boxed{\dot{\mathbf{x}}^*(t) = \left[\mathbf{A} - \mathbf{B}\mathbf{R}^{-1}\mathbf{B}'\bar{\mathbf{P}}\right]\mathbf{x}^*(t)} \qquad (3.5.16)$$

and the optimal cost is given by

$$\boxed{J^* = \frac{1}{2}\mathbf{x}^{*\prime}(t)\bar{\mathbf{P}}\mathbf{x}^*(t).} \qquad (3.5.17)$$

The entire procedure is now summarized in Table 3.3 and the implementation of the closed-loop optimal control (CLOC) is shown in Figure 3.8

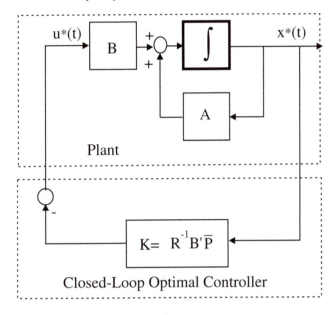

Figure 3.8   Implementation of the Closed-Loop Optimal Control:
Infinite Final Time

Next, an example is given to illustrate the infinite interval regulator
system and the associated matrix algebraic Riccati equation. Let us
reconsider the same Example 3.1 with final time $t_f \to \infty$ and $\mathbf{F} = 0$.

Example 3.2

Given a second order plant

$$\dot{x}_1(t) = x_2(t), \qquad x_1(0) = 2$$
$$\dot{x}_2(t) = -2x_1(t) + x_2(t) + u(t), \qquad x_2(0) = -3 \qquad (3.5.18)$$

and the performance index

$$J = \frac{1}{2} \int_0^\infty \left[2x_1^2(t) + 6x_1(t)x_2(t) + 5x_2^2(t) + 0.25u^2(t) \right] dt, \quad (3.5.19)$$

obtain the feedback optimal control law.

Solution: Comparing the plant (3.5.18) and PI (3.5.19) of the
present system with the corresponding general formulation of plant
(3.5.12) and PI (3.5.13), respectively, let us first identify the various

Table 3.3 Procedure Summary of Infinite-Interval Linear
Quadratic Regulator System: Time-Invariant Case

A. Statement of the Problem
Given the plant as $\dot{\mathbf{x}}(t) = \mathbf{A}\mathbf{x}(t) + \mathbf{B}\mathbf{u}(t)$, the performance index as $J = \frac{1}{2} \int_0^\infty [\mathbf{x}'(t)\mathbf{Q}\mathbf{x}(t) + \mathbf{u}'(t)\mathbf{R}\mathbf{u}(t)] \, dt$, and the boundary conditions as $\mathbf{x}(t_0) = \mathbf{x}_0; \quad \mathbf{x}(\infty) = 0$, find the optimal control, state and index.

B. Solution of the Problem	
Step 1	Solve the matrix algebraic Riccati equation (ARE) $-\bar{\mathbf{P}}\mathbf{A} - \mathbf{A}'\bar{\mathbf{P}} - \mathbf{Q} + \bar{\mathbf{P}}\mathbf{B}\mathbf{R}^{-1}\mathbf{B}'\bar{\mathbf{P}} = 0..$
Step 2	Solve the optimal state $\mathbf{x}^*(t)$ from $\dot{\mathbf{x}}^*(t) = \left[\mathbf{A} - \mathbf{B}\mathbf{R}^{-1}\mathbf{B}'\bar{\mathbf{P}}\right]\mathbf{x}^*(t)$ with initial condition $\mathbf{x}(t_0) = \mathbf{x}_0$.
Step 3	Obtain the optimal control $\mathbf{u}^*(t)$ from $\mathbf{u}^*(t) = -\mathbf{R}^{-1}\mathbf{B}'\bar{\mathbf{P}}\mathbf{x}^*(t)$.
Step 4	Obtain the optimal performance index from $J^* = \frac{1}{2}\mathbf{x}^{*\prime}(t)\bar{\mathbf{P}}\mathbf{x}^*(t)$.

matrices as

$$\mathbf{A} = \begin{bmatrix} 0 & 1 \\ -2 & 1 \end{bmatrix}; \quad \mathbf{B} = \begin{bmatrix} 0 \\ 1 \end{bmatrix}; \tag{3.5.20}$$

$$\mathbf{Q} = \begin{bmatrix} 2 & 3 \\ 3 & 5 \end{bmatrix}; \quad \mathbf{R} = r = \frac{1}{4}; \quad t_0 = 0; \quad t_f = \infty. \tag{3.5.21}$$

Let $\bar{\mathbf{P}}$ be the 2x2 symmetric matrix

$$\bar{\mathbf{P}} = \begin{bmatrix} \bar{p}_{11} & \bar{p}_{12} \\ \bar{p}_{12} & \bar{p}_{22} \end{bmatrix}. \tag{3.5.22}$$

Then, the optimal control (3.5.14) is given by

$$u^*(t) = -4\begin{bmatrix} 0 & 1 \end{bmatrix} \begin{bmatrix} \bar{p}_{11} & \bar{p}_{12} \\ \bar{p}_{12} & \bar{p}_{22} \end{bmatrix} \begin{bmatrix} x_1^*(t) \\ x_2^*(t) \end{bmatrix},$$

$$= -4[\bar{p}_{12}x_1^*(t) + \bar{p}_{22}x_2^*(t)], \tag{3.5.23}$$

where, $\bar{\mathbf{P}}$, the 2x2 symmetric, positive definite matrix, is the solution of the matrix algebraic Riccati equation (3.5.15)

$$
\begin{bmatrix} 0\,0 \\ 0\,0 \end{bmatrix} = - \begin{bmatrix} \bar{p}_{11} & \bar{p}_{12} \\ \bar{p}_{12} & \bar{p}_{22} \end{bmatrix} \begin{bmatrix} 0 & 1 \\ -2 & 1 \end{bmatrix} - \begin{bmatrix} 0 & -2 \\ 1 & 1 \end{bmatrix} \begin{bmatrix} \bar{p}_{11} & \bar{p}_{12} \\ \bar{p}_{12} & \bar{p}_{22} \end{bmatrix} +
$$
$$
\begin{bmatrix} \bar{p}_{11} & \bar{p}_{12} \\ \bar{p}_{12} & \bar{p}_{22} \end{bmatrix} \begin{bmatrix} 0 \\ 1 \end{bmatrix} 4 \begin{bmatrix} 0 & 1 \end{bmatrix} \begin{bmatrix} \bar{p}_{11} & \bar{p}_{12} \\ \bar{p}_{12} & \bar{p}_{22} \end{bmatrix} - \begin{bmatrix} 2 & 3 \\ 3 & 5 \end{bmatrix}.
$$

$$\text{(3.5.24)}$$

Simplifying the equation (3.5.24), we get

$$
4\bar{p}_{12}^2 + 4\bar{p}_{12} - 2 = 0
$$
$$
-\bar{p}_{11} - \bar{p}_{12} + 2\bar{p}_{22} + 4\bar{p}_{12}\bar{p}_{22} - 3 = 0
$$
$$
-2\bar{p}_{12} - 2\bar{p}_{22} + 4\bar{p}_{22}^2 - 5 = 0. \qquad (3.5.25)
$$

Solving the previous equations for positive definiteness of $\bar{\mathbf{P}}$ is easy in this particular case. Thus, solve the first equation in (3.5.25) for $\bar{p}_{12}$, using this value of $\bar{p}_{12}$ in the third equation solve for $\bar{p}_{22}$ and finally using the values of $\bar{p}_{12}$ and $\bar{p}_{22}$ in the second equation, solve for $\bar{p}_{11}$. In general, we have to solve the nonlinear algebraic equations and pick up the positive definite values for $\bar{\mathbf{P}}$. Hence, we get

$$
\bar{\mathbf{P}} = \begin{bmatrix} 1.7363 & 0.3660 \\ 0.3660 & 1.4729 \end{bmatrix}. \qquad (3.5.26)
$$

Using these Riccati coefficients (gains), the closed-loop optimal control (3.5.23) is given by

$$
u^*(t) = -4[0.366x_1^*(t) + 1.4729x_2^*(t)]
$$
$$
= -[1.464x_1^*(t) + 5.8916x_2^*(t)]. \qquad (3.5.27)
$$

Using the closed-loop optimal control $u^*(t)$ from (3.5.27) in the original open-loop system (3.5.18), the closed-loop optimal system becomes

$$
\dot{x}_1^*(t) = x_2^*(t)
$$
$$
\dot{x}_2^*(t) = -2x_1^*(t) + x_2^*(t) - 4[0.366x_1^*(t) + 1.4729x_2^*(t)] \quad (3.5.28)
$$

and the implementation of the closed-loop optimal control is shown in Figure 3.9.

Using the initial conditions and the Riccati coefficient matrix (3.5.26), the optimal cost (3.5.17) is obtained as

$$
J^* = \frac{1}{2}\mathbf{x}'(0)\bar{\mathbf{P}}\mathbf{x}(0) = \frac{1}{2}\begin{bmatrix} 2 & -3 \end{bmatrix} \begin{bmatrix} 1.7363 & 0.3660 \\ 0.3660 & 1.4729 \end{bmatrix} \begin{bmatrix} 2 \\ -3 \end{bmatrix},
$$
$$
= 7.9047. \qquad (3.5.29)
$$

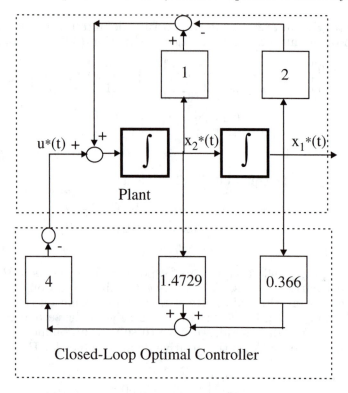

Figure 3.9   Closed-Loop Optimal Control System

The previous results can also easily obtained using Control System Toolbox of the MATLAB©, Version 6 as shown below.

```
************************************************
%% Solution using Control System Toolbox in
%% The  MATLAB. Version 6
%% For Example:4-3
%%
 x10=2; %% initial condition on state x1
 x20=-3; %% initial condition on state x2
 X0=[x10;x20];
 A=[0 1;-2 1]; %% system matrix A
 B=[0;1]; %% system matrix B
 Q=[2 3;3 5]; %% performance index weighted matrix
 R=[0.25]; %% performance index weighted matrix
 [K,P,EV]=lqr(A,B,Q,R) %% K = feedback matrix;
%% P = Riccati matrix;
%% EV = eigenvalues of closed loop system A - B*K
K =
```

```
      1.4641    5.8916
P =
      1.7363    0.3660
      0.3660    1.4729
EV =
     -4.0326
     -0.8590
BIN=[0;0]; % dummy BIN for "initial" command
C=[1 1];
D=[1];
tfinal=10;
t=0:0.05:10;
[Y,X,t]=initial(A-B*K,BIN,C,D,X0,tfinal);
x1t=[1 0]*X'; %% extracting x1 from vector X
x2t=[0 1]*X'; %% extracting x2 from vector X
ut=-K*X';
plot(t,x1t,'k',t,x2t,'k')
xlabel('t')
gtext('x_1(t)')
gtext('x_2(t)')
plot(t,ut,'k')
xlabel('t')
gtext('u(t)')
****************************************************
```

Using the optimal control $u^*(t)$ given by (3.5.23), the plant equations (3.5.18) are solved using MATLAB© to obtain the optimal states $x_1^*(t)$ and $x_2^*(t)$ and the optimal control $u^*(t)$ as shown in Figure 3.10 and Figure 3.11. Note that

1. the values of $\bar{\mathbf{P}}$ obtained in the example, are exactly the steady-state values of Example 3.1 and

2. the original plant (3.5.18) is *unstable* (eigenvalues at $2 \pm j1$) whereas the optimal closed-loop system (3.5.28) is *stable* (eigenvalues at $-4.0326, -0.8590$).

3.5.4 *Stability Issues of Time-Invariant Regulator*

Let us consider the previous result for linear time-invariant system with infinite-time horizon from relations (3.5.12) to (3.5.17) and Table 3.3. We address briefly some stability remarks of the infinite-time regulator system [3, 89].

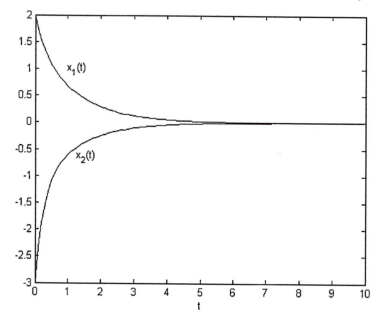

Figure 3.10   Optimal States for Example 3.2

1. The closed-loop optimal system (3.5.16) is not *always stable* especially when the original plant is unstable and these unstable states are not weighted in the PI (3.5.13). In order to prevent such a situation, we need the assumption that the pair $[\mathbf{A}, \mathbf{C}]$ is detectable, where C is any matrix such that $\mathbf{C}'\mathbf{C} = \mathbf{Q}$, which guarantees the stability of closed-loop optimal system. This assumption essentially ensures that all the potentially unstable states will show up in the $\mathbf{x}'(t)\mathbf{Q}\mathbf{x}(t)$ part of the performance measure.

2. The Riccati coefficient matrix $\bar{\mathbf{P}}$ is positive definite if and only if $[\mathbf{A}, \mathbf{C}]$ is completely observable.

3. The *detectability* condition is necessary for stability of the closed-loop optimal system.

4. Thus both *detectability* and *stabilizability* conditions are necessary for the existence of a stable closed-loop system.

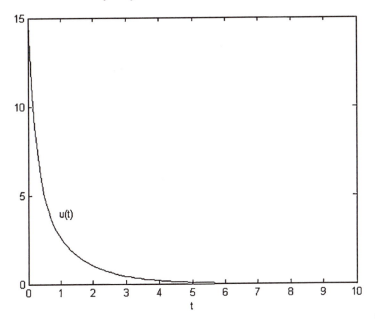

Figure 3.11   Optimal Control for Example 3.2

3.5.5   *Equivalence of Open-Loop and Closed-Loop Optimal Controls*

Next, we present a simple example to show an interesting property that an optimal control system can be solved and implemented as an *open-loop* optimal control (OLOC) configuration or a *closed-loop* optimal control (CLOC) configuration. We will also demonstrate the simplicity of the CLOC.

Example 3.3

Consider a simple first order system

$$\dot{x}(t) = -3x(t) + u(t) \tag{3.5.30}$$

and the cost function (CF) as

$$J = \int_0^\infty [x^2(t) + u^2(t)]dt \tag{3.5.31}$$

where, $x(0) = 1$ and the final state $x(\infty) = 0$. Find the open-loop and closed-loop optimal controllers.

Solution: (a) Open-Loop Optimal Control: We use the Pontryagin procedure given in Chapter 2 (see Table 2.1). First of all, comparing the given plant (3.5.30) and the CF (3.5.31) with the general formulations (see Table 2.1), identify that

$$V(x(t), u(t)) = x^2(t) + u^2(t),$$
$$f(x(t), u(t)) = -3x(t) + u(t). \tag{3.5.32}$$

Now, we use the step-by-step procedure given in Table 2.1.

- **Step 1:** Formulate the Hamiltonian as

$$\mathcal{H} = V(x(t), u(t)) + \lambda(t)f(x(t), u(t))$$
$$= x^2(t) + u^2(t) + \lambda(t)[-3x(t) + u(t)]. \tag{3.5.33}$$

- **Step 2:** The optimal control $u^*(t)$ is obtained by minimizing the previous Hamiltonian w.r.t. u as

$$\frac{\partial \mathcal{H}}{\partial u} = 0 \longrightarrow 2u^*(t) + \lambda^*(t) = 0 \longrightarrow u^*(t) = -\frac{1}{2}\lambda^*(t). \tag{3.5.34}$$

- **Step 3:** Using optimal control (3.5.34) in the Hamiltonian function (3.5.33), find the optimal Hamiltonian function as

$$\mathcal{H}^* = x^{*2}(t) - \frac{1}{4}\lambda^{*2}(t) - 3\lambda^*(t)x^*(t). \tag{3.5.35}$$

- **Step 4:** Using the previous optimal $\mathcal{H}^*$, obtain the set of state and costate equations

$$\dot{x}^*(t) = \frac{\partial \mathcal{H}^*}{\partial \lambda} \longrightarrow \dot{x}^*(t) = -\frac{1}{2}\lambda^*(t) - 3x^*(t), \tag{3.5.36}$$

$$\dot{\lambda}^*(t) = -\frac{\partial \mathcal{H}^*}{\partial x} \longrightarrow \dot{\lambda}^*(t) = -2x^*(t) + 3\lambda^*(t), \tag{3.5.37}$$

yielding

$$\ddot{x}^*(t) - 10x^*(t) = 0, \tag{3.5.38}$$

the solution of which becomes

$$x^*(t) = C_1 e^{\sqrt{10}t} + C_2 e^{-\sqrt{10}t}. \tag{3.5.39}$$

Using the optimal state (3.5.39) in (3.5.36), we have the costate as

$$\lambda^*(t) = 2[-\dot{x}^*(t) - 3x^*(t)]$$
$$= -2C_1(\sqrt{10} + 3)e^{\sqrt{10}t} + 2C_2(\sqrt{10} - 3)e^{-\sqrt{10}t}. \tag{3.5.40}$$

Using the initial condition $x(0) = 1$ in the optimal state (3.5.39), and the final condition (for δx_f being free) $\lambda(t_f = \infty) = 0$ in the optimal costate (3.5.40), we get

$$x(0) = 1 \longrightarrow C_1 + C_2 = 1$$
$$\lambda(\infty) = 0 \longrightarrow C_1 = 0. \tag{3.5.41}$$

Then, the previous optimal state and costate are given as

$$x^*(t) = e^{-\sqrt{10}t}; \qquad \lambda^*(t) = 2(\sqrt{10} - 3)e^{-\sqrt{10}t}. \tag{3.5.42}$$

- **Step 5:** Using the previous costate solution (3.5.34) of Step 2, we get the open-loop optimal control as

$$u^*(t) = -(\sqrt{10} - 3)e^{-\sqrt{10}t}. \tag{3.5.43}$$

(b) Closed-Loop Optimal Control: Here, we use the matrix algebraic Riccati equation (ARE) to find the closed-loop optimal control, as summarized in Table 3.3. First of all, comparing the present plant (3.5.30) and the PI (3.5.31) with the general formulation of the plant (3.5.12) and the PI (3.5.13), respectively, we identify the various coefficients and matrices as

$$\mathbf{A} = a = -3; \quad \mathbf{B} = b = 1;$$
$$\mathbf{Q} = q = 2; \quad \mathbf{R} = r = 2; \quad \bar{\mathbf{P}} = \bar{p}. \tag{3.5.44}$$

Note the PI (3.5.31) does not contain the factor $1/2$ as in the general PI (3.5.13) and accordingly, we have $\mathbf{Q} = q = 2$ and $\mathbf{R} = r = 2$.

- **Step 1:** With the previous values, the ARE (3.5.15) becomes

$$\bar{p}(-3) + (-3)\bar{p} - \bar{p}(1)\left(\frac{1}{2}\right)(1)\bar{p} + 2 = 0 \longrightarrow$$
$$\bar{p}^2 + 12\bar{p} - 4 = 0, \tag{3.5.45}$$

the solution of which is

$$\bar{p} = -6 \pm 2\sqrt{10}. \tag{3.5.46}$$

- **Step 2:** Using the positive value of the Riccati coefficient (3.5.46), the closed-loop optimal control (3.5.14) becomes

$$u^*(t) = -r^{-1}b\bar{p}x^*(t) = -\frac{1}{2}(-6 + 2\sqrt{10})x^*(t)$$
$$= -(-3 + \sqrt{10})x^*(t). \tag{3.5.47}$$

- **Step 3:** Using the optimal control (3.5.47), the optimal state is solved from (3.5.16) as

$$\dot{x}(t) = -3x^*(t) - (-3 + \sqrt{10})x^*(t) = -\sqrt{10}x^*(t). \quad (3.5.48)$$

Solving the previous along with the initial condition $x(0) = 1$, we get the optimal state as

$$x^*(t) = e^{-\sqrt{10}t} \qquad (3.5.49)$$

with which the optimal control (3.5.47) becomes

$$u^*(t) = -(\sqrt{10} - 3)e^{-\sqrt{10}t}. \qquad (3.5.50)$$

Thus, we note that the optimal control (3.5.50) and optimal state (3.5.49) obtained from using the closed-loop optimal control are *identical* to those of (3.5.43) and (3.5.42), respectively. We can easily extend this analysis for the general case. Intuitively, this equivalence should exist as the optimal control being *unique* should be same by any method.

The implementation of this open-loop optimal controller (OLOC) is shown in Figure 3.12(a), and that of the closed-loop optimal controller (CLOC) is shown in Figure 3.12(b).

From the previous example, it is clear that

1. from the implementation point of view, the closed-loop optimal controller ($\sqrt{10} - 3$) is much *simpler* than the open-loop optimal controller (($\sqrt{10}-3)e^{-\sqrt{10}t}$) which is an exponential time function and

2. with a closed-loop configuration, all the advantages of conventional feedback are incorporated.

3.6   *Notes and Discussion*

We know that linear quadratic optimal control is concerned with *linear* plants, performance measures *quadratic* in controls and states, and regulation and tracking errors. In particular, the resulting optimal controller is closed-loop and *linear* in state. Note that linear quadratic optimal control is a special class of the general optimal control which includes *nonlinear* systems and *nonlinear* performance measures. There are many useful advantages and attractive features of linear quadratic optimal control systems which are enumerated below [3].

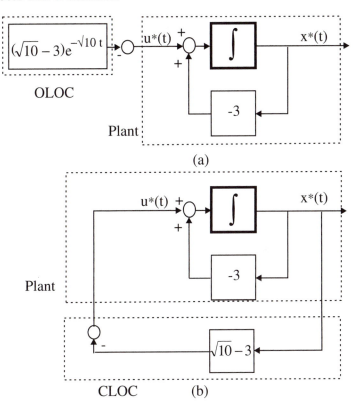

Figure 3.12   (a) Open-Loop Optimal Controller (OLOC) and (b)
Closed-Loop Optimal Controller (CLOC)

1. Many engineering and physical systems operate in *linear* range
   during normal operations.

2. There is a *wealth* of theoretical results available for *linear* systems
   which can be useful for linear quadratic methods.

3. The resulting optimal controller is *linear* in state and thus easy
   and simple for implementation purposes in real application of the
   LQ results.

4. Many (nonlinear) optimal control systems do not have solutions
   which can be easily computed. On the other hand, LQ optimal
   control systems have easily *computable* solutions.

5. As is well known, *nonlinear* systems can be examined for *small*
   variations from their normal operating conditions. For example,

assume that after a heavy computational effort, we obtained an optimal solution for a nonlinear plant and there is a small change from an operating point. Then, one can easily use to a first approximation a *linear model* and obtain linear optimal control to drive the original nonlinear system to its operating point.

6. Many of the concepts, techniques and computational procedures that are developed for *linear* optimal control systems in many cases many be carried on to *nonlinear* optimal control systems.

7. Linear optimal control designs for plants whose states are measurable possess a number of desirable *robustness* properties (such as good gain and phase margins and a good tolerance to nonlinearities) of classical control designs.

3.7 Problems

1. Make reasonable assumptions wherever necessary.

2. Use MATLAB© wherever possible to solve the problems and plot all the optimal controls and states for all problems. Provide the relevant MATLAB© m files.

Problem 3.1 A first order system is given by

$$\dot{x}(t) = x(t) + u(t).$$

(a) Find the unconstrained optimal control law which minimizes the performance index

$$J = \int_0^{t_f} [2x^2(t) + 0.25u^2(t)]dt,$$

such that the final time t_f is fixed and the final state $x(t_f)$ is free.
(b) Find the optimal control law as $t_f \to \infty$.

Problem 3.2 A system is described by

$$\ddot{x}(t) + x(t) = u(t)$$

with initial conditions $x(0) = 0$ and $\dot{x}(0) = 1$ and the performance index

$$J = \int_0^\infty \left[2x^2(t) + 0.25u^2(t) \right] dt.$$

Find the closed-loop optimal control in terms of x and $\dot{x}$ and the optimal cost function.

Problem 3.3 A first order system is described by

$$\dot{x}(t) = ax(t) + bu(t)$$

with performance index

$$J = \frac{1}{2} \int_0^{t_f} [qx^2(t) + ru^2(t)]dt$$

and with a fixed initial state $x(0)$ and final state $x(t_f) = 0$, where t_f is fixed. Show that the solution of the Riccati equation is given by

$$p(t) = \frac{r}{b^2}[a - \beta coth\{\beta(t - t_f)\}]$$

and the solution of the optimal state $x^*(t)$ is given by

$$x^*(t) = x(0)\frac{sinh\{\beta(t_f - t)\}}{sinh\beta}$$

where, $\beta = \sqrt{a^2 + b^2 q/r}$.

Problem 3.4 Find the optimal feedback control for the plant

$$\dot{x}_1(t) = x_2(t)$$
$$\dot{x}_2(t) = -2x_1(t) + 4x_2(t) + 5u(t)$$

with performance criterion

$$J = \frac{1}{2}f_{11}[x_1(t_f) - 4]^2 + \frac{1}{2}f_{22}[x_2(t_f) - 2]^2 +$$
$$\frac{1}{2}\int_0^{t_f}\left[5x_1^2(t) + 2x_2^2(t) + 4u^2(t)\right]dt$$

and initial conditions as $x(0) = [1\ 2]'$ and the final state $x(t_f)$ is free, where t_f is specified.

Problem 3.5 Find the closed-loop, unconstrained, optimal control for the system

$$\dot{x}_1(t) = x_2(t)$$
$$\dot{x}_2(t) = -2x_1(t) - 3x_2(t) + u(t)$$

and the performance index

$$J = \int_0^\infty[x_1^2(t) + x_2^2(t) + u^2(t)]dt.$$

Problem 3.6 Find the optimal feedback control law for the plant

$$\dot{x}_1(t) = x_2(t) + u(t)$$
$$\dot{x}_2(t) = -x_1(t) - x_2(t) + u(t)$$

and the cost function

$$J = \int_0^\infty[2x_1^2(t) + 4x_2^2(t) + 0.5u^2(t)]dt.$$

Problem 3.7 Consider a second order system

$$\ddot{x}(t) + b\dot{x}(t) + cx(t) = u(t)$$

and the performance index to be minimized as

$$J = \int_0^\infty [qx^2(t) + ru^2(t)]dt.$$

Determine the closed-loop optimal control in terms of the state $x(t)$ and its derivative $\dot{x}(t)$.

Problem 3.8 Given a third order plant,

$$\dot{x}_1(t) = x_2(t)$$
$$\dot{x}_2(t) = x_3(t)$$
$$\dot{x}_3(t) = -5x_1(t) + -7x_2(t) - 10x_3(t) + 4u(t)$$

and the performance index

$$J = \int_0^\infty [q_{11}x_1^2(t) + q_{22}x_2^2(t) + q_{33}x_3^2(t) + ru^2(t)]dt,$$

for

1. $q_{11} = q_{22} = q_{33} = 1, r = 1,$

2. $q_{11} = 10, q_{22} = 1, q_{33} = 1, r = 1,$ and

3. $q_{11} = q_{22} = q_{33} = 1, r = 10,$

find the positive definite solution for Riccati coefficient matrix $\bar{\mathbf{P}}$, optimal feedback gain matrix $\bar{\mathbf{K}}$ and the eigenvalues of the closed-loop system matrix $\mathbf{A} - \mathbf{B}\bar{\mathbf{K}}$.

Problem 3.9 Determine the optimal feedback coefficients and the optimal control law for the multi-input, multi-output (MIMO) system

$$\dot{\mathbf{x}}(t) = \begin{bmatrix} 0 & 1 \\ 1 & 1 \end{bmatrix} \mathbf{x}(t) + \begin{bmatrix} 1 & 1 \\ 0 & 1 \end{bmatrix} \mathbf{u}(t)$$

and the cost function

$$J = \int_0^\infty [2x_1^2(t) + 4x_2^2(t) + 0.5u_1^2(t) + 0.25u_2^2(t)]dt.$$

Problem 3.10 For the D.C. motor speed control system described in Problem 1.1, find the closed-loop optimal control to keep the speed constant at a particular value and the system to respond for any disturbances from the regulated value.

Problem 3.11 For the liquid-level control system described in Problem 1.2, find the closed-loop optimal control to keep the liquid level constant at a reference value and the system to act only if there is a change in the liquid level.

Problem 3.12 [35] For the inverted pendulum control system described in Problem 1.3, find the closed-loop optimal control to keep the pendulum in a vertical position.

Problem 3.13 For the mechanical control system described in Problem 1.4, find the closed-loop optimal control to keep the system at equilibrium condition and act only if there is a disturbance.

Problem 3.14 For the automobile suspension system described in Problem 1.5, find the closed-loop optimal control to keep the system at equilibrium condition and act only if there is a disturbance.

Problem 3.15 For the chemical control system described in Problem 1.6, find the closed-loop optimal control to keep the system at equilibrium condition and act only if there is a disturbance.

@@@@@@@@@@@@@@

Chapter 4

Linear Quadratic Optimal Control Systems II

In the previous chapter, we addressed the linear quadratic *regulator* system, where the aim was to obtain the optimal control to regulate (or keep) the state around zero. In this chapter, we discuss linear quadratic tracking (LQT) systems, and some related topics in linear quadratic regulator theory. It is suggested that the student reviews the material in Appendices A and B given at the end of the book. This chapter is based on [6, 89, 3][1].

Trajectory Following Systems

In tracking (trajectory following) systems, we require that the output of a system *track* or *follow* a desired trajectory in some optimal sense. Thus, we see that this is a generalization of *regulator* system in the sense that the *desired* trajectory for the regulator is simply the zero state.

---

[1]The permissions given by John Wiley for F. L. Lewis, *Optimal Control*, John Wiley & Sons, Inc., New York, NY, 1986 and McGraw-Hill for M. Athans and P. L. Falb, *Optimal Control: An Introduction to the Theory and Its Applications*, McGraw-Hill Book Company, New York, NY, 1966, are hereby acknowledged.

4.1 Linear Quadratic Tracking System: Finite-Time Case

In this section, we discuss the linear quadratic tracking (LQT)system to maintain the output as close as possible to the *desired* output with minimum control energy [6]. We are given a linear, observable system

$$\dot{\mathbf{x}}(t) = \mathbf{A}(t)\mathbf{x}(t) + \mathbf{B}(t)\mathbf{u}(t)$$
$$\mathbf{y}(t) = \mathbf{C}(t)\mathbf{x}(t) \tag{4.1.1}$$

where, $\mathbf{x}(t)$ is the nth order state vector, $\mathbf{u}(t)$ is the rth order control vector, and $\mathbf{y}(t)$ is the mth order output vector. Let $\mathbf{z}(t)$ be the mth order *desired output* and the various matrices $\mathbf{A}(t), \mathbf{B}(t)$ and $\mathbf{C}(t)$ be of appropriate dimensionality. Our objective is to control the system (4.1.1) in such a way that the output $\mathbf{y}(t)$ *tracks* the *desired* output $\mathbf{z}(t)$ as close as possible during the interval $[t_0, \ t_f]$ with minimum expenditure of control effort. For this, let us define the *error vector* as

$$\mathbf{e}(t) = \mathbf{z}(t) - \mathbf{y}(t) \tag{4.1.2}$$

and choose the performance index as

$$J = \frac{1}{2}\mathbf{e}'(t_f)\mathbf{F}(t_f)\mathbf{e}(t_f)$$
$$+ \frac{1}{2}\int_{t_0}^{t_f} \left[\mathbf{e}'(t)\mathbf{Q}(t)\mathbf{e}(t) + \mathbf{u}'(t)\mathbf{R}(t)\mathbf{u}(t)\right] dt \tag{4.1.3}$$

with t_f *specified* and $\mathbf{x}(t_f)$ *not specified*. In this way we are dealing with *free-final state system*. Also, we assume that $\mathbf{F}(t_f)$ and $\mathbf{Q}(t)$ are $m \times m$ symmetric, positive *semidefinite* matrices, and $\mathbf{R}(t)$ is $r \times r$ symmetric, positive *definite* matrix. We now use the Pontryagin Minimum Principle in the following order.

- **Step 1:** *Hamiltonian*

- **Step 2:** *Open-Loop Optimal Control*

- **Step 3:** *State and Costate System*

- **Step 4:** *Riccati and Vector Equations*

- **Step 5:** *Closed-Loop Optimal Control*

- **Step 6:** *Optimal State*

- **Step 7:** *Optimal Cost*

Now we discuss these steps in detail. Also, note that we heavily draw upon the results of the previous Chapters 2 and 3. First of all, let us note from (4.1.1) and (4.1.2) that the error $\mathbf{e}(t)$ can be expressed as a function of $\mathbf{z}(t)$ and $\mathbf{x}(t)$ as

$$\mathbf{e}(t) = \mathbf{z}(t) - \mathbf{C}(t)\mathbf{x}(t). \tag{4.1.4}$$

- **Step 1:** *Hamiltonian:* Formulate the Hamiltonian as (see Table 3.1)

$$\mathcal{H}(\mathbf{x}(t), \mathbf{u}(t), \boldsymbol{\lambda}(t)) = \frac{1}{2}\left[\mathbf{z}(t) - \mathbf{C}(t)\mathbf{x}(t)\right]' \mathbf{Q}(t) \left[\mathbf{z}(t) - \mathbf{C}(t)\mathbf{x}(t)\right]$$
$$+ \frac{1}{2}\mathbf{u}'(t)\mathbf{R}(t)\mathbf{u}(t)$$
$$+ \boldsymbol{\lambda}'(t)\left[\mathbf{A}(t)\mathbf{x}(t) + \mathbf{B}(t)\mathbf{u}(t)\right]. \tag{4.1.5}$$

- **Step 2:** *Open-Loop Optimal Control:* Using the Hamiltonian (4.1.5), obtain the control equation from

$$\frac{\partial \mathcal{H}}{\partial \mathbf{u}} = 0 \longrightarrow \mathbf{R}(t)\mathbf{u}(t) + \mathbf{B}'(t)\boldsymbol{\lambda}(t) = 0 \tag{4.1.6}$$

from which we have the optimal control as

$$\boxed{\mathbf{u}^*(t) = -\mathbf{R}^{-1}(t)\mathbf{B}'(t)\boldsymbol{\lambda}^*(t).} \tag{4.1.7}$$

Since the second partial of $\mathcal{H}$ in (4.1.5) w.r.t. $\mathbf{u}^*(t)$ is just $\mathbf{R}(t)$, and we chose $\mathbf{R}(t)$ to be positive *definite*, we are dealing with a control which *minimizes* the cost functional (4.1.3).

- **Step 3:** *State and Costate System:* The state is given in terms of the Hamiltonian (4.1.5) as

$$\dot{\mathbf{x}}(t) = \frac{\partial \mathcal{H}}{\partial \boldsymbol{\lambda}} = \mathbf{A}(t)\mathbf{x}(t) + \mathbf{B}(t)\mathbf{u}(t) \tag{4.1.8}$$

and with the optimal control (4.1.7), the optimal state equation (4.1.8) becomes

$$\dot{\mathbf{x}}^*(t) = \mathbf{A}(t)\mathbf{x}^*(t) - \mathbf{B}(t)\mathbf{R}^{-1}(t)\mathbf{B}'(t)\boldsymbol{\lambda}^*(t). \tag{4.1.9}$$

Using the Hamiltonian (4.1.5), the optimal costate equation becomes

$$\dot{\boldsymbol{\lambda}}^*(t) = -\frac{\partial \mathcal{H}}{\partial \mathbf{x}}$$
$$= -\mathbf{C}'(t)\mathbf{Q}(t)\mathbf{C}(t)\mathbf{x}^*(t) - \mathbf{A}'(t)\boldsymbol{\lambda}^*(t)$$
$$+\mathbf{C}'(t)\mathbf{Q}(t)\mathbf{z}(t). \tag{4.1.10}$$

For the sake of simplicity, let us define

$$\mathbf{E}(t) = \mathbf{B}(t)\mathbf{R}^{-1}(t)\mathbf{B}'(t), \quad \mathbf{V}(t) = \mathbf{C}'(t)\mathbf{Q}(t)\mathbf{C}(t),$$
$$\mathbf{W}(t) = \mathbf{C}'(t)\mathbf{Q}(t). \tag{4.1.11}$$

Using the relation (4.1.11) and combining the state (4.1.8) and costate (4.1.10) equations, we obtain the Hamiltonian canonical system as

$$\begin{bmatrix} \dot{\mathbf{x}}^*(t) \\ \dot{\boldsymbol{\lambda}}^*(t) \end{bmatrix} = \begin{bmatrix} \mathbf{A}(t) & -\mathbf{E}(t) \\ -\mathbf{V}(t) & -\mathbf{A}'(t) \end{bmatrix} \begin{bmatrix} \mathbf{x}^*(t) \\ \boldsymbol{\lambda}^*(t) \end{bmatrix} + \begin{bmatrix} \mathbf{0} \\ \mathbf{W}(t) \end{bmatrix} \mathbf{z}(t). \tag{4.1.12}$$

This canonical system of $2n$ differential equations is linear, time-varying, but *nonhomogeneous* with $\mathbf{W}(t)\mathbf{z}(t)$ as forcing function. The boundary conditions for this state and costate equations are given by the initial condition on the state as

$$\mathbf{x}(t = t_0) = \mathbf{x}(t_0) \tag{4.1.13}$$

and the final condition on the costate (for the final time t_f *specified* and state $\mathbf{x}(t_f)$ being *free*) given by (3.2.10), which along with (4.1.4) become

$$\boldsymbol{\lambda}(t_f) = \frac{\partial}{\partial \mathbf{x}(t_f)} \left[\frac{1}{2}\mathbf{e}'(t_f)\mathbf{F}(t_f)\mathbf{e}(t_f) \right]$$
$$= \frac{\partial}{\partial \mathbf{x}(t_f)} \left[\frac{1}{2}[\mathbf{z}(t_f) - \mathbf{C}(t_f)\mathbf{x}(t_f)]'\mathbf{F}(t_f)[\mathbf{z}(t_f) - \mathbf{C}(t_f)\mathbf{x}(t_f)] \right]$$
$$= \mathbf{C}'(t_f)\mathbf{F}(t_f)\mathbf{C}(t_f)\mathbf{x}(t_f) - \mathbf{C}'(t_f)\mathbf{F}(t_f)\mathbf{z}(t_f). \tag{4.1.14}$$

- **Step 4:** *Riccati and Vector Equations:* The boundary condition (4.1.14) and the solution of the system (4.1.12) indicate that the state and costate are linearly related as

$$\boldsymbol{\lambda}^*(t) = \mathbf{P}(t)\mathbf{x}^*(t) - \mathbf{g}(t) \tag{4.1.15}$$

where, the *nxn matrix* $\mathbf{P}(t)$ and *n vector* $\mathbf{g}(t)$ are yet to be determined so as to satisfy the canonical system (4.1.12). This is done by substituting the linear (Riccati) transformation (4.1.15) in the Hamiltonian system (4.1.12) and eliminating the costate function $\boldsymbol{\lambda}^*(t)$. Thus, we first differentiate (4.1.15) to get

$$\dot{\boldsymbol{\lambda}}^*(t) = \dot{\mathbf{P}}(t)\mathbf{x}^*(t) + \mathbf{P}(t)\dot{\mathbf{x}}^*(t) - \dot{\mathbf{g}}(t). \qquad (4.1.16)$$

Now, substituting for $\dot{\mathbf{x}}^*(t)$ and $\dot{\boldsymbol{\lambda}}^*(t)$ from (4.1.12) and eliminating $\boldsymbol{\lambda}^*(t)$ with (4.1.15), we get

$$-\mathbf{V}(t)\mathbf{x}^*(t) - \mathbf{A}'(t)\left[\mathbf{P}(t)\mathbf{x}^*(t) - \mathbf{g}(t)\right] + \mathbf{W}(t)\mathbf{z}(t) = \dot{\mathbf{P}}(t)\mathbf{x}^*(t)$$
$$+\mathbf{P}(t)[\mathbf{A}(t)\mathbf{x}(t) - \mathbf{E}(t)\left\{\mathbf{P}(t)\mathbf{x}^*(t) - \mathbf{g}(t)\right\}] - \dot{\mathbf{g}}(t). \qquad (4.1.17)$$

Rearranging the above, we get

$$\left[\dot{\mathbf{P}}(t) + \mathbf{P}(t)\mathbf{A}(t) + \mathbf{A}'(t)\mathbf{P}(t) - \mathbf{P}(t)\mathbf{E}(t)\mathbf{P}(t) + \mathbf{V}(t)\right]\mathbf{x}^*(t) -$$
$$[\dot{\mathbf{g}}(t) + \mathbf{A}'(t)\mathbf{g}(t) - \mathbf{P}(t)\mathbf{E}(t)\mathbf{g}(t) + \mathbf{W}(t)\mathbf{z}(t)] = 0. \qquad (4.1.18)$$

Now, this relation (4.1.18) must satisfy for all $\mathbf{x}^*(t)$, $\mathbf{z}(t)$ and t, which leads us to the *nxn* matrix $\mathbf{P}(t)$ to satisfy the *matrix differential Riccati equation* (DRE)

$$\boxed{\dot{\mathbf{P}}(t) = -\mathbf{P}(t)\mathbf{A}(t) - \mathbf{A}'(t)\mathbf{P}(t) + \mathbf{P}(t)\mathbf{E}(t)\mathbf{P}(t) - \mathbf{V}(t)}$$
$$(4.1.19)$$

and the *n* vector $\mathbf{g}(t)$ to satisfy the *vector* differential equation

$$\boxed{\dot{\mathbf{g}}(t) = [\mathbf{P}(t)\mathbf{E}(t) - \mathbf{A}'(t)]\,\mathbf{g(t)} - \mathbf{W}(t)\mathbf{z(t)}.} \qquad (4.1.20)$$

Since $\mathbf{P}(t)$ is *nxn* symmetric matrix, and $\mathbf{g}(t)$ is of *n* vector, the equations (4.1.19) and (4.1.20) are a set of $n(n+1)/2 + n$ first-order differential equations. The boundary conditions are obtained from (4.1.15) as

$$\boldsymbol{\lambda}(t_f) = \mathbf{P}(t_f)\mathbf{x}(t_f) - \mathbf{g}(t_f), \qquad (4.1.21)$$

which compared with the boundary condition (4.1.14) gives us for all $\mathbf{x}(t_f)$ and $\mathbf{z}(t_f)$,

$$\boxed{\mathbf{P}(t_f) = \mathbf{C}'(t_f)\mathbf{F}(t_f)\mathbf{C}(t_f),} \qquad (4.1.22)$$

$$\boxed{\mathbf{g}(t_f) = \mathbf{C}'(t_f)\mathbf{F}(t_f)\mathbf{z}(t_f).} \qquad (4.1.23)$$

Thus, the matrix DRE (4.1.19) and the vector equation (4.1.20) are to be solved *backward* using the boundary conditions (4.1.22) and (4.1.23).

- **Step 5:** *Closed-Loop Optimal Control:* The optimal control (4.1.7) is now given in terms of the state using the linear transformation (4.1.15)

$$\mathbf{u}^*(t) = -\mathbf{R}^{-1}(t)\mathbf{B}'(t)\left[\mathbf{P}(t)\mathbf{x}^*(t) - \mathbf{g}(t)\right]$$
$$= -\mathbf{K}(t)\mathbf{x}^*(t) + \mathbf{R}^{-1}(t)\mathbf{B}'(t)\mathbf{g}(t) \qquad (4.1.24)$$

where, $\mathbf{K}(t) = \mathbf{R}^{-1}(t)\mathbf{B}'(t)\mathbf{P}(t)$, is the *Kalman gain*.

- **Step 6:** *Optimal State:* Using this optimal control $\mathbf{u}^*(t)$ from (4.1.24) in the original plant (4.1.1), we have the optimal state obtained from

$$\dot{\mathbf{x}}^*(t) = \left[\mathbf{A}(t) - \mathbf{B}(t)\mathbf{R}^{-1}(t)\mathbf{B}'(t)\mathbf{P}(t)\right]\mathbf{x}^*(t)$$
$$+\mathbf{B}(t)\mathbf{R}^{-1}(t)\mathbf{B}'(t)\mathbf{g}(t)$$
$$= [\mathbf{A}(t) - \mathbf{E}(t)\mathbf{P}(t)]\mathbf{x}^*(t) + \mathbf{E}(t)\mathbf{g}(t). \qquad (4.1.25)$$

- **Step 7:** *Optimal Cost:* The optimal cost $J^*(t)$ for any time t can be obtained as (see [6])

$$J^*(t) = \frac{1}{2}\mathbf{x}^{*'}(t)\mathbf{P}(t)\mathbf{x}^*(t) - \mathbf{x}^{*'}(t)\mathbf{g}(t) + \mathbf{h}(t) \qquad (4.1.26)$$

where, the new function $\mathbf{h}(t)$ satisfies [3, 6]

$$\dot{\mathbf{h}}(t) = -\frac{1}{2}\mathbf{g}'(t)\mathbf{B}(t)\mathbf{R}^{-1}(t)\mathbf{B}'(t)\mathbf{g}(t) - \frac{1}{2}\mathbf{z}'(t)\mathbf{Q}(t)\mathbf{z}(t)$$
$$= -\frac{1}{2}\mathbf{g}'(t)\mathbf{E}(t)\mathbf{g}(t) - \frac{1}{2}\mathbf{z}'(t)\mathbf{Q}(t)\mathbf{z}(t) \qquad (4.1.27)$$

with final condition

$$\mathbf{h}(t_f) = -\mathbf{z}'(t_f)\mathbf{P}(t_f)\mathbf{z}(t_f). \qquad (4.1.28)$$

For further details on this, see [3, 6, 89, 90]. We now summarize the tracking system.

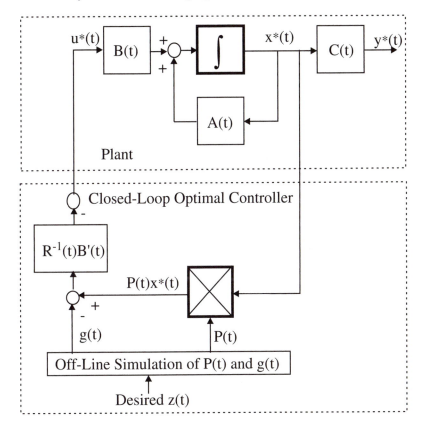

Figure 4.1 Implementation of the Optimal Tracking System

4.1.1 *Linear Quadratic Tracking System: Summary*

Given the linear, observable system (see Figure 4.1)

$$\dot{\mathbf{x}}(t) = \mathbf{A}(t)\mathbf{x}(t) + \mathbf{B}(t)\mathbf{u}(t)$$
$$\mathbf{y}(t) = \mathbf{C}(t)\mathbf{x}(t) \tag{4.1.29}$$

the desired output $\mathbf{z}(t)$, the error $\mathbf{e}(t) = \mathbf{z}(t) - \mathbf{y}(t)$, and the performance index

$$J = \frac{1}{2}\mathbf{e}'(t_f)\mathbf{F}(t_f)\mathbf{e}(t_f) + \frac{1}{2}\int_{t_0}^{t_f} \left[\mathbf{e}'(t)\mathbf{Q}(t)\mathbf{e}(t) + \mathbf{u}'(t)\mathbf{R}(t)\mathbf{u}(t)\right] dt$$

$$\tag{4.1.30}$$

then the optimal control $\mathbf{u}^*(t)$ is given by

$$\mathbf{u}^*(t) = -\mathbf{R}^{-1}(t)\mathbf{B}'(t)\left[\mathbf{P}(t)\mathbf{x}^*(t) - \mathbf{g}(t)\right]$$
$$= -\mathbf{K}(t)\mathbf{x}^*(t) + \mathbf{R}^{-1}(t)\mathbf{B}'(t)\mathbf{g}(t) \qquad (4.1.31)$$

where, the nxn symmetric, positive definite matrix $\mathbf{P}(t)$ is the solution of the nonlinear, matrix differential Riccati equation (DRE)

$$\dot{\mathbf{P}}(t) = -\mathbf{P}(t)\mathbf{A}(t) - \mathbf{A}'(t)\mathbf{P}(t) + \mathbf{P}(t)\mathbf{E}(t)\mathbf{P}(t) - \mathbf{V}(t) \quad (4.1.32)$$

with final condition

$$\mathbf{P}(t_f) = \mathbf{C}'(t_f)\mathbf{F}(t_f)\mathbf{C}(t_f), \qquad (4.1.33)$$

and the nth order $\mathbf{g}(t)$ is the solution of the linear, nonhomogeneous vector differential equation

$$\dot{\mathbf{g}}(t) = -\left[\mathbf{A}(t) - \mathbf{E}(t)\mathbf{P}(t)\right]' \mathbf{g}(t) - \mathbf{W}(t)\mathbf{z}(t) \qquad (4.1.34)$$

with final condition

$$\mathbf{g}(t_f) = \mathbf{C}'(t_f)\mathbf{F}(t_f)\mathbf{z}(t_f) \qquad (4.1.35)$$

where, $\mathbf{E}(t)$, $\mathbf{V}(t)$ and $\mathbf{W}(t)$ are defined in (4.1.11), the optimal state (trajectory) is the solution of the linear state equation

$$\dot{\mathbf{x}}^*(t) = \left[\mathbf{A}(t) - \mathbf{E}'(t)\mathbf{P}(t)\right]\mathbf{x}^*(t) + \mathbf{E}(t)\mathbf{g}(t), \qquad (4.1.36)$$

and the optimal cost J^*

$$J^*(t_0) = \frac{1}{2}\mathbf{x}^{*'}(t_0)\mathbf{P}(t_0)\mathbf{x}^*(t_0) - \mathbf{x}^*(t_0)\mathbf{g}(t_0) + \mathbf{h}(t_0). \quad (4.1.37)$$

The implementation of the tracking system is shown in Figure 4.1. The entire procedure is now summarized in Table 4.1.

4.1.2 Salient Features of Tracking System

1. *Riccati Coefficient Matrix* $\mathbf{P}(t)$: We note that the *desired output* $\mathbf{z}(t)$ has no influence on the matrix differential Riccati equation (4.1.32) and its boundary condition (4.1.33). This means that once the problem is specified in terms of the final time t_f, the plant matrices $\mathbf{A}(t)$, $\mathbf{B}(t)$, and $\mathbf{C}(t)$, and the cost functional matrices $\mathbf{F}(t_f)$, $\mathbf{Q}(t)$, and $\mathbf{R}(t)$, the matrix function $\mathbf{P}(t)$ is completely determined.

Table 4.1   Procedure Summary of Linear Quadratic Tracking System

A. Statement of the Problem
Given the plant as $\dot{\mathbf{x}}(t) = \mathbf{A}(t)\mathbf{x}(t) + \mathbf{B}(t)\mathbf{u}(t)$, $\mathbf{y}(t) = \mathbf{C}(t)\mathbf{x}(t)$, $\mathbf{e}(t) = \mathbf{z}(t) - \mathbf{y}(t)$, the performance index as $J = \frac{1}{2}\mathbf{e}'(t_f)\mathbf{F}(t_f)\mathbf{e}(t_f) + \frac{1}{2}\int_{t_0}^{t_f} [\mathbf{e}'(t)\mathbf{Q}(t)\mathbf{e}(t) + \mathbf{u}'(t)\mathbf{R}(t)\mathbf{u}(t)] \, dt$, and the boundary conditions as $\mathbf{x}(t_0) = \mathbf{x}_0$, $\mathbf{x}(t_f)$ is free, find the optimal control, state and performance index.

	B. Solution of the Problem
Step 1	Solve the matrix differential Riccati equation $\dot{\mathbf{P}}(t) = -\mathbf{P}(t)\mathbf{A}(t) - \mathbf{A}'(t)\mathbf{P}(t) + \mathbf{P}(t)\mathbf{E}(t)\mathbf{P}(t) - \mathbf{V}(t)$, with final condition $\mathbf{P}(t_f) = \mathbf{C}'(t_f)\mathbf{F}(t_f)\mathbf{C}(t_f)$, and the non-homogeneous vector differential equation $\dot{\mathbf{g}}(t) = - [\mathbf{A}(t) - \mathbf{E}(t)\mathbf{P}(t)]' \, \mathbf{g}(t) - \mathbf{W}(t)\mathbf{z}(t)$, with final condition $\mathbf{g}(t_f) = \mathbf{C}'(t_f)\mathbf{F}(t_f)\mathbf{z}(t_f)$ where $\mathbf{E}(t) = \mathbf{B}(t)\mathbf{R}^{-1}(t)\mathbf{B}'(t)$, $\mathbf{V}(t) = \mathbf{C}'(t)\mathbf{Q}(t)\mathbf{C}(t)$, $\mathbf{W}(t) = \mathbf{C}'(t)\mathbf{Q}(t)$.
Step 2	Solve the optimal state $\mathbf{x}^*(t)$ from $\dot{\mathbf{x}}^*(t) = [\mathbf{A}(t) - \mathbf{E}(t)\mathbf{P}(t)] \, \mathbf{x}^*(t) + \mathbf{E}(t)\mathbf{g}(t)$ with initial condition $\mathbf{x}(t_0) = \mathbf{x}_0$.
Step 3	Obtain optimal control $\mathbf{u}^*(t)$ from $\mathbf{u}^*(t) = -\mathbf{K}(t)\mathbf{x}^*(t) + \mathbf{R}^{-1}(t)\mathbf{B}'(t)\mathbf{g}(t)$, where, $\mathbf{K}(t) = \mathbf{R}^{-1}(t)\mathbf{B}'(t)\mathbf{P}(t)$.
Step 4	The optimal cost $J^*(t_0)$ is $J^*(t_0) = \frac{1}{2}\mathbf{x}^{*'}(t_0)\mathbf{P}(t_0)\mathbf{x}^*(t_0) - \mathbf{x}^*(t_0)\mathbf{g}(t_0) + \mathbf{h}(t_0)$ where $\mathbf{h}(t)$ is the solution of $\dot{\mathbf{h}}(t) = -\frac{1}{2}\mathbf{g}'(t)\mathbf{E}(t)\mathbf{g}(t) - \frac{1}{2}\mathbf{z}'(t)\mathbf{Q}(t)\mathbf{z}(t)$ with final condition $\mathbf{h}(t_f) = -\mathbf{z}'(t_f)\mathbf{P}(t_f)\mathbf{z}(t_f)$.

2. *Closed Loop Eigenvalues:* From the costate relation (4.1.36), we see the closed-loop system matrix $[\mathbf{A}(t) - \mathbf{B}(t)\mathbf{R}^{-1}(t)\mathbf{B}'(t)\mathbf{P}(t)]$ is again independent of the desired output $\mathbf{z}(t)$. This means the eigenvalues of the closed-loop, optimal tracking system are *independent* of the desired output $\mathbf{z}(t)$.

3. *Tracking and Regulator Systems:* The main difference between the optimal *output tracking* system and the optimal *state regulator* system is in the vector $\mathbf{g}(t)$. As shown in Figure 4.1, one can think of the desired output $\mathbf{z}(t)$ as the forcing function of the closed-loop optimal system which generates the signal $\mathbf{g}(t)$.

4. Also, note that if we make $\mathbf{C}(t) = \mathbf{I}(t)$, then in (4.1.11), $\mathbf{V}(t) = \mathbf{Q}(t)$. Thus, the matrix DRE (4.1.19) becomes the same matrix DRE (3.2.34) that was obtained in LQR system in Chapter 3.

Let us consider a second order example to illustrate the linear quadratic tracking system.

Example 4.1

A second order plant

$$\dot{x}_1(t) = x_2(t),$$
$$\dot{x}_2(t) = -2x_1(t) - 3x_2(t) + u(t)$$
$$\mathbf{y}(t) = \mathbf{x}(t) \tag{4.1.38}$$

is to be controlled to minimize the performance index

$$J = [1 - x_1(t_f)]^2$$
$$+ \int_{t_0}^{t_f} \left[[1 - x_1(t)]^2 + 0.002u^2(t) \right] dt. \tag{4.1.39}$$

The final time t_f is specified at 20, the final state $\mathbf{x}(t_f)$ is free and the admissible controls and states are unbounded. It is required to keep the state $x_1(t)$ close to 1. Obtain the feedback control law and plot all the time histories of Riccati coefficients, $\mathbf{g}$ vector components, optimal states and control.

Solution: The performance index indicates that the state $x_1(t)$ is to be kept close to the reference input $z_1(t) = 1$ and since there is no condition on state $x_2(t)$, one can choose arbitrarily as $z_2(t) = 0$. Now, in our present case, with $\mathbf{e}(t) = \mathbf{z}(t) - \mathbf{C}\mathbf{x}(t)$, we have $e_1(t) = z_1(t) - x_1(t)$, $e_2(t) = z_2(t) - x_2(t)$ and $\mathbf{C} = \mathbf{I}$.

Next, let us identify the various matrices in the present tracking system by comparing the state (4.1.38) and the PI (4.1.39) of the present system (note the absence of the factor $1/2$ in PI) with the corresponding state (4.1.29) and the PI (4.1.30), respectively, of the general formulation of the problem, we get

$$\mathbf{A} = \begin{bmatrix} 0 & 1 \\ -2 & -3 \end{bmatrix}; \quad \mathbf{B} = \begin{bmatrix} 0 \\ 1 \end{bmatrix}; \quad \mathbf{C} = \mathbf{I}; \quad \mathbf{z}(t) = \begin{bmatrix} 1 \\ 0 \end{bmatrix};$$

$$\mathbf{Q} = \begin{bmatrix} 2 & 0 \\ 0 & 0 \end{bmatrix} = \mathbf{F}(t_f); \quad \mathbf{R} = r = 0.004. \tag{4.1.40}$$

Let $\mathbf{P}(t)$ be the 2x2 symmetric matrix and $\mathbf{g}(t)$ be the 2x1 vector as

$$\mathbf{P}(t) = \begin{bmatrix} p_{11}(t) & p_{12}(t) \\ p_{12}(t) & p_{22}(t) \end{bmatrix}; \quad \mathbf{g}(t) = \begin{bmatrix} g_1(t) \\ g_2(t) \end{bmatrix}. \tag{4.1.41}$$

Then, the optimal control given by (4.1.31) becomes

$$u^*(t) = -250 \left[p_{12} x_1^*(t) + p_{22} x_2^*(t) - g_2(t) \right] \tag{4.1.42}$$

where, $\mathbf{P}(t)$, the positive definite matrix, is the solution of the matrix differential Riccati equation (4.1.32)

$$\begin{bmatrix} \dot{p}_{11}(t) & \dot{p}_{12}(t) \\ \dot{p}_{12}(t) & \dot{p}_{22}(t) \end{bmatrix} = - \begin{bmatrix} p_{11}(t) & p_{12}(t) \\ p_{12}(t) & p_{22}(t) \end{bmatrix} \begin{bmatrix} 0 & 1 \\ -2 & -3 \end{bmatrix}$$

$$- \begin{bmatrix} 0 & -2 \\ 1 & -3 \end{bmatrix} \begin{bmatrix} p_{11}(t) & p_{12}(t) \\ p_{12}(t) & p_{22}(t) \end{bmatrix}$$

$$+ \begin{bmatrix} p_{11}(t) & p_{12}(t) \\ p_{12}(t) & p_{22}(t) \end{bmatrix} \begin{bmatrix} 0 \\ 1 \end{bmatrix} \frac{1}{0.004} \begin{bmatrix} 0 & 1 \end{bmatrix} \text{x}$$

$$\begin{bmatrix} p_{11}(t) & p_{12}(t) \\ p_{12}(t) & p_{22}(t) \end{bmatrix} - \begin{bmatrix} 2 & 0 \\ 0 & 0 \end{bmatrix} \tag{4.1.43}$$

and $\mathbf{g}(t)$, is the solution of the nonhomogeneous vector differential equation obtained from (4.1.34) as

$$\begin{bmatrix} \dot{g}_1(t) \\ \dot{g}_2(t) \end{bmatrix} = - \left\{ \begin{bmatrix} 0 & 1 \\ -2 & -3 \end{bmatrix} \right.$$

$$- \begin{bmatrix} 0 \\ 1 \end{bmatrix} \frac{1}{0.004} \begin{bmatrix} 0 & 1 \end{bmatrix} \begin{bmatrix} p_{11}(t) & p_{12}(t) \\ p_{12}(t) & p_{22}(t) \end{bmatrix} \right\}' \begin{bmatrix} g_1(t) \\ g_2(t) \end{bmatrix}$$

$$- \begin{bmatrix} 1 & 0 \\ 0 & 1 \end{bmatrix}' \begin{bmatrix} 2 & 0 \\ 0 & 0 \end{bmatrix} \begin{bmatrix} 1 \\ 0 \end{bmatrix}. \tag{4.1.44}$$

Simplifying the equations (4.1.43) and (4.1.44), we get

$$\dot{p}_{11}(t) = 250p_{12}^2(t) + 4p_{12}(t) - 2$$
$$\dot{p}_{12}(t) = 250p_{12}(t)p_{22}(t) - p_{11}(t) + 3p_{12}(t) + 2p_{22}(t)$$
$$\dot{p}_{22}(t) = 250p_{22}^2(t) - 2p_{12}(t) + 6p_{22}(t) \qquad (4.1.45)$$

with final condition (4.1.33) as

$$\begin{bmatrix} p_{11}(t_f) & p_{12}(t_f) \\ p_{12}(t_f) & p_{22}(t_f) \end{bmatrix} = \begin{bmatrix} 2 & 0 \\ 0 & 0 \end{bmatrix} \qquad (4.1.46)$$

and

$$\dot{g}_1(t) = [250p_{12}(t) + 2]\, g_2(t) - 2$$
$$\dot{g}_2(t) = -g_1(t) + [3 + 250p_{22}(t)]\, g_2(t) \qquad (4.1.47)$$

with final condition

$$\begin{bmatrix} g_1(t_f) \\ g_2(t_f) \end{bmatrix} = \begin{bmatrix} 2 \\ 0 \end{bmatrix}. \qquad (4.1.48)$$

Note: One has to try various values of the matrix **R** in order to get a better tracking of the states. Solutions for the functions $p_{11}(t), p_{12}(t)$, and $p_{22}(t)$ (Figure 4.2), functions $g_1(t)$ and $g_2(t)$ (Figure 4.3), optimal states (Figures 4.4) and control (Figure 4.5) for initial conditions $\mathbf{x}(0) = [-0.5 \ 0]$ and the final time $t_f = 20$ are obtained using MATLAB© routines given in Appendix C under continuous-time tracking system.

Example 4.2

Consider the same Example 4.1 with a different PI as

$$J = \int_{t_0}^{t_f} \left[[2t - x_1(t)]^2 + 0.02u^2(t) \right] dt \qquad (4.1.49)$$

where, t_f is specified and $\mathbf{x}(t_f)$ is free. Find the optimal control in order that the state $x_1(t)$ *track* a ramp function $z_1(t) = 2t$ and without much expenditure of control energy. Plot all the variables (Riccati coefficients, optimal states and control) for initial conditions $\mathbf{x}(0) = [-1 \ 0]'$.

Solution: The performance index (4.1.49) indicates that the state $x_1(t)$ is to be kept close to the reference input $z_1(t) = 2t$ and since there is no condition on state $x_2(t)$, one can choose arbitrarily as $z_2(t) = 0$. Now, in our present case, with $\mathbf{e}(t) = \mathbf{z}(t) - \mathbf{C}\mathbf{x}(t)$, we have $e_1(t) = z_1(t) - x_1(t)$, $e_2(t) = z_2(t) - x_2(t)$ and $\mathbf{C} = \mathbf{I}$.

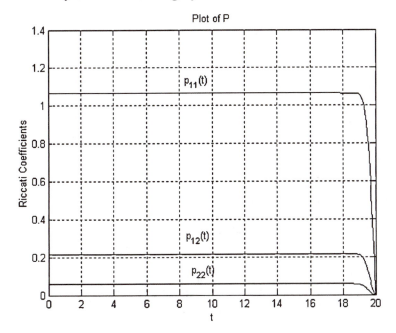

Figure 4.2   Riccati Coefficients for Example 4.1

Next, let us identify the various matrices in the present tracking system by comparing the state (4.1.38) and the PI (4.1.39) of the present problem (note the absence of the factor $1/2$ in PI) with the corresponding state (4.1.29) and the PI (4.1.30), respectively, of the general formulation of the problem, we get

$$\mathbf{A} = \begin{bmatrix} 0 & 1 \\ -2 & -3 \end{bmatrix}; \quad \mathbf{B} = \begin{bmatrix} 0 \\ 1 \end{bmatrix}; \quad \mathbf{C} = \mathbf{I}; \quad \mathbf{z}(t) = \begin{bmatrix} 2t \\ 0 \end{bmatrix};$$

$$\mathbf{Q} = \begin{bmatrix} 2 & 0 \\ 0 & 0 \end{bmatrix} \quad \mathbf{R} = r = 0.04. \tag{4.1.50}$$

Let $\mathbf{P}(t)$ be the 2x2 symmetric matrix and $\mathbf{g}(t)$ be the 2x1 vector as

$$\mathbf{P}(t) = \begin{bmatrix} p_{11}(t) & p_{12}(t) \\ p_{12}(t) & p_{22}(t) \end{bmatrix}; \quad \mathbf{g}(t) = \begin{bmatrix} g_1(t) \\ g_2(t) \end{bmatrix}. \tag{4.1.51}$$

Then, the optimal control given by (4.1.31) becomes

$$u^*(t) = -250 \left[p_{12}x_1^*(t) + p_{22}x_2^*(t) - g_2(t) \right] \tag{4.1.52}$$

where, $\mathbf{P}(t)$, the positive definite matrix, is the solution of the

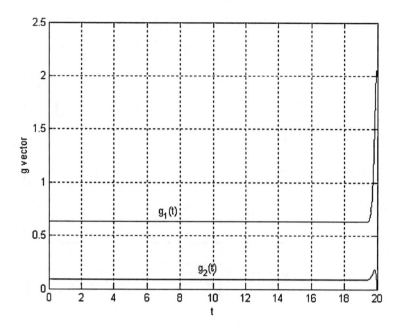

Figure 4.3   Coefficients $g_1(t)$ and $g_2(t)$ for Example 4.1

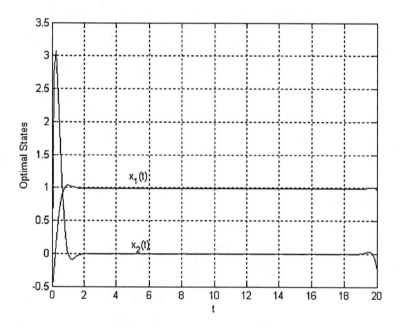

Figure 4.4   Optimal States for Example 4.1

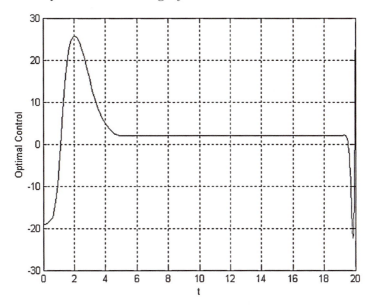

Figure 4.5   Optimal Control for Example 4.1

matrix differential Riccati equation (4.1.32)

$$\begin{bmatrix} \dot{p}_{11}(t) & \dot{p}_{12}(t) \\ \dot{p}_{12}(t) & \dot{p}_{22}(t) \end{bmatrix} = - \begin{bmatrix} p_{11}(t) & p_{12}(t) \\ p_{12}(t) & p_{22}(t) \end{bmatrix} \begin{bmatrix} 0 & 1 \\ -2 & -3 \end{bmatrix}$$

$$- \begin{bmatrix} 0 & -2 \\ 1 & -3 \end{bmatrix} \begin{bmatrix} p_{11}(t) & p_{12}(t) \\ p_{12}(t) & p_{22}(t) \end{bmatrix}$$

$$+ \begin{bmatrix} p_{11}(t) & p_{12}(t) \\ p_{12}(t) & p_{22}(t) \end{bmatrix} \begin{bmatrix} 0 \\ 1 \end{bmatrix} \frac{1}{0.04} \begin{bmatrix} 0 & 1 \end{bmatrix}.$$

$$\begin{bmatrix} p_{11}(t) & p_{12}(t) \\ p_{12}(t) & p_{22}(t) \end{bmatrix} - \begin{bmatrix} 2 & 0 \\ 0 & 0 \end{bmatrix} \qquad (4.1.53)$$

and $\mathbf{g}(t)$, is the solution of the nonhomogeneous vector differential equation obtained from (4.1.34) as

$$\begin{bmatrix} \dot{g}_1(t) \\ \dot{g}_2(t) \end{bmatrix} = - \left\{ \begin{bmatrix} 0 & 1 \\ -2 & -3 \end{bmatrix} \right.$$

$$\left. - \begin{bmatrix} 0 \\ 1 \end{bmatrix} \frac{1}{0.04} \begin{bmatrix} 0 & 1 \end{bmatrix} \begin{bmatrix} p_{11}(t) & p_{12}(t) \\ p_{12}(t) & p_{22}(t) \end{bmatrix} \right\}' \begin{bmatrix} g_1(t) \\ g_2(t) \end{bmatrix}$$

$$- \begin{bmatrix} 1 & 0 \\ 0 & 1 \end{bmatrix}' \begin{bmatrix} 2 & 0 \\ 0 & 0 \end{bmatrix} \begin{bmatrix} 2t \\ 0 \end{bmatrix}. \qquad (4.1.54)$$

Simplifying the equations (4.1.53) and (4.1.54), we get

$$\dot{p}_{11}(t) = 25p_{12}^2(t) + 4p_{12}(t) - 2$$
$$\dot{p}_{12}(t) = 25p_{12}(t)p_{22}(t) - p_{11}(t) + 3p_{12}(t) + 2p_{22}(t)$$
$$\dot{p}_{22}(t) = 25p_{22}^2(t) - 2p_{12}(t) + 6p_{22}(t) \tag{4.1.55}$$

with final condition (4.1.33) as

$$\begin{bmatrix} p_{11}(t_f) & p_{12}(t_f) \\ p_{12}(t_f) & p_{22}(t_f) \end{bmatrix} = \begin{bmatrix} 0 & 0 \\ 0 & 0 \end{bmatrix} \tag{4.1.56}$$

and

$$\dot{g}_1(t) = [25p_{12}(t) + 2]\, g_2(t) - 4t$$
$$\dot{g}_2(t) = -g_1(t) + [3 + 25p_{22}(t)]\, g_2(t) \tag{4.1.57}$$

with final condition

$$\begin{bmatrix} g_1(t_f) \\ g_2(t_f) \end{bmatrix} = \begin{bmatrix} 0 \\ 0 \end{bmatrix}. \tag{4.1.58}$$

See the plots of the Riccati coefficients $p_{11}(t), p_{12}(t)$ and $p_{22}(t)$ in Figure 4.6 and coefficients $g_1(t)$ and $g_2(t)$ in Figure 4.7. Also see the plots of the optimal control $u^*(t)$ in Figure 4.9 and optimal states $x_1^*(t)$ and $x_2^*(t)$ in Figure 4.8.

4.2   LQT System: Infinite-Time Case

In Chapter 3, in the case of linear quadratic regulator system, we extended the results of finite-time case to infinite-time (limiting or steady-state) case. Similarly, we now extend the results of *finite-time* case of the linear quadratic tracking system to the case of *infinite time* [3]. Thus, we restrict our treatment to *time-invariant* matrices in the plant and the performance index. Consider a linear *time-invariant* plant as

$$\dot{\mathbf{x}}(t) = \mathbf{A}\mathbf{x}(t) + \mathbf{B}\mathbf{u}(t) \tag{4.2.1}$$
$$\mathbf{y}(t) = \mathbf{C}\mathbf{x}(t). \tag{4.2.2}$$

The error is

$$\mathbf{e}(t) = \mathbf{z}(t) - \mathbf{y}(t), \tag{4.2.3}$$

and choose the performance index as

$$\lim_{t_f \to \infty} J = \lim_{t_f \to \infty} \frac{1}{2} \int_0^\infty [\mathbf{e}'(t)\mathbf{Q}\mathbf{e}(t) + \mathbf{u}'(t)\mathbf{R}\mathbf{u}(t)]\, dt \tag{4.2.4}$$

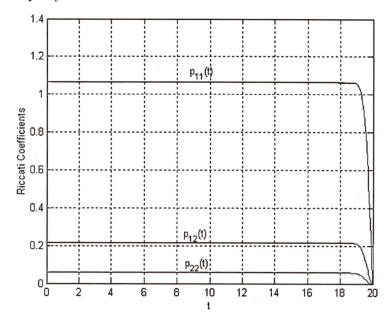

Figure 4.6   Riccati Coefficients for Example 4.2

to track the desired signal $\mathbf{z}(t)$. Also, we assume that $\mathbf{Q}$ is an $n{\times}n$ symmetric, positive *semidefinite matrix*, and $\mathbf{R}$ is a $r{\times}r$ symmetric, positive *definite* matrix. Note that there is no terminal cost function in the PI (4.2.4) and hence $\mathbf{F} = 0$.

An obvious way of getting results for infinite-time (steady-state) case is to write down the results of finite-time case and then simply let $t_f \to \infty$. Thus, as $t_f \to \infty$, the matrix function $\mathbf{P}(t)$ in (4.1.19) tends to the steady-state value $\bar{\mathbf{P}}$ as the solution of

$$\boxed{-\bar{\mathbf{P}}\mathbf{A} - \mathbf{A}'\bar{\mathbf{P}} + \bar{\mathbf{P}}\mathbf{B}\mathbf{R}^{-1}\mathbf{B}'\bar{\mathbf{P}} - \mathbf{C}'\mathbf{Q}\mathbf{C} = 0.} \qquad (4.2.5)$$

Also, the vector function $\mathbf{g}(t)$ in (4.1.20) tends to a finite function $\bar{\mathbf{g}}(t)$ as the solution of

$$\dot{\bar{\mathbf{g}}}(t) = \left[\bar{\mathbf{P}}\mathbf{E} - \mathbf{A}'\right]\bar{\mathbf{g}}(t) - \mathbf{W}\mathbf{z}(t) \qquad (4.2.6)$$

where, $\mathbf{E} = \mathbf{B}\mathbf{R}^{-1}\mathbf{B}'$ and $\mathbf{W} = \mathbf{C}'\mathbf{Q}$. The optimal control (4.1.31) becomes

$$\boxed{\mathbf{u}(t) = -\mathbf{R}^{-1}\mathbf{B}' \left[\bar{\mathbf{P}}\mathbf{x}(t) - \bar{\mathbf{g}}(t)\right].} \qquad (4.2.7)$$

Further details on this are available in Anderson and Moore [3].

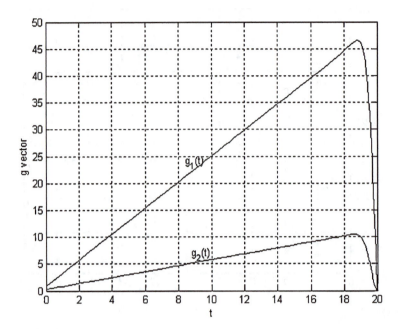

Figure 4.7   Coefficients $g_1(t)$ and $g_2(t)$ for Example 4.2

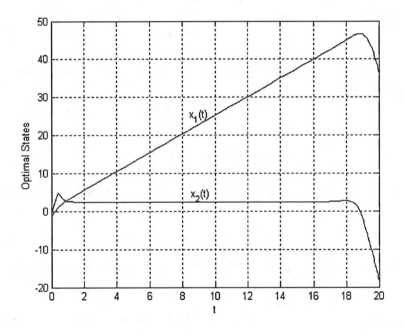

Figure 4.8   Optimal Control and States for Example 4.2

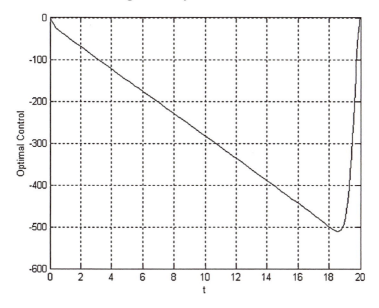

Figure 4.9   Optimal Control and States for Example 4.2

4.3   *Fixed-End-Point Regulator System*

In this section, we discuss the *fixed-end-point* state regulator system, where the final state $\mathbf{x}(t_f)$ is zero and the final time t_f is *fixed* [5]. This is different from the conventional *free-end-point* state regulator system with the final time t_f being *free*, leading to the matrix Riccati differential equation that was discussed in Chapter 3. Following the procedure similar to the free-end-point system, we will arrive at the same matrix differential Riccati equation (3.2.18). But, if we use the earlier transformation (3.2.11) to find the corresponding boundary condition for (3.2.18), we see that

$$\boldsymbol{\lambda}(t_f) = \mathbf{F}(t_f)\mathbf{x}(t_f) = \mathbf{P}(t_f)\mathbf{x}(t_f) \tag{4.3.1}$$

and for the fixed final condition $\mathbf{x}(t_f) = 0$, and for arbitrary $\boldsymbol{\lambda}(t_f)$, we have

$$\mathbf{P}(t_f) = \infty. \tag{4.3.2}$$

This means that for the fixed-end-point regulator system, we solve the matrix DRE (3.2.18) using the final condition (4.3.2). In practice, we may start with a very large value of $\mathbf{P}(t_f)$ instead of ∞.

   Alternatively, we present a different procedure to find *closed-loop optimal control* for the fixed-end-point system [5]. In fact, we will use what is known as *inverse* Riccati transformation between the state and costate variables and arrive at matrix *inverse Riccati* equation.

As before, consider a linear, time-varying system

$$\dot{\mathbf{x}}(t) = \mathbf{A}(t)\mathbf{x}(t) + \mathbf{B}(t)\mathbf{u}(t) \tag{4.3.3}$$

with a cost functional

$$J(\mathbf{u}) = \frac{1}{2}\int_{t_0}^{t_f} \left[\mathbf{x}'(t)\mathbf{Q}(t)\mathbf{x}(t) + \mathbf{u}'(t)\mathbf{R}(t)\mathbf{u}(t) \right] dt \tag{4.3.4}$$

where, $\mathbf{x}(t)$ is *n*th *state* vector, $\mathbf{u}(t)$ is *r*th *control* vector, $\mathbf{A}(t)$ is *nxn* *state* matrix, and $\mathbf{B}(t)$ is *nxr* *control* matrix. We assume that the control is *unconstrained*. The boundary conditions are given as

$$\mathbf{x}(t = t_0) = \mathbf{x}_0; \qquad \mathbf{x}(t = t_f) = \mathbf{x}_f = 0 \tag{4.3.5}$$

where, t_f is *fixed* or given *a priori*. Here, we can easily see that for *fixed* final condition, there is no meaning in having a terminal cost term in the cost function (4.3.4).

We develop the procedure for free-end-point regulator system under the following steps (see Table 2.1).

- **Step 1:** *Hamiltonian*

- **Step 2:** *Optimal Control*

- **Step 3:** *State and Costate System*

- **Step 4:** *Closed-Loop Optimal Control*

- **Step 5:** *Boundary Conditions*

Now, we address these steps in detail.

- **Step 1:** *Hamiltonian:* Formulate the Hamiltonian as

$$\mathcal{H}(\mathbf{x}(t), \mathbf{u}(t), \boldsymbol{\lambda}(t)) = \frac{1}{2}\mathbf{x}'(t)\mathbf{Q}(t)\mathbf{x}(t) + \frac{1}{2}\mathbf{u}'(t)\mathbf{R}(t)\mathbf{u}(t)$$
$$+ \boldsymbol{\lambda}'(t)\left[\mathbf{A}(t)\mathbf{x}(t) + \mathbf{B}(t)\mathbf{u}(t)\right] \tag{4.3.6}$$

- **Step 2:** *Optimal Control:* Taking the partial of $\mathcal{H}$ w.r.t. $\mathbf{u}$, we have

$$\frac{\partial \mathcal{H}}{\partial \mathbf{u}} = 0 \longrightarrow \mathbf{R}(t)\mathbf{u}(t) + \mathbf{B}'(t)\boldsymbol{\lambda}(t) = 0 \tag{4.3.7}$$

which gives optimal control $\mathbf{u}^*(t)$ as

$$\mathbf{u}^*(t) = -\mathbf{R}^{-1}(t)\mathbf{B}'(t)\boldsymbol{\lambda}^*(t). \tag{4.3.8}$$

- **Step 3:** *State and Costate System:* Obtain the state and costate equations as

$$\dot{\mathbf{x}}^*(t) = +\frac{\partial \mathcal{H}}{\partial \boldsymbol{\lambda}} \longrightarrow \dot{\mathbf{x}}^*(t) = \mathbf{A}(t)\mathbf{x}^*(t) + \mathbf{B}(t)\mathbf{u}^*(t), \qquad (4.3.9)$$

$$\dot{\boldsymbol{\lambda}}^*(t) = -\frac{\partial \mathcal{H}}{\partial \mathbf{x}} \longrightarrow \dot{\boldsymbol{\lambda}}^*(t) = -\mathbf{Q}(t)\mathbf{x}^*(t) - \mathbf{A}'(t)\boldsymbol{\lambda}^*(t). \quad (4.3.10)$$

Eliminating control $\mathbf{u}^*(t)$ from (4.3.8) and (4.3.9) to obtain the canonical system of equations

$$\begin{bmatrix} \dot{\mathbf{x}}^*(t) \\ \dot{\boldsymbol{\lambda}}^*(t) \end{bmatrix} = \begin{bmatrix} \mathbf{A}(t) & -\mathbf{E}(t) \\ -\mathbf{Q}(t) & -\mathbf{A}'(t) \end{bmatrix} \begin{bmatrix} \mathbf{x}^*(t) \\ \boldsymbol{\lambda}^*(t) \end{bmatrix} \qquad (4.3.11)$$

where, $\mathbf{E}(t) = \mathbf{B}(t)\mathbf{R}^{-1}(t)\mathbf{B}'(t)$. This state and costate system, along with the given boundary conditions (4.3.5), constitutes a two-point boundary value problem (TPBVP), which when solved gives optimal state $\mathbf{x}^*(t)$ and costate $\boldsymbol{\lambda}^*(t)$ functions. This optimal costate function $\boldsymbol{\lambda}^*(t)$ substituted in (4.3.8) gives optimal control $\mathbf{u}^*(t)$. This leads us to *open-loop* optimal control as discussed in Chapter 2. But our interest here is to obtain *closed-loop* optimal control for the fixed-end-point regulator system.

- **Step 4:** *Closed-Loop Optimal Control:* Now if this were a free-end-point system ($\mathbf{x}(t_f)$ free), using transversality conditions, we would be able to obtain a final condition on the costate $\boldsymbol{\lambda}(t_f)$,which lets us assume a Riccati transformation between the state and costate function as

$$\boldsymbol{\lambda}^*(t) = \mathbf{P}(t)\mathbf{x}^*(t). \qquad (4.3.12)$$

In the absence of any knowledge on the final condition of the costate function $\boldsymbol{\lambda}^*(t)$, we are led to assume a kind of *inverse* Riccati transformation as [104, 113]

$$\mathbf{x}^*(t) = \mathbf{M}(t)\boldsymbol{\lambda}^*(t) \qquad (4.3.13)$$

where, the $n \times n$ matrix $\mathbf{M}(t)$ is yet to be determined. Note the difference between the transformations (4.3.12) and (4.3.13). Now as before in the case of free-end-point system, by simple manipulation of the state and costate system (4.3.11) and (4.3.13) (that is, eliminating $\mathbf{x}^*(t)$), we obtain

$$\dot{\mathbf{x}}^*(t) = \dot{\mathbf{M}}(t)\boldsymbol{\lambda}^*(t) + \mathbf{M}(t)\dot{\boldsymbol{\lambda}}^*(t) \longrightarrow$$

$$\left[\dot{\mathbf{M}}(t) - \mathbf{A}(t)\mathbf{M}(t) - \mathbf{M}(t)\mathbf{A}'(t) - \mathbf{M}(t)\mathbf{Q}(t)\mathbf{M}(t) + \right.$$

$$\left. \mathbf{B}(t)\mathbf{R}^{-1}\mathbf{B}'(t)\right]\boldsymbol{\lambda}^*(t) = 0. \quad (4.3.14)$$

Now, if the previous equation should be valid for all $t \in [t_0, t_f]$, and for any arbitrary $\boldsymbol{\lambda}^*(t)$, we then have

$$\boxed{\dot{\mathbf{M}}(t) = \mathbf{A}(t)\mathbf{M}(t) + \mathbf{M}(t)\mathbf{A}'(t) + \mathbf{M}(t)\mathbf{Q}(t)\mathbf{M}(t) - \mathbf{B}(t)\mathbf{R}^{-1}\mathbf{B}'(t).}$$

$$(4.3.15)$$

Let us call this the *inverse* matrix differential Riccati equation (DRE) just to distinguish from the normal DRE (3.2.34).

- **Step 5:** *Boundary Conditions*: Now the boundary condition for (4.3.15) is obtained as follows. Here, we have different cases to be discussed.

  1. $\mathbf{x}(t_f) = 0$ *and* $\mathbf{x}(t_0) \neq 0$: We know from the given boundary conditions (4.3.5) that $\mathbf{x}(t_f) = 0$ and using this in (4.3.13), we get

$$\mathbf{x}(t_f) = 0 = \mathbf{M}(t_f)\boldsymbol{\lambda}(t_f). \quad (4.3.16)$$

     For arbitrary $\boldsymbol{\lambda}(t_f)$, (4.3.16) becomes

$$\boxed{\mathbf{M}(t_f) = 0.} \quad (4.3.17)$$

  2. $\mathbf{x}(t_f) \neq 0$ *and* $\mathbf{x}(t_0) = 0$: Here, at $t = t_0$, (4.3.13) becomes

$$\mathbf{x}(t_0) = 0 = \mathbf{M}(t_0)\boldsymbol{\lambda}(t_0) \quad (4.3.18)$$

     and for arbitrary $\boldsymbol{\lambda}(t_0)$, (4.3.18) becomes

$$\boxed{\mathbf{M}(t_0) = 0.} \quad (4.3.19)$$

  Thus, we solve the inverse matrix DRE (4.3.15) *backward* using the final condition (4.3.17) or *forward* using the initial condition (4.3.19).

  The optimal control (4.3.8) with the transformation (4.3.13) becomes

$$\boxed{\mathbf{u}^*(t) = -\mathbf{R}^{-1}(t)\mathbf{B}'(t)\mathbf{M}^{-1}(t)\mathbf{x}^*(t).} \quad (4.3.20)$$

3. *General Boundary Conditions:* $\mathbf{x}(t_0) \neq 0$ and $\mathbf{x}(t_f) \neq 0$. Here, both the given boundary conditions are not zero, and we assume a transformation as

$$\mathbf{x}^*(t) = \mathbf{M}(t)\boldsymbol{\lambda}^*(t) + \mathbf{v}(t). \qquad (4.3.21)$$

As before, we substitute the transformation (4.3.21) in the state and costate system (4.3.11) and eliminate $\mathbf{x}^*(t)$ to get

$$\dot{\mathbf{x}}^*(t) = \dot{\mathbf{M}}(t)\boldsymbol{\lambda}^*(t) + \mathbf{M}(t)\dot{\boldsymbol{\lambda}}^*(t) + \dot{\mathbf{v}}(t) \qquad (4.3.22)$$

leading to

$$\mathbf{A}(t)\left[\mathbf{M}(t)\boldsymbol{\lambda}^*(t) + \mathbf{v}(t)\right] - \mathbf{B}(t)\mathbf{R}^{-1}(t)\mathbf{B}'(t)\boldsymbol{\lambda}^*(t)$$
$$= \dot{\mathbf{M}}(t)\boldsymbol{\lambda}^*(t) + \mathbf{M}(t)\left[-\mathbf{Q}(t)\left[\mathbf{M}(t)\boldsymbol{\lambda}^*(t) + \mathbf{v}(t)\right]\right.$$
$$\left. -\mathbf{A}'(t)\boldsymbol{\lambda}^*(t)\right] + \dot{\mathbf{v}}(t) \qquad (4.3.23)$$

further leading to

$$\left[\dot{\mathbf{M}}(t) - \mathbf{A}(t)\mathbf{M}(t) - \mathbf{M}(t)\mathbf{A}'(t) - \mathbf{M}(t)\mathbf{Q}(t)\mathbf{M}(t) + \right.$$
$$\left. \mathbf{B}(t)\mathbf{R}^{-1}(t)\mathbf{B}'(t)\right]\boldsymbol{\lambda}^*(t) +$$
$$\left[\dot{\mathbf{v}}(t) - \mathbf{M}(t)\mathbf{Q}(t)\mathbf{v}(t) - \mathbf{A}(t)\mathbf{v}(t)\right] = 0.$$
$$(4.3.24)$$

This should be valid for any orbit ray value of $\boldsymbol{\lambda}^*(t)$, which leads us to a set of equations

$$\dot{\mathbf{M}}(t) = \mathbf{A}(t)\mathbf{M}(t) + \mathbf{M}(t)\mathbf{A}'(t) + \mathbf{M}(t)\mathbf{Q}(t)\mathbf{M}(t)$$
$$-\mathbf{B}(t)\mathbf{R}^{-1}(t)\mathbf{B}'(t) \qquad (4.3.25)$$
$$\dot{\mathbf{v}}(t) = \mathbf{M}(t)\mathbf{Q}(t)\mathbf{v}(t) + \mathbf{A}(t)\mathbf{v}(t). \qquad (4.3.26)$$

At $t = t_0$, (4.3.21) becomes

$$\mathbf{x}^*(t_0) = \mathbf{M}(t_0)\boldsymbol{\lambda}^*(t_0) + \mathbf{v}(t_0). \qquad (4.3.27)$$

Since $\boldsymbol{\lambda}^*(t_0)$ is arbitrary, (4.3.27) gives us

$$\mathbf{M}(t_0) = 0; \qquad \mathbf{v}(t_0) = \mathbf{x}(t_0). \qquad (4.3.28)$$

At $t = t_f$, (4.3.21) becomes

$$\mathbf{x}^*(t_f) = \mathbf{M}(t_f)\boldsymbol{\lambda}^*(t_f) + \mathbf{v}(t_f). \qquad (4.3.29)$$

Again, since $\boldsymbol{\lambda}^*(t_f)$ is arbitrary, (4.3.29) becomes

$$\mathbf{M}(t_f) = 0; \qquad \mathbf{v}(t_f) = \mathbf{x}(t_f). \qquad (4.3.30)$$

Thus, the set of the equations (4.3.25) and (4.3.26) are solved using either the *initial* conditions (4.3.28) or *final* conditions (4.3.30).

Finally, using the transformation (4.3.21) in the optimal control (4.3.8), the closed-loop optimal control is given by

$$\mathbf{u}^*(t) = -\mathbf{R}^{-1}(t)\mathbf{B}'(t)\mathbf{M}^{-1}(t)[\mathbf{x}^*(t) - \mathbf{v}(t)] \qquad (4.3.31)$$

where, it is assumed that $\mathbf{M}(t)$ is invertible.

Now to illustrate the previous method and to be able to get analytical solutions, we present a first order example.

Example 4.3

Given the plant as

$$\dot{x}(t) = ax(t) + bu(t), \qquad (4.3.32)$$

and the performance index as

$$J = \frac{1}{2}\int_{t_0}^{t_f}[qx^2(t) + ru^2(t)]dt, \qquad (4.3.33)$$

and, the boundary conditions as

$$x(t=0) = x_0; \qquad x(t=t_f) = 0, \qquad (4.3.34)$$

find the closed-loop optimal control.

Solution: Follow the procedure of the inverse matrix DRE described in the last section. We see that with the boundary conditions (4.3.34), we need to use the scalar version of the inverse matrix DRE (4.3.15) having the boundary condition (4.3.17). The optimal control (4.3.20) is given by

$$u^*(t) = -r^{-1}bm^{-1}(t)x^*(t) \qquad (4.3.35)$$

where, $m(t)$ is the solution of the scalar DRE (4.3.15)

$$\dot{m}(t) = 2am(t) + m^2(t)q - \frac{b^2}{r} \qquad (4.3.36)$$

with the boundary condition (4.3.17) as $m(t_f) = 0$. Solving (4.3.36) with this boundary condition, we get

$$m(t) = \frac{b^2}{r} \left[\frac{e^{-\beta(t-t_f)} - e^{\beta(t-t_f)}}{(a + \beta)e^{-\beta(t-t_f)} - (a - \beta)e^{\beta(t-t_f)}} \right] \quad (4.3.37)$$

where, $\beta = \sqrt{a^2 + q\frac{b^2}{r}}$. Then the optimal control (4.3.35) becomes

$$u^*(t) = \frac{1}{b} \left[\frac{(a + \beta)e^{-\beta(t-t_f)} - (a - \beta)e^{\beta(t-t_f)}}{e^{-\beta(t-t_f)} - e^{\beta(t-t_f)}} \right] x^*(t). \quad (4.3.38)$$

4.4   LQR with a Specified Degree of Stability

In this section, we examine the state regulator system with infinite time interval and with a specified degree of stability for a time-invariant system [3, 2]. Let us consider a linear time-invariant plant as

$$\dot{\mathbf{x}}(t) = \mathbf{A}\mathbf{x}(t) + \mathbf{B}\mathbf{u}(t); \quad \mathbf{x}(t = t_0) = \mathbf{x}(0), \quad (4.4.1)$$

and the cost functional as

$$J = \frac{1}{2} \int_{t_0}^{\infty} e^{2\alpha t} \left[\mathbf{x}'(t)\mathbf{Q}\mathbf{x}(t) + \mathbf{u}'(t)\mathbf{R}\mathbf{u}(t) \right] dt \quad (4.4.2)$$

where, α is a positive parameter. Here, we first assume that the pair $[A + \alpha\mathbf{I}, B]$ is completely *stabilizable* and $\mathbf{R}$ and $\mathbf{Q}$ are constant, symmetric, positive *definite* and positive *semidefinite* matrices, respectively. The problem is to find the optimal control which minimizes the performance index (4.4.2) under the dynamical constraint (4.4.1).

This can be solved by modifying the previous system to fit into the standard infinite-time regulator system discussed earlier in Chapter 3. Thus, we make the following transformations

$$\hat{\mathbf{x}}(t) = e^{\alpha t}\mathbf{x}(t); \quad \hat{\mathbf{u}}(t) = e^{\alpha t}\mathbf{u}(t). \quad (4.4.3)$$

Then, using the transformations (4.4.3), it is easy to see that the *modified* system becomes

$$\dot{\hat{\mathbf{x}}}(t) = \frac{d}{dt}\{e^{\alpha t}\mathbf{x}(t)\} = \alpha e^{\alpha t}\mathbf{x}(t) + e^{\alpha t}\dot{\mathbf{x}}(t)$$

$$= \alpha\hat{\mathbf{x}}(t) + e^{\alpha t}[\mathbf{A}\mathbf{x}(t) + \mathbf{B}\mathbf{u}(t)]$$

$$\dot{\hat{\mathbf{x}}}(t) = (\mathbf{A} + \alpha\mathbf{I})\hat{\mathbf{x}}(t) + \mathbf{B}\hat{\mathbf{u}}(t). \quad (4.4.4)$$

We note that the initial conditions for the *original* system (4.4.1) and the *modified* system (4.4.4) are simply related as $\hat{\mathbf{x}}(t_0) = e^{\alpha t_0}\mathbf{x}(t_0)$ and in particular, if $t_0 = 0$ the initial conditions are the same for the original and modified systems. Also, using the transformations (4.4.3), the original performance measure (4.4.2) can be modified to

$$\hat{J} = \frac{1}{2}\int_{t_0}^{\infty} \left[\hat{\mathbf{x}}'(t)\mathbf{Q}\hat{\mathbf{x}}(t) + \hat{\mathbf{u}}'(t)\mathbf{R}\hat{\mathbf{u}}(t)\right] dt. \tag{4.4.5}$$

Considering the minimization of the modified system defined by (4.4.4) and (4.4.5), we see that the optimal control is given by (see Chapter 3, Table 3.3)

$$\hat{\mathbf{u}}^*(t) = -\mathbf{R}^{-1}\mathbf{B}'\bar{\mathbf{P}}\hat{\mathbf{x}}^*(t) = -\bar{\mathbf{K}}\hat{\mathbf{x}}^*(t) \tag{4.4.6}$$

where, $\bar{\mathbf{K}} = \mathbf{R}^{-1}\mathbf{B}'\bar{\mathbf{P}}$ and the matrix $\bar{\mathbf{P}}$ is the positive definite, symmetric solution of the algebraic Riccati equation

$$\bar{\mathbf{P}}(\mathbf{A} + \alpha\mathbf{I}) + (\mathbf{A}' + \alpha\mathbf{I})\bar{\mathbf{P}} - \bar{\mathbf{P}}\mathbf{B}\mathbf{R}^{-1}\mathbf{B}'\bar{\mathbf{P}} + \mathbf{Q} = 0. \tag{4.4.7}$$

Using the optimal control (4.4.6) in the modified system (4.4.4), we get the optimal closed-loop system as

$$\dot{\hat{\mathbf{x}}}^*(t) = (\mathbf{A} + \alpha\mathbf{I} - \mathbf{B}\mathbf{R}^{-1}\mathbf{B}'\bar{\mathbf{P}})\hat{\mathbf{x}}^*(t). \tag{4.4.8}$$

Now, we can simply apply these results to the original system. Thus, using the transformations (4.4.3) in (4.4.6), the optimal control of the original system (4.4.1) and the associated performance measure (4.4.2) is given by

$$\mathbf{u}^*(t) = e^{-\alpha t}\hat{\mathbf{u}}^*(t) = -e^{-\alpha t}\mathbf{R}^{-1}\mathbf{B}'\bar{\mathbf{P}}e^{\alpha t}\mathbf{x}^*(t)$$
$$= -\mathbf{R}^{-1}\mathbf{B}'\bar{\mathbf{P}}\mathbf{x}^*(t) = -\bar{\mathbf{K}}\mathbf{x}^*(t). \tag{4.4.9}$$

Interestingly, this desired (original) optimal control (4.4.9) has the same structure as the optimal control (4.4.6) of the modified system. The optimal performance index for original system or modified system is the same and equals to

$$\hat{J}^* = \frac{1}{2}\hat{\mathbf{x}}^{*\prime}(t_0)\bar{\mathbf{P}}\hat{\mathbf{x}}^*(t_0)$$
$$J^* = \frac{1}{2}e^{2\alpha t_0}\mathbf{x}^{*\prime}(t_0)\bar{\mathbf{P}}\mathbf{x}^*(t_0). \tag{4.4.10}$$

We see that the closed-loop optimal control system (4.4.8) has eigenvalues with real parts less than $-\alpha$. In other words, the state $\mathbf{x}^*(t)$ approaches zero at least as fast as $e^{-\alpha t}$. Then, we say that the closed-loop optimal system (4.4.8) has a degree of stability of at least α.

4.4.1 *Regulator System with Prescribed Degree of Stability: Summary*

For a controllable, linear, time-invariant plant

$$\dot{\mathbf{x}}(t) = \mathbf{A}\mathbf{x}(t) + \mathbf{B}\mathbf{u}(t), \tag{4.4.11}$$

and the infinite interval cost functional

$$J = \frac{1}{2} \int_{t_0}^{\infty} e^{2\alpha t} \left[\mathbf{x}'(t)\mathbf{Q}\mathbf{x}(t) + \mathbf{u}'(t)\mathbf{R}\mathbf{u}(t) \right] dt, \tag{4.4.12}$$

the optimal control is given by

$$\mathbf{u}^*(t) = -\mathbf{R}^{-1}\mathbf{B}'\bar{\mathbf{P}}\mathbf{x}^*(t) = -\bar{\mathbf{K}}\mathbf{x}^*(t) \tag{4.4.13}$$

where, $\bar{\mathbf{K}} = \mathbf{R}^{-1}\mathbf{B}'\bar{\mathbf{P}}$ and $\bar{\mathbf{P}}$, the nxn constant, positive *definite*, symmetric matrix, is the solution of the nonlinear, matrix algebraic Riccati equation (ARE)

$$\bar{\mathbf{P}}(\mathbf{A} + \alpha\mathbf{I}) + (\mathbf{A}' + \alpha\mathbf{I})\bar{\mathbf{P}} - \bar{\mathbf{P}}\mathbf{B}\mathbf{R}^{-1}\mathbf{B}'\bar{\mathbf{P}} + \mathbf{Q} = 0, \tag{4.4.14}$$

the optimal trajectory is the solution of

$$\dot{\mathbf{x}}^*(t) = \left(\mathbf{A} - \mathbf{B}\mathbf{R}^{-1}\mathbf{B}'\bar{\mathbf{P}} \right)\mathbf{x}^*(t), \tag{4.4.15}$$

and the optimal cost is given by

$$J^* = \frac{1}{2}e^{2\alpha t_0}\mathbf{x}^{*\prime}(t_0)\bar{\mathbf{P}}\mathbf{x}^*(t_0). \tag{4.4.16}$$

The entire procedure is now summarized in Table 4.2. Consider a first-order system example to illustrate the previous method.

Example 4.4

Consider a first-order system

$$\dot{x}(t) = -x(t) + u(t), \quad x(0) = 1 \tag{4.4.17}$$

and a performance measure

$$J = \frac{1}{2} \int_0^{\infty} e^{2\alpha t}[x^2(t) + u^2(t)]dt. \tag{4.4.18}$$

Find the optimal control law and show that the closed-loop optimal system has a degree of stability of at least α.

Table 4.2   Procedure Summary of Regulator System with Prescribed Degree of Stability

A. Statement of the Problem
Given the plant as $\dot{\mathbf{x}}(t) = \mathbf{A}\mathbf{x}(t) + \mathbf{B}\mathbf{u}(t)$, the performance index as $J = \frac{1}{2} \int_{t_0}^{\infty} e^{2\alpha t} \left[\mathbf{x}'(t)\mathbf{Q}\mathbf{x}(t) + \mathbf{u}'(t)\mathbf{R}\mathbf{u}(t) \right] dt$, and the boundary conditions as $\mathbf{x}(t_0) = \mathbf{x}_0; \quad \mathbf{x}(\infty) = 0$, find the optimal control, state and index.

B. Solution of the Problem	
Step 1	Solve the matrix algebraic Riccati equation $\bar{\mathbf{P}}(\mathbf{A} + \alpha\mathbf{I}) + (\mathbf{A}' + \alpha\mathbf{I})\bar{\mathbf{P}} + \mathbf{Q} - \bar{\mathbf{P}}\mathbf{B}\mathbf{R}^{-1}\mathbf{B}'\bar{\mathbf{P}} = 0$.
Step 2	Solve the optimal state $\mathbf{x}^*(t)$ from $\dot{\mathbf{x}}^*(t) = \left(\mathbf{A} - \mathbf{B}\mathbf{R}^{-1}\mathbf{B}'\bar{\mathbf{P}} \right) \mathbf{x}^*(t)$ with initial condition $\mathbf{x}(t_0) = \mathbf{x}_0$.
Step 3	Obtain the optimal control $\mathbf{u}^*(t)$ from $\mathbf{u}^*(t) = -\mathbf{R}^{-1}\mathbf{B}'\bar{\mathbf{P}}\mathbf{x}^*(t)$.
Step 4	Obtain the optimal performance index from $J^* = \frac{1}{2}e^{2\alpha t_0}\mathbf{x}^{*\prime}(t_0)\bar{\mathbf{P}}\mathbf{x}^*(t_0)$.

Solution: Essentially, we show that the eigenvalue of this closed-loop optimal system is less than or equal to $-\alpha$. First of all, in the above, we note that $A = a = -1, B = b = 1, Q = q = 1$ and $R = r = 1$. Then, the algebraic Riccati equation (4.4.14) becomes

$$2\bar{p}(\alpha - 1) - \bar{p}^2 + 1 = 0 \quad \longrightarrow \quad \bar{p}^2 - 2\bar{p}(\alpha - 1) - 1. \qquad (4.4.19)$$

Solving the previous for positive value of $\bar{p}$, we have

$$\bar{p} = -1 + \alpha + \sqrt{(\alpha - 1)^2 + 1}. \qquad (4.4.20)$$

The optimal control (4.4.15) becomes

$$u^*(t) = -\bar{p}x^*(t). \qquad (4.4.21)$$

The optimal system (4.4.22) becomes

$$\dot{x}^*(t) = \left(-\alpha - \sqrt{(\alpha - 1)^2 + 1} \right) x^*(t). \qquad (4.4.22)$$

It is easy to see that the eigenvalue for the system (4.4.22) is related as

$$-\alpha - \sqrt{(\alpha - 1)^2 + 1} < -\alpha. \qquad (4.4.23)$$

This shows the desired result that the optimal system has the eigenvalue less than α.

4.5   *Frequency-Domain Interpretation*

In this section, we use frequency domain to derive some results from the classical control point of view for a linear, time-invariant, continuous-time, optimal control system with infinite-time horizon case. For this, we know that the closed-loop optimal control involves the solution of matrix algebraic Riccati equation [89, 3]. For ready reference, we repeat here some of the results of Chapter 3.

Consider a controllable, linear, time-invariant plant

$$\dot{\mathbf{x}}(t) = \mathbf{A}\mathbf{x}(t) + \mathbf{B}\mathbf{u}(t), \qquad (4.5.1)$$

and the infinite-time interval cost functional

$$J = \frac{1}{2} \int_0^\infty \left[\mathbf{x}'(t)\mathbf{Q}\mathbf{x}(t) + \mathbf{u}'(t)\mathbf{R}\mathbf{u}(t) \right] dt. \qquad (4.5.2)$$

The optimal control is given by

$$\mathbf{u}^*(t) = -\mathbf{R}^{-1}\mathbf{B}'\bar{\mathbf{P}}\mathbf{x}^*(t) = -\bar{\mathbf{K}}\mathbf{x}^*(t), \qquad (4.5.3)$$

where, $\bar{\mathbf{K}} = \mathbf{R}^{-1}\mathbf{B}'\bar{\mathbf{P}}$, and $\bar{\mathbf{P}}$, the nxn constant, positive definite, symmetric matrix, is the solution of the nonlinear, matrix ARE

$$-\bar{\mathbf{P}}\mathbf{A} - \mathbf{A}'\bar{\mathbf{P}} + \bar{\mathbf{P}}\mathbf{B}\mathbf{R}^{-1}\mathbf{B}'\bar{\mathbf{P}} - \mathbf{Q} = 0. \qquad (4.5.4)$$

The optimal trajectory (state) is the solution of

$$\dot{\mathbf{x}}^*(t) = \left(\mathbf{A} - \mathbf{B}\mathbf{R}^{-1}\mathbf{B}'\bar{\mathbf{P}} \right) \mathbf{x}^*(t) = \left(\mathbf{A} - \mathbf{B}\bar{\mathbf{K}} \right) \mathbf{x}^*(t), \qquad (4.5.5)$$

which is asymptotically stable. Here, we assume that $[\mathbf{A}, \mathbf{B}]$ is stabilizable and $[\mathbf{A}, \sqrt{\mathbf{Q}}]$ is observable. Then, the open-loop characteristic polynomial of the system is [89]

$$\boldsymbol{\Delta}_o(s) = |s\mathbf{I} - \mathbf{A}|, \qquad (4.5.6)$$

where, s is the Laplace variable and the optimal closed-loop characteristic polynomial is

$$\begin{aligned}
\mathbf{\Delta}_c(s) &= |s\mathbf{I} - \mathbf{A} + \mathbf{B}\bar{\mathbf{K}}| \\
&= |\mathbf{I} + \mathbf{B}\bar{\mathbf{K}}[s\mathbf{I} - \mathbf{A}]^{-1}|.|s\mathbf{I} - \mathbf{A}|, \\
&= |\mathbf{I} + \bar{\mathbf{K}}[s\mathbf{I} - \mathbf{A}]^{-1}\mathbf{B}|\mathbf{\Delta}_o(s).
\end{aligned} \qquad (4.5.7)$$

This is a relation between the open-loop $\mathbf{\Delta}_o(s)$ and closed-loop $\mathbf{\Delta}_c(s)$ characteristic polynomials. From Figure 4.10, we note that

1. $-\bar{\mathbf{K}}[s\mathbf{I} - \mathbf{A}]^{-1}\mathbf{B}$ is called the *loop gain matrix*, and

2. $\mathbf{I} + \bar{\mathbf{K}}[s\mathbf{I} - \mathbf{A}]^{-1}\mathbf{B}$ is termed *return difference matrix*.

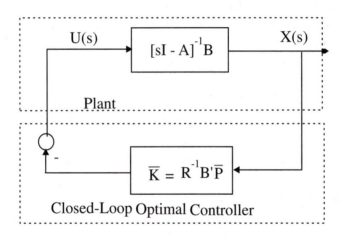

Figure 4.10   Optimal Closed-Loop Control in Frequency Domain

To derive the desired factorization result, we use the matrix ARE (4.5.4). Let us rewrite the ARE as

$$-\bar{\mathbf{P}}\mathbf{A} - \mathbf{A}'\bar{\mathbf{P}} + \bar{\mathbf{P}}\mathbf{B}\mathbf{R}^{-1}\mathbf{B}'\bar{\mathbf{P}} = \mathbf{Q}. \qquad (4.5.8)$$

First adding and subtracting $s\bar{\mathbf{P}}, s = j\omega$ to the previous ARE, we get

$$\bar{\mathbf{P}}[s\mathbf{I} - \mathbf{A}] + [-s\mathbf{I} - \mathbf{A}']\bar{\mathbf{P}} + \bar{\mathbf{K}}'\mathbf{R}\bar{\mathbf{K}} = \mathbf{Q}. \qquad (4.5.9)$$

Next, premultiplying by $\mathbf{B}'\mathbf{\Phi}'(-s)$ and post multiplying by $\mathbf{\Phi}(s)\mathbf{B}$, the previous equation becomes

$$\begin{aligned}
\mathbf{B}'\mathbf{\Phi}'(-s)\bar{\mathbf{P}}\mathbf{B} + \mathbf{B}'\bar{\mathbf{P}}\mathbf{\Phi}(s)\mathbf{B} + \mathbf{B}'\mathbf{\Phi}'(-s)\bar{\mathbf{K}}'\mathbf{R}\bar{\mathbf{K}}\mathbf{\Phi}(s)\mathbf{B} \\
= \mathbf{B}'\mathbf{\Phi}'(-s)\mathbf{Q}\mathbf{\Phi}(s)\mathbf{B} \quad (4.5.10)
\end{aligned}$$

where, we used

$$\boldsymbol{\Phi}(s) = [s\mathbf{I} - \mathbf{A}]^{-1}; \quad \boldsymbol{\Phi}'(-s) = [-s\mathbf{I} - \mathbf{A}']^{-1}. \qquad (4.5.11)$$

Finally, using $\bar{\mathbf{K}} = \mathbf{R}^{-1}\mathbf{B}'\bar{\mathbf{P}} \longrightarrow \bar{\mathbf{K}}' = \bar{\mathbf{P}}\mathbf{B}\mathbf{R}^{-1} \longrightarrow \bar{\mathbf{P}}\mathbf{B} = \bar{\mathbf{K}}'\mathbf{R}$ and adding $\mathbf{R}$ to both sides of (4.5.10), we have the desired factorization result as

$$\boxed{\mathbf{B}'\boldsymbol{\Phi}'(-s)\mathbf{Q}\boldsymbol{\Phi}(s)\mathbf{B} + \mathbf{R} = \left[\mathbf{I} + \bar{\mathbf{K}}\boldsymbol{\Phi}(-s)\mathbf{B}\right]'\mathbf{R}\left[\mathbf{I} + \bar{\mathbf{K}}\boldsymbol{\Phi}(s)\mathbf{B}\right]}$$

$$(4.5.12)$$

or equivalently,

$$\mathbf{B}'[-s\mathbf{I} - \mathbf{A}']^{-1}\mathbf{Q}[s\mathbf{I} - \mathbf{A}]^{-1}\mathbf{B} + \mathbf{R}$$
$$= \left[\mathbf{I} + \bar{\mathbf{K}}[-s\mathbf{I} - \mathbf{A}]^{-1}\mathbf{B}\right]'\mathbf{R}\left[\mathbf{I} + \bar{\mathbf{K}}[s\mathbf{I} - \mathbf{A}]^{-1}\mathbf{B}\right].$$

$$(4.5.13)$$

The previous relation is also called the *Kalman equation in frequency domain.*

4.5.1   Gain Margin and Phase Margin

We know that in classical control theory, the features of gain and phase margins are important in evaluating the system performance with respect to robustness to plant parameter variations and uncertainties. The engineering specifications often place lower bounds on the phase and gain margins. Here, we interpret some of the classical control features such as gain margin and phase margin for the closed-loop optimal control system [3]. For ready reference let us rewrite the return-difference result (4.5.13)) with $s = j\omega$ here as

$$\mathbf{B}'[-j\omega\mathbf{I} - \mathbf{A}']^{-1}\mathbf{Q}[j\omega\mathbf{I} - \mathbf{A}]^{-1}\mathbf{B} + \mathbf{R}$$
$$= \left[\mathbf{I} + \bar{\mathbf{K}}[-j\omega\mathbf{I} - \mathbf{A}]^{-1}\mathbf{B}\right]'\mathbf{R}[\mathbf{I} + \bar{\mathbf{K}}[j\omega\mathbf{I} - \mathbf{A}]^{-1}\mathbf{B}].$$

$$(4.5.14)$$

The previous result can be viewed as

$$\mathbf{M}(j\omega) = \mathbf{W}'(-j\omega)\mathbf{W}(j\omega) \qquad (4.5.15)$$

where,

$$\mathbf{W}(j\omega) = \mathbf{R}^{1/2}\left[\mathbf{I} + \bar{\mathbf{K}}[j\omega\mathbf{I} - \mathbf{A}]^{-1}\mathbf{B}\right]$$
$$\mathbf{M}(j\omega) = \mathbf{R} + \mathbf{B}'[-j\omega\mathbf{I} - \mathbf{A}']^{-1}\mathbf{Q}[j\omega\mathbf{I} - \mathbf{A}]^{-1}\mathbf{B}. \qquad (4.5.16)$$

Note that $\mathbf{M}(j\omega) \geq \mathbf{R} > 0$. Using $\mathbf{Q} = \mathbf{CC}'$, $\mathbf{R} = \mathbf{DD}' = \mathbf{I}$ and the notation

$$\mathbf{W}'(-j\omega)\mathbf{W}(j\omega) = ||\mathbf{W}(j\omega)||^2, \qquad (4.5.17)$$

the factorization result (4.5.14) can be written in neat form as

$$\boxed{||\mathbf{I} + \bar{\mathbf{K}}'[j\omega\mathbf{I} - \mathbf{A}]^{-1}\mathbf{B}||^2 = \mathbf{I} + ||\mathbf{C}'[j\omega\mathbf{I} - \mathbf{A}]^{-1}\mathbf{B}||^2.} \quad (4.5.18)$$

This result can be used to find the optimal feedback matrix $\bar{\mathbf{K}}$ given the other quantities $\mathbf{A}, \mathbf{B}, \mathbf{Q}, \mathbf{R} = \mathbf{I}$. Note that in (4.5.18), we need not solve for the Riccati coefficient matrix $\bar{\mathbf{P}}$, instead we directly obtain the feedback matrix $\bar{\mathbf{K}}$.

In the single-input case, the various matrices become scalars or vectors as $\mathbf{B} = \mathbf{b}, \mathbf{R} = r, \bar{\mathbf{K}} = \bar{\mathbf{k}}$. Then, the factorization result (4.5.14) boils down to

$$r + \mathbf{b}'[-j\omega\mathbf{I} - \mathbf{A}']^{-1}\mathbf{Q}[j\omega\mathbf{I} - \mathbf{A}]^{-1}\mathbf{b}$$
$$= r|1 + \bar{\mathbf{k}}[j\omega\mathbf{I} - \mathbf{A}]^{-1}\mathbf{b}|^2. \qquad (4.5.19)$$

In case $\mathbf{Q} = \mathbf{cc}'$, we can write (4.5.19) as

$$r + |\mathbf{c}'[j\omega\mathbf{I} - \mathbf{A}']^{-1}\mathbf{b}|^2 = r|(1 + \bar{\mathbf{k}}[j\omega\mathbf{I} - \mathbf{A}]^{-1}\mathbf{b})|^2. \quad (4.5.20)$$

The previous result may be called another version of the *Kalman equation in frequency domain.* The previous relation (also from (4.5.18) for a scalar case) implies that

$$\boxed{|(1 + \bar{\mathbf{k}}[j\omega\mathbf{I} - \mathbf{A}]^{-1}\mathbf{b})|^2 \geq 1.} \qquad (4.5.21)$$

Thus, the return difference is lower bounded by 1 for all ω.

Example 4.5

Consider a simple example where we can verify the analytical solutions by another known method. Find the optimal feedback coefficients for the system

$$\dot{x}_1(t) = x_2(t)$$
$$\dot{x}_2(t) = u(t) \qquad (4.5.22)$$

and the performance measure

$$J = \frac{1}{2} \int_0^\infty \left[x_1^2(t) + x_2^2(t) + u^2(t) \right] dt. \qquad (4.5.23)$$

Solution: First it is easy to identify the various matrices as

$$\mathbf{A} = \begin{bmatrix} 0 & 1 \\ 0 & 0 \end{bmatrix}; \quad \mathbf{B} = \mathbf{b} = \begin{bmatrix} 0 \\ 1 \end{bmatrix}; \quad \mathbf{Q} = \begin{bmatrix} 1 & 0 \\ 0 & 1 \end{bmatrix}; \quad \mathbf{R} = r = 1. \quad (4.5.24)$$

Also, note since $\mathbf{Q} = \mathbf{R} = \mathbf{I}$, we have $\mathbf{C} = \mathbf{D} = \mathbf{I}$ and $\mathbf{B} = \mathbf{I}$. Thus, the Kalman equation (4.5.18) with $j\omega = s$ becomes

$$||\mathbf{I} + \bar{\mathbf{K}}'[s\mathbf{I} - \mathbf{A}]^{-1}\mathbf{B}||^2 = \mathbf{I} + ||\mathbf{C}'[s\mathbf{I} - \mathbf{A}]^{-1}\mathbf{B}||^2. \quad (4.5.25)$$

Further, we have

$$[s\mathbf{I} - \mathbf{A}]^{-1} = \begin{bmatrix} \frac{1}{s} & 0 \\ \frac{1}{s^2} & \frac{1}{s} \end{bmatrix}; \quad \bar{\mathbf{K}} = \bar{\mathbf{k}} = [k_{11} \ k_{12}]. \quad (4.5.26)$$

Then, the Kalman equation (4.5.25) becomes

$$\left[1 + \begin{bmatrix} k_{11} & k_{12} \end{bmatrix} \begin{bmatrix} \frac{1}{-s} & \frac{1}{(-s)^2} \\ 0 & \frac{1}{-s} \end{bmatrix} \begin{bmatrix} 0 \\ 1 \end{bmatrix} \right] \left[1 + \begin{bmatrix} k_{11} & k_{12} \end{bmatrix} \begin{bmatrix} \frac{1}{s} & \frac{1}{s^2} \\ 0 & \frac{1}{s} \end{bmatrix} \begin{bmatrix} 0 \\ 1 \end{bmatrix} \right] =$$
$$1 + \begin{bmatrix} 0 & 1 \end{bmatrix} \begin{bmatrix} \frac{1}{-s} & \frac{1}{(-s)^2} \\ 0 & \frac{1}{-s} \end{bmatrix} \begin{bmatrix} 1 & 0 \\ 0 & 1 \end{bmatrix} \begin{bmatrix} \frac{1}{s} & \frac{1}{s^2} \\ 0 & \frac{1}{s} \end{bmatrix} \begin{bmatrix} 0 \\ 1 \end{bmatrix}.$$
$$(4.5.27)$$

By simple matrix multiplication and equating the coefficients of like powers of s on either side, we get a set of algebraic equations in general, and in particular in this example we have a single scalar equation as

$$1 + (2k_{11} - k_{12}^2)\frac{1}{s^2} + k_{11}^2\frac{1}{s^4} = 1 - \frac{1}{s^2} + \frac{1}{s^4} \quad (4.5.28)$$

giving us

$$k_{11} = 1, \quad k_{12} = \sqrt{3} \quad (4.5.29)$$

and the optimal feedback control as

$$u^*(t) = -\bar{\mathbf{K}}\mathbf{x}^*(t) = -\begin{bmatrix} 1 & \sqrt{3} \end{bmatrix}\mathbf{x}^*(t). \quad (4.5.30)$$

Note: This example can be easily verified by using the algebraic Riccati equation (3.5.15) (of Chapter 3)

$$\bar{\mathbf{P}}\mathbf{A} + \mathbf{A}'\bar{\mathbf{P}} - \bar{\mathbf{P}}\mathbf{B}\mathbf{R}^{-1}\mathbf{B}'\bar{\mathbf{P}} + \mathbf{Q} = 0 \quad (4.5.31)$$

discussed in Chapter 3. Using the previous relation, we get

$$\bar{\mathbf{P}} = \begin{bmatrix} \sqrt{3} & 1 \\ 1 & \sqrt{3} \end{bmatrix} \tag{4.5.32}$$

and the optimal control (3.5.14) as

$$\mathbf{u}^*(t) = -\mathbf{R}^{-1}\mathbf{B}'\bar{\mathbf{P}}\mathbf{x}^*(t) = -\begin{bmatrix} 1 & \sqrt{3} \end{bmatrix}\mathbf{x}^*(t) \tag{4.5.33}$$

which is the same as (4.5.30).

Let us redraw the closed-loop optimal control system in Figure 4.10 as unity feedback system shown in Figure 4.11.

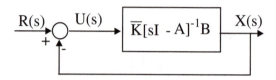

Figure 4.11   Closed-Loop Optimal Control System with Unity Feedback

Here, we can easily recognize that for a single-input, single-output case, the optimal feedback control system is exactly like a classical feedback control system with unity negative feedback and transfer function as $G_o(s) = \bar{\mathbf{k}}[s\mathbf{I} - \mathbf{A}]^{-1}\mathbf{b}$. Thus, the frequency domain interpretation in terms of gain margin, phase margin can be easily done using Nyquist, Bode, or some other plot of the transfer function $G_o(s)$.

Gain Margin

We recall that the gain margin of a feedback control system is the amount of loop gain (usually in decibels) that can be changed before the closed-loop system becomes unstable. Let us now apply the well-known Nyquist criterion to the unity feedback, optimal control system depicted in Figure 4.11. Here, we assume that the *Nyquist path* is clockwise (CW) and the corresponding Nyquist plot makes counter-clockwise (CCW) encirclements around the critical point $-1 + j0$. According to Nyquist stability criterion, for closed-loop stability, the *Nyquist plot* (or diagram) makes CCW encirclements as many times as there are poles of the transfer function $G_o(s)$ lying in the *right half* of the s-plane.

From Figure 4.11 and the return difference relation (4.5.21), we note that

$$\left|(1 + \bar{\mathbf{k}}[j\omega\mathbf{I} - \mathbf{A}]^{-1}\mathbf{b})\right| \geq 1 \qquad (4.5.34)$$

implies that the distance between the critical point $-1 + j0$ and any point on the Nyquist plot is at least 1 and the resulting Nyquist plot is shown in Figure 4.12 for all positive values of ω (i.e., 0 to ∞). This

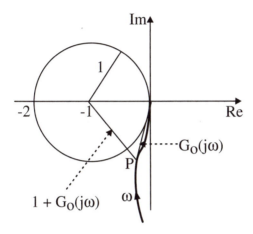

Figure 4.12 Nyquist Plot of $G_o(j\omega)$

means that the Nyquist plot of $G_o(j\omega)$ is constrained to avoid all the points inside the unit circle (centered at $-1 + j0$). Thus, it is clear that the closed-loop optimal system has *infinite gain margin*. Let us proceed further to see if there is a lower limit on the gain factor.

Now, if we multiply the open-loop gain with some constant factor β, the closed-loop system will be asymptotically stable if the Nyquist plot of $\beta G_0(j\omega)$ encircles $-1 + j0$ in CCW direction as many times as there are poles of $\beta G_o(s)$ in the right-half plane. This means that the closed-loop system will be stable if the Nyquist diagram of $G_o(j\omega)$ encircles the critical point $-(1/\beta) + j0$ the same number of times as there are open-loop poles in the right-half plane. But the set of points $-(1/\beta)$ for all real $\beta > \frac{1}{2}$ lies inside the critical unit circle and thus are encircled CCW the same number of times as the original point $-1 + j0$. Consequently, there is a lower limit as $\beta > \frac{1}{2}$. In other words, for values of $\beta < \frac{1}{2}$, the set of points $-1/\beta + j0$ lies outside the unit circle and contradicts the Nyquist criterion for stability of closed-loop

optimal control system. Thus, we have an infinite gain margin on the
upper side and a lower gain margin of $\beta = 1/2$.

Phase Margin

Let us first recall that the phase margin is the amount of phase shift in
CW direction (without affecting the gain) through which the Nyquist
plot can be rotated about the origin so that the gain crossover (unit
distance from the origin) passes through the $-1 + j0$ point. Simply,
it is the amount by which Nyquist plot can be rotated CW to make
the system unstable. Consider a point P at unit distance from the
origin on the Nyquist plot (see Figure 4.12). Since we know that the
Nyquist plot of an optimal regulator must avoid the unit circle centered
at $-1 + j0$, the set of points which are at unit distance *from the origin*
and lying on Nyquist diagram of an optimal regulator are constrained
to lie on the portion marked X on the circumference of the circle with
unit radius and centered at the origin as shown in Figure 4.13. Here,

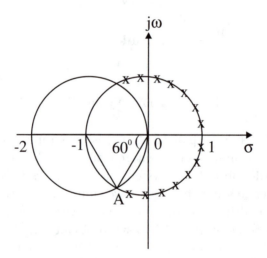

Figure 4.13   Intersection of Unit Circles Centered at Origin and
$-1 + j0$

we notice that the *smallest angle* through which one of the admissible
points A on (the circumference of the circle centered at origin) the

Nyquist plot could be shifted in a CW direction to reach $-1 + j0$ point is 60 degrees. Thus, the closed-loop optimal system or LQR system has a phase margin of *at least* 60 degrees.

4.6   Problems

1. Make reasonable assumptions wherever necessary.

2. Use MATLAB© wherever possible to solve the problems and plot all the optimal controls and states for all problems. Provide the relevant MATLAB© m files.

Problem 4.1 A second order plant

$$\dot{x}_1(t) = x_2(t),$$
$$\dot{x}_2(t) = -2x_1(t) - 3x_2(t) + u(t)$$
$$\mathbf{y}(t) = \mathbf{x}(t)$$

is to be controlled to minimize a performance index and to keep the state $x_1(t)$ close to a ramp function $2t$. The final time t_f is specified, the final state $\mathbf{x}(t_f)$ is free and the admissible controls and states are unbounded. Formulate the performance index, obtain the feedback control law and plot all the time histories of Riccati coefficients, optimal states and control.

Problem 4.2 A second order plant

$$\dot{x}_1(t) = x_2(t),$$
$$\dot{x}_2(t) = -2x_1(t) - 4x_2(t) + 0.5u(t)$$
$$\mathbf{y}(t) = \mathbf{x}(t)$$

is to be controlled to minimize the performance index

$$J = \int_{t_0}^{t_f} \left[4x_1^2(t) + 6x_2^2(t) + 0.02u^2(t) \right] dt.$$

The final time t_f is specified, the final state $\mathbf{x}(t_f)$ is fixed and the admissible controls and states are unbounded. Obtain the feedback control law and plot all the time histories of inverse Riccati coefficients, optimal states and control.

Problem 4.3 For a linear, time-varying system (3.2.48) given as

$$\dot{\mathbf{x}}(t) = \mathbf{A}(t)\mathbf{x}(t) + \mathbf{B}(t)\mathbf{u}(t),$$
$$\mathbf{y}(t) = \mathbf{C}(t)\mathbf{x}(t)$$

with a general cost functional (3.2.49) as

$$J = \frac{1}{2}\mathbf{x}'(t_f)\mathbf{F}(t_f)\mathbf{x}(t_f)$$
$$+ \frac{1}{2}\int_{t_0}^{t_f} \begin{bmatrix} \mathbf{x}'(t) & \mathbf{u}'(t) \end{bmatrix} \begin{bmatrix} \mathbf{Q}(t) & \mathbf{S}(t) \\ \mathbf{S}'(t) & \mathbf{R}(t) \end{bmatrix} \begin{bmatrix} \mathbf{x}(t) \\ \mathbf{u}(t) \end{bmatrix} dt,$$

where, the various vectors and matrices are defined in Chapter 3, formulate a tracking problem and obtain the results similar to those obtained in Chapter 4.

Problem 4.4 Using the frequency-domain results, determine the optimal feedback coefficients and the closed-loop optimal control for the multi-input, multi-output system

$$\dot{\mathbf{x}}(t) = \begin{bmatrix} 0 & 1 \\ -2 & -3 \end{bmatrix} \mathbf{x}(t) + \begin{bmatrix} 1 & 0 \\ 0 & 1 \end{bmatrix} \mathbf{u}(t)$$

and the cost function

$$J = \int_0^\infty [4x_1^2(t) + 4x_2^2(t) + 0.5u_1^2(t) + u_2^2(t)]dt.$$

Problem 4.5 For the D.C. motor speed control system described in Problem 1.1, find the closed-loop optimal control to track the speed at a particular value.

Problem 4.6 For the liquid-level control system described in Problem 1.2, find the closed-loop optimal control to track the liquid level along a ramp function $0.25t$.

Problem 4.7 For the mechanical control system described in Problem 1.4, find the closed-loop optimal control to track the system along (i) a constant value and (ii) a ramp function.

Problem 4.8 For the chemical control system described in Problem 1.6, find the closed-loop optimal control to track the system along (i) a constant value and (ii) a ramp function.

@@@@@@@@@@@@@

Chapter 5

Discrete-Time Optimal Control Systems

In previous chapters, the optimal control of *continuous-time* systems has been presented. In this chapter, the optimal control of *discrete-time* systems is presented. We start with the basic calculus of variations and then touch upon all the topics discussed in the previous chapters with respect to the continuous-time systems such as open-loop optimal control, linear quadratic regulator system, tracking system, etc. It is suggested that the student reviews the material in Appendices A and B given at the end of the book. This chapter is inspired by [84, 89, 120][1].

5.1 Variational Calculus for Discrete-Time Systems

In earlier chapters, we discussed the optimal control of *continuous-time* systems described by *differential* equations. There, we minimized cost functionals which are essentially *integrals* of scalar functions. Now, we know that *discrete-time* systems are characterized by *difference* equations, and we focus on minimizing the cost functionals which are *summations* of some scalar functions.

---

[1]The permission given by John Wiley for F. L. Lewis, *Optimal Control*, John Wiley & Sons, Inc., New York, NY, 1986, is hereby acknowledged.

5.1.1   Extremization of a Functional

In this section, we obtain the *necessary* conditions for optimization of cost functionals which are *summations* such as

$$J(x(k_0), k_0) = J = \sum_{k=k_0}^{k_f-1} V(x(k), x(k+1), k) \qquad (5.1.1)$$

where, the discrete instant $k = k_0, k_1, \cdots, k_f - 1$. Note the following points.

1. For a given interval $k \in [k_0, k_f]$ and a given function $V(x(k), x(k+1), k)$, the summation interval in (5.1.1) needs to be $[k_0, k_f - 1]$.

2. We consider first a *scalar* case for simplicity and then we generalize for the *vector* case.

3. We are also given the initial condition $x(k = k_0) = x(k_0)$.

4. Consider the case of a *free-final point system*, such that k is *fixed* or *specified* and $x(k_f)$ is *free* or *unspecified*.

5. Also, if T is the sampling period, then $x(k) = x(kT)$.

6. Let us note that if we are directly considering the discrete-time version of the continuous-time cost functionals (such as (2.3.1) addressed in Chapter 2), we have the sampling period T multiplying the cost functional (5.1.1).

For extremization (maximization or minimization) of functionals, analogous to the case of continuous-time systems addressed in Chapter 2, we use the fundamental theorem of the calculus of variations (CoV) which states that the *first variation must be equal to zero*. The methodology for this simple case of optimization of a functional is carried out briefly under the following steps.

- **Step 1:** *Variations*

- **Step 2:** *Increment*

- **Step 3:** *First Variation*

- **Step 4:** *Euler-Lagrange Equation*

- **Step 5:** *Boundary Conditions*

Now consider these items in detail.

- **Step 1:** *Variations:* We first let $x(k)$ and $x(k+1)$ take on *variations* $\delta x(k)$ and $\delta x(k+1)$ from their optimal values $x^*(k)$ and $x^*(k+1)$, respectively, such that

$$x(k) = x^*(k) + \delta x(k); \quad x(k+1) = x^*(k+1) + \delta x(k+1). \quad (5.1.2)$$

   Now with these variations, the performance index (5.1.1) becomes

$$
\begin{aligned}
J^* &= J(x^*(k_0), k_0) \\
&= \sum_{k=k_0}^{k_f-1} V(x^*(k), x^*(k+1), k) \quad (5.1.3)
\end{aligned}
$$

$$
\begin{aligned}
J &= J(x(k_0), k_0) \\
&= \sum_{k=k_0}^{k_f-1} V\left(x^*(k) + \delta x(k), x^*(k+1) + \delta x(k+1), k\right). \quad (5.1.4)
\end{aligned}
$$

- **Step 2:** *Increment:* The *increment* of the functionals defined by (5.1.2) and (5.1.3) is defined as

$$\Delta J = J - J^*. \quad (5.1.5)$$

- **Step 3:** *First Variation:* The *first variation* δJ is the *first* order approximation of the increment ΔJ. Thus, using the Taylor series expansion of (5.1.4) along with (5.1.3), we have

$$
\begin{aligned}
\delta J = \sum_{k=k_0}^{k_f-1} \Bigg[&\frac{\partial V(x^*(k), x^*(k+1), k)}{\partial x^*(k)} \delta x(k) \\
&+ \frac{\partial V(x^*(k), x^*(k+1), k)}{\partial x^*(k+1)} \delta x(k+1) \Bigg]. \quad (5.1.6)
\end{aligned}
$$

Now in order to express the coefficient $\delta x(k+1)$ also in terms of $\delta x(k)$, consider the second expression in (5.1.6).

$$\sum_{k=k_0}^{k_f-1} \frac{\partial V(x^*(k), x^*(k+1), k)}{\partial x^*(k+1)} \delta x(k+1)$$

$$= \frac{\partial V(x^*(k_0), x^*(k_0+1), k_0)}{\partial x^*(k_0+1)} \delta x(k_0+1)$$

$$+ \frac{\partial V(x^*(k_0+1), x^*(k_0+2), k_0+1)}{\partial x^*(k_0+2)} \delta x(k_0+2)$$

$$+ \cdots \cdots \cdots \cdots$$

$$+ \frac{\partial V(x^*(k_f-2), x^*(k_f-1), k_f-2)}{\partial x^*(k_f-1)} \delta x(k_f-1)$$

$$+ \frac{\partial V(x^*(k_f-1), x^*(k_f), k_f-1)}{\partial x^*(k_f)} \delta x(k_f)$$

$$+ \frac{\partial V(x^*(k_0-1), x^*(k_0), k_0-1)}{\partial x^*(k_0)} \delta x(k_0)$$

$$- \frac{\partial V(x^*(k_0-1), x^*(k_0), k_0-1)}{\partial x^*(k_0)} \delta x(k_0) \qquad (5.1.7)$$

where, the last two terms in (5.1.7) are added without affecting the rest of the equation. The entire equation (5.1.7) (except the last term and the last but the two terms) is rewritten as

$$\sum_{k=k_0}^{k_f-1} \frac{\partial V(x^*(k), x^*(k+1), k)}{\partial x^*(k+1)} \delta x(k+1)$$

$$= \sum_{k=k_0}^{k_f-1} \frac{\partial V(x^*(k-1), x^*(k), k-1)}{\partial x^*(k)} \delta x(k)$$

$$+ \frac{\partial V(x^*(k_f-1), x^*(k_f), k_f-1)}{\partial x^*(k_f)} \delta x(k_f)$$

$$- \frac{\partial V(x^*(k_0-1), x^*(k_0), k_0-1)}{\partial x^*(k_0)} \delta x(k_0)$$

$$= \sum_{k=k_0}^{k_f-1} \frac{\partial V(x^*(k-1), x^*(k), k-1)}{\partial x^*(k)} \delta x(k)$$

$$+ \left[\frac{\partial V(x^*(k-1), x^*(k), k-1)}{\partial x^*(k)} \delta x(k) \right] \Big|_{k=k_0}^{k=k_f}. \qquad (5.1.8)$$

Substituting (5.1.8) in (5.1.6) and noting that the first variation should be zero, we have

$$
\sum_{k=k_0}^{k_f-1} \left[\frac{\partial V(x^*(k), x^*(k+1), k)}{\partial x^*(k)} + \frac{\partial V(x^*(k-1), x^*(k), k-1)}{\partial x^*(k)} \right] \delta x(k)
$$

$$
+ \left[\frac{\partial V(x^*(k-1), x^*(k), k-1)}{\partial x^*(k)} \delta x(k) \right] \Bigg|_{k=k_0}^{k=k_f} = 0.
$$

$$(5.1.9)$$

- **Step 4:** *Euler-Lagrange Equation:* For (5.1.9) to be satisfied for *arbitrary* variations $\delta x(k)$, we have the condition that the coefficient of $\delta x(k)$ in the first term in (5.1.9) be zero. That is

$$
\boxed{\frac{\partial V(x^*(k), x^*(k+1), k)}{\partial x^*(k)} + \frac{\partial V(x^*(k-1), x^*(k), k-1)}{\partial x^*(k)} = 0.}
$$

$$(5.1.10)$$

This may very well be called the *discrete-time* version of the *Euler-Lagrange (EL) equation.*

- **Step 5:** *Boundary Conditions:* The *boundary* or *transversality* condition is obtained by setting the second term in (5.1.9) equal to zero. That is

$$
\boxed{\left[\frac{\partial V(x^*(k-1), x^*(k), k-1)}{\partial x^*(k)} \delta x(k) \right] \Bigg|_{k=k_0}^{k=k_f} = 0.} \quad (5.1.11)
$$

Now, we discuss two important cases:

1. For a *fixed-end point system*, we have the boundary conditions $x(k_0)$ and $x(k_f)$ *fixed* and hence $\delta x(k_0) = \delta x(k_f) = 0$. The additional (or derived) boundary condition (5.1.11) does not exist.

2. For a *free-final point system*, we are given the initial condition $x(k_0)$ and hence $\delta x(k_0) = 0$ in (5.1.11). Next, at the final point, k_f is *specified*, and $x(k_f)$ is not specified or is *free*,

and hence $\delta x(k_f)$ is *arbitrary*. Thus, the coefficient of $\delta x(k)$ at $k = k_f$ is zero in the condition (5.1.11) which reduces to

$$\left[\frac{\partial V(x^*(k-1), x^*(k), k-1)}{\partial x^*(k)}\right]\bigg|_{k=k_f} = 0. \quad (5.1.12)$$

Let us note that the *necessary* condition (5.1.10) and the associated boundary or *transversality* condition (5.1.12) are derived for the *scalar* function $x(k)$ only.

The previous analysis can be easily extended to the *vector* function $\mathbf{x}(k)$ of nth order. Thus, consider a functional which is the vector version of the scalar functional (5.1.1) as

$$J(\mathbf{x}(k_0), k_0) = J = \sum_{k=k_0}^{k_f-1} V(\mathbf{x}(k), \mathbf{x}(k+1), k). \quad (5.1.13)$$

We will only give the corresponding final Euler-Lagrange equation and the transversality condition, respectively, as

$$\frac{\partial V(\mathbf{x}^*(k), \mathbf{x}^*(k+1), k)}{\partial \mathbf{x}^*(k)} + \frac{\partial V(\mathbf{x}^*(k-1), \mathbf{x}^*(k), k-1)}{\partial \mathbf{x}^*(k)} = 0 \quad (5.1.14)$$

and

$$\left[\frac{\partial V(\mathbf{x}^*(k-1), \mathbf{x}^*(k), k-1)}{\partial \mathbf{x}^*(k)}\right]\bigg|_{k=k_f} = 0. \quad (5.1.15)$$

Note in the Euler-Lagrange equation (5.1.10) or (5.1.14),

1. the first term involves taking the partial derivative of the *given* function $V(\mathbf{x}^*(k), \mathbf{x}^*(k+1), k)$ w.r.t. $\mathbf{x}(k)$ and

2. the second term considers taking the partial derivative of $V(\mathbf{x}^*(k-1), \mathbf{x}^*(k), k-1)$ (one step behind) w.r.t. the same function $\mathbf{x}(k)$.

The second function $V(\mathbf{x}^*(k-1), \mathbf{x}^*(k), k-1)$ can be easily obtained from the given function $V(\mathbf{x}^*(k), \mathbf{x}^*(k+1), k)$ just by replacing k by $k-1$. Also, compare the previous results with the corresponding results for continuous-time systems in Chapter 2.

5.1.2   Functional with Terminal Cost

Let us formulate the cost functional with terminal cost (in addition to summation cost) as

$$J = J(\mathbf{x}(k_0), k_0)$$

$$= S(\mathbf{x}(k_f), k_f) + \sum_{k=k_0}^{k_f-1} V(\mathbf{x}(k), \mathbf{x}(k+1), k) \qquad (5.1.16)$$

given the initial condition $\mathbf{x}(k_0)$ and the final time k_f as *fixed*, and the final state $\mathbf{x}(k_f)$ as *free*. First, assume optimal (*) condition and then consider the variations as

$$\mathbf{x}(k) = \mathbf{x}^*(k) + \delta\mathbf{x}(k)$$

$$\mathbf{x}(k+1) = \mathbf{x}^*(k+1) + \delta\mathbf{x}(k+1). \qquad (5.1.17)$$

Then, the corresponding functionals J and J^* become

$$J^* = S(\mathbf{x}^*(k_f), k_f) + \sum_{k=k_0}^{k_f-1} V(\mathbf{x}^*(k), \mathbf{x}^*(k+1), k), \qquad (5.1.18)$$

$$J = S(\mathbf{x}^*(k_f) + \delta\mathbf{x}(k_f), k_f)$$

$$+ \sum_{k=k_0}^{k_f-1} V(\mathbf{x}^*(k) + \delta\mathbf{x}(k), \mathbf{x}^*(k+1) + \delta\mathbf{x}(k+1), k). \qquad (5.1.19)$$

Following the same procedure as given previously for a functional without terminal cost, we get the *first variation* as

$$\delta J = \sum_{k=k_0}^{k_f-1} \left[\frac{\partial V(\mathbf{x}^*(k), \mathbf{x}^*(k+1), k)}{\partial \mathbf{x}^*(k)} + \frac{\partial V[\mathbf{x}^*(k-1), \mathbf{x}^*(k), k-1)}{\partial \mathbf{x}^*(k)} \right]' \delta\mathbf{x}(k)$$

$$+ \left[\frac{\partial V(\mathbf{x}^*(k-1), \mathbf{x}^*(k), k-1)}{\partial \mathbf{x}^*(k)} \delta\mathbf{x}(k) \right] \Bigg|_{k=k_0}^{k=k_f}$$

$$+ \frac{\partial S(\mathbf{x}^*(k_f), k_f)}{\partial \mathbf{x}^*(k_f)} \delta\mathbf{x}(k_f). \qquad (5.1.20)$$

For extremization, the *first variation* δJ must be zero. Hence, from (5.1.20) the Euler-Lagrange equation becomes

$$\boxed{\frac{\partial V(\mathbf{x}^*(k), \mathbf{x}^*(k+1), k)}{\partial \mathbf{x}^*(k)} + \frac{\partial V(\mathbf{x}^*(k-1), \mathbf{x}^*(k), k-1)}{\partial \mathbf{x}^*(k)} = 0.}$$

$$(5.1.21)$$

and the *transversality* condition for the *free-final point* becomes

$$\left[\left[\frac{\partial V(\mathbf{x}^*(k-1), \mathbf{x}^*(k), k-1)}{\partial \mathbf{x}^*(k)} + \frac{\partial S(\mathbf{x}^*(k_f), k_f)}{\partial \mathbf{x}^*(k_f)}\right]\right]_{k=k_f} = 0.$$

(5.1.22)

Let us now illustrate the application of the Euler-Lagrange equation for discrete-time functionals.

Example 5.1

Consider the minimization of a functional

$$J(x(k_0), k_0) = J = \sum_{k=k_0}^{k_f-1} [x(k)x(k+1) + x^2(k)]$$

(5.1.23)

subject to the boundary conditions x(0) = 2, and x(10) = 5.

Solution: Let us identify in (5.1.23) that

$$V(x(k), x(k+1)) = x(k)x(k+1) + x^2(k)$$

(5.1.24)

and hence

$$V(x(k-1), x(k)) = x(k-1)x(k) + x^2(k-1).$$

(5.1.25)

Then using the Euler-Lagrange equation (5.1.10), which is the same as the scalar version of (5.1.21), we get

$$x(k+1) + 2x(k) + x(k-1) = 0$$

(5.1.26)

or

$$x(k+2) + 2x(k+1) + x(k) = 0$$

(5.1.27)

which upon solving with the given boundary conditions $x(0) = 2$ and $x(10) = 5$, becomes

$$x(k) = 2(-1)^k + 0.3k(-1)^k.$$

(5.1.28)

5.2   *Discrete-Time Optimal Control Systems*

We develop the Minimum Principle for discrete-time control systems analogous to that for continuous-time control systems addressed in previous Chapters 2, 3, and 4. Instead of repeating all the topics of the continuous-time systems for the discrete-time systems, we focus on *linear quadratic optimal control problem*. We essentially approach the problem using the Lagrangian and Hamiltonian (or Pontryagin) functions.

   Consider a linear, time-varying, discrete-time control system described by

$$\mathbf{x}(k+1) = \mathbf{A}(k)\mathbf{x}(k) + \mathbf{B}(k)\mathbf{u}(k) \tag{5.2.1}$$

where, $k = k_0, k_1, \ldots, k_f - 1$, $\mathbf{x}(k)$ is *n*th order *state* vector, $\mathbf{u}(k)$ is *r*th order *control* vector, and $\mathbf{A}(k)$ and $\mathbf{B}(k)$ are matrices of *nxn* and *nxr* dimensions, respectively. Note that we used $\mathbf{A}$ and $\mathbf{B}$ for the state space representation for discrete-time case as well as for the continuous-time case as shown in the previous chapters. One can alternatively use, say $\mathbf{G}$ and $\mathbf{E}$ for the discrete-time case so that the case of discretization of a continuous-time system with $\mathbf{A}$ and $\mathbf{B}$ will result in $\mathbf{G}$ and $\mathbf{E}$ in the discrete-time representation. However, the present notation should not cause any confusion once we redefine the matrices in the discrete-time case. We are given the initial condition as

$$\mathbf{x}(k = k_0) = \mathbf{x}(k_0). \tag{5.2.2}$$

We will discuss later the final state condition and the resulting relations. We are also given a general performance index (PI) with terminal cost as

$$\begin{aligned}
J &= J(\mathbf{x}(k_0), \mathbf{u}(k_0), k_0) \\
&= \frac{1}{2}\mathbf{x}'(k_f)\mathbf{F}(k_f)\mathbf{x}(k_f) \\
&\quad + \frac{1}{2}\sum_{k=k_0}^{k_f-1} [\mathbf{x}'(k)\mathbf{Q}(k)\mathbf{x}(k) + \mathbf{u}'(k)\mathbf{R}(k)\mathbf{u}(k)]
\end{aligned} \tag{5.2.3}$$

where, $\mathbf{F}(k_f)$ and $\mathbf{Q}(k)$ are each *nxn* order symmetric, positive *semi-definite* matrices, and $\mathbf{R}(k)$ is *rxr* symmetric, positive *definite* matrix.

   The methodology for *linear quadratic optimal control* problem is carried out under the following steps.

- **Step 1:** *Augmented Performance Index*

- **Step 2:** *Lagrangian*

- **Step 3:** *Euler-Lagrange Equation*

- **Step 4:** *Hamiltonian*

- **Step 5:** *Open-Loop Optimal Control*

- **Step 6:** *State and Costate System*

Now these steps are described in detail.

- **Step 1:** *Augmented Performance Index:* First, we formulate an *augmented* cost functional by adjoining the original cost functional (5.2.3) with the condition or plant relation (5.2.1) using Lagrange multiplier (later to be called as costate function) $\lambda(k+1)$ as

$$J_a = \frac{1}{2}\mathbf{x}'(k_f)\mathbf{F}(k_f)\mathbf{x}(k_f)$$

$$+\frac{1}{2}\sum_{k=k_0}^{k_f-1}\left[\mathbf{x}'(k)\mathbf{Q}(k)\mathbf{x}(k) + \mathbf{u}'(k)\mathbf{R}(k)\mathbf{u}(k)\right]$$

$$+\lambda(k+1)\left[\mathbf{A}(k)\mathbf{x}(k) + \mathbf{B}(k)\mathbf{u}(k) - \mathbf{x}(k+1)\right]. \quad (5.2.4)$$

Minimization of the augmented cost functional (5.2.4) is the same as that of the original cost functional (5.2.3), since $J = J_a$. The reason for associating the stage $(k+1)$ with the Lagrange multiplier $\lambda(k+1)$ is mainly the simplicity of the final result as will be apparent later.

- **Step 2:** *Lagrangian:* Let us now define a new function called Lagrangian as

$$\mathcal{L}(\mathbf{x}(k), \mathbf{u}(k), \mathbf{x}(k+1), \lambda(k+1))$$
$$= \frac{1}{2}\mathbf{x}'(k)\mathbf{Q}(k)\mathbf{x}(k) + \frac{1}{2}\mathbf{u}'(k)\mathbf{R}(k)\mathbf{u}(k)$$
$$+\lambda'(k+1)\left[\mathbf{A}(k)\mathbf{x}(k) + \mathbf{B}(k)\mathbf{u}(k) - \mathbf{x}(k+1)\right]. \quad (5.2.5)$$

- **Step 3:** *Euler-Lagrange Equations:* We now apply the Euler-Lagrange (EL) equation (5.1.21) to this new function $\mathcal{L}$ with respect to the variables $\mathbf{x}(k)$, $\mathbf{u}(k)$, and $\boldsymbol{\lambda}(k+1)$. Thus, we get

$$\frac{\partial \mathcal{L}(\mathbf{x}^*(k), \mathbf{x}^*(k+1), \mathbf{u}^*(k), \boldsymbol{\lambda}^*(k+1))}{\partial \mathbf{x}^*(k)}$$
$$+ \frac{\partial \mathcal{L}(\mathbf{x}^*(k-1), \mathbf{x}^*(k), \mathbf{u}^*(k-1), \boldsymbol{\lambda}^*(k))}{\partial \mathbf{x}^*(k)} = 0 \qquad (5.2.6)$$

$$\frac{\partial \mathcal{L}(\mathbf{x}^*(k), \mathbf{x}^*(k+1), \mathbf{u}^*(k), \boldsymbol{\lambda}^*(k+1))}{\partial \mathbf{u}^*(k)}$$
$$+ \frac{\partial \mathcal{L}(\mathbf{x}^*(k-1), \mathbf{x}^*(k), \mathbf{u}^*(k-1), \boldsymbol{\lambda}^*(k))}{\partial \mathbf{u}^*(k)} = 0 \qquad (5.2.7)$$

$$\frac{\partial \mathcal{L}(\mathbf{x}^*(k), \mathbf{x}^*(k+1), \mathbf{u}^*(k), \boldsymbol{\lambda}^*(k+1))}{\partial \boldsymbol{\lambda}^*(k)}$$
$$+ \frac{\partial \mathcal{L}(\mathbf{x}^*(k-1), \mathbf{x}^*(k), \mathbf{u}^*(k-1), \boldsymbol{\lambda}^*(k))}{\partial \boldsymbol{\lambda}^*(k)} = 0 \qquad (5.2.8)$$

and the boundary (final) condition (5.1.22) becomes

$$\left[\frac{\partial \mathcal{L}(\mathbf{x}(k-1), \mathbf{x}(k), \mathbf{u}(k-1), \boldsymbol{\lambda}(k))}{\partial \mathbf{x}(k)} + \frac{\partial S(\mathbf{x}(k), k)}{\partial \mathbf{x}(k)} \right]'_* \delta \mathbf{x}(k) \Bigg|_{k=k_0}^{k=k_f} = 0$$
$$(5.2.9)$$

where, from (5.2.3),

$$S(\mathbf{x}(k_f), k_f) = \frac{1}{2} \mathbf{x}'(k_f) \mathbf{F}(k_f) \mathbf{x}(k_f). \qquad (5.2.10)$$

- **Step 4:** *Hamiltonian:* Although relations (5.2.6) to (5.2.10) give the required conditions for optimum, we proceed to get the results in a more elegant manner in terms of the Hamiltonian which is defined as

$$\mathcal{H}(\mathbf{x}^*(k), \mathbf{u}^*(k), \boldsymbol{\lambda}^*(k+1)) = \frac{1}{2} \mathbf{x}^{*\prime}(k) \mathbf{Q}(k) \mathbf{x}^*(k)$$
$$+ \frac{1}{2} \mathbf{u}^{*\prime}(k) \mathbf{R}(k) \mathbf{u}^*(k)$$
$$+ \boldsymbol{\lambda}^{*\prime}(k+1) \left[\mathbf{A}(k) \mathbf{x}^*(k) + \mathbf{B}(k) \mathbf{u}^*(k) \right]. \qquad (5.2.11)$$

Thus, the Lagrangian (5.2.5) and the Hamiltonian (5.2.11) are related as

$$
\begin{aligned}
\mathcal{L}(\mathbf{x}^*(k), \mathbf{x}^*(k+1), \mathbf{u}^*(k), \boldsymbol{\lambda}^*(k+1)) = \\
\mathcal{H}(\mathbf{x}^*(k), \mathbf{u}^*(k), \boldsymbol{\lambda}^*(k+1)) \\
- \boldsymbol{\lambda}^*(k+1)\mathbf{x}^*(k+1).
\end{aligned} \tag{5.2.12}
$$

Now, using the relation (5.2.12) in the set of Euler-Lagrange equations (5.2.6) to (5.2.8), we get the required conditions for extremum in terms of the Hamiltonian as

$$
\boxed{\boldsymbol{\lambda}^*(k) = \frac{\partial \mathcal{H}(\mathbf{x}^*(k), \mathbf{u}^*(k), \boldsymbol{\lambda}^*(k+1))}{\partial \mathbf{x}^*(k)}}, \tag{5.2.13}
$$

$$
\boxed{0 = \frac{\partial \mathcal{H}(\mathbf{x}^*(k), \mathbf{u}^*(k), \boldsymbol{\lambda}^*(k+1))}{\partial \mathbf{u}^*(k)}}, \tag{5.2.14}
$$

$$
\boxed{\mathbf{x}^*(k) = \frac{\partial \mathcal{H}(\mathbf{x}^*(k-1), \mathbf{u}^*(k-1), \boldsymbol{\lambda}^*(k))}{\partial \boldsymbol{\lambda}^*(k)}}. \tag{5.2.15}
$$

Note that the relation (5.2.15) can also be written in a more appropriate way by considering the whole relation at the next stage as

$$
\mathbf{x}^*(k+1) = \frac{\partial \mathcal{H}(\mathbf{x}^*(k), \mathbf{u}^*(k), \boldsymbol{\lambda}^*(k+1))}{\partial \boldsymbol{\lambda}^*(k+1)}. \tag{5.2.16}
$$

For the present system described by the plant (5.2.1) and the performance index (5.2.3) we have the relations (5.2.16), (5.2.13), and (5.2.14) for the state, costate, and control, transforming respectively, to

$$
\mathbf{x}^*(k+1) = \mathbf{A}(k)\mathbf{x}^*(k) + \mathbf{B}(k)\mathbf{u}^*(k) \tag{5.2.17}
$$
$$
\boldsymbol{\lambda}^*(k) = \mathbf{Q}(k)\mathbf{x}^*(k) + \mathbf{A}'(k)\boldsymbol{\lambda}^*(k+1) \tag{5.2.18}
$$
$$
0 = \mathbf{R}(k)\mathbf{u}^*(k) + \mathbf{B}'(k)\boldsymbol{\lambda}^*(k+1). \tag{5.2.19}
$$

- **Step 5:** *Open-Loop Optimal Control:* The optimal control is then given by (5.2.19) as

$$
\boxed{\mathbf{u}^*(k) = -\mathbf{R}^{-1}(k)\mathbf{B}'(k)\boldsymbol{\lambda}^*(k+1)} \tag{5.2.20}
$$

where,the positive definiteness of $\mathbf{R}(k)$ ensures its invertibility. Using the optimal control (5.2.20) in the state equation (5.2.17) we get

$$\begin{aligned}
\mathbf{x}^*(k+1) &= \mathbf{A}(k)\mathbf{x}^*(k) - \mathbf{B}(k)\mathbf{R}^{-1}(k)\mathbf{B}'(k)\boldsymbol{\lambda}^*(k+1) \\
&= \mathbf{A}(k)\mathbf{x}^*(k) - \mathbf{E}(k)\boldsymbol{\lambda}^*(k+1) \quad\quad (5.2.21)
\end{aligned}$$

where, $\mathbf{E}(k) = \mathbf{B}(k)\mathbf{R}^{-1}(k)\mathbf{B}'(k)$.

- **Step 6:** *State and Costate System:* The canonical (state and costate) system of (5.2.21) and (5.2.18) becomes

$$\begin{bmatrix} \mathbf{x}^*(k+1) \\ \boldsymbol{\lambda}^*(k) \end{bmatrix} = \begin{bmatrix} \mathbf{A}(k) & -\mathbf{E}(k) \\ \mathbf{Q}(k) & \mathbf{A}'(k) \end{bmatrix} \begin{bmatrix} \mathbf{x}^*(k) \\ \boldsymbol{\lambda}^*(k+1) \end{bmatrix}. \quad (5.2.22)$$

The state and costate (or Hamiltonian) system (5.2.22) is shown in Figure 5.1. Note that the preceding Hamiltonian system (5.2.22) is not symmetrical in the sense that $\mathbf{x}^*(k+1)$ and $\boldsymbol{\lambda}^*(k)$ are related in terms of $\mathbf{x}^*(k)$ and $\boldsymbol{\lambda}^*(k+1)$.

5.2.1   *Fixed-Final State and Open-Loop Optimal Control*

Let us now discuss the boundary condition and the associated control configurations. For the given or *fixed-initial* condition (5.2.2) and the *fixed-final state* as

$$\mathbf{x}(k = k_f) = \mathbf{x}(k_f), \quad\quad (5.2.23)$$

the terminal cost term in the performance index (5.2.3) makes no sense and hence we can set $\mathbf{F}(k_f) = 0$. Also, in view of the fixed-final state condition (5.2.23), the variation $\delta\mathbf{x}(k_f) = 0$ and hence the boundary condition (5.2.9) does not exist for this case. Thus, the state and costate system (5.2.22) along with the initial condition (5.2.2) and the fixed-final condition (5.2.23) constitute a two-point boundary value problem (TPBVP). The solution of this TPBVP, gives $\mathbf{x}^*(k)$ and $\boldsymbol{\lambda}^*(k)$ or $\boldsymbol{\lambda}^*(k+1)$ which along with the control relation (5.2.20) leads us to the so-called *open-loop optimal control.* The entire procedure is now summarized in Table 5.1.

We now illustrate the previous procedure by considering a simple system.

Table 5.1　Procedure Summary of Discrete-Time Optimal Control System: Fixed-End Points Condition

A. Statement of the Problem
Given the plant as $\mathbf{x}(k+1) = \mathbf{A}(k)\mathbf{x}(k) + \mathbf{B}(k)\mathbf{u}(k)$, the performance index as $J(k_0) = \frac{1}{2} \sum_{k=k_0}^{k_f-1} [\mathbf{x}'(k)\mathbf{Q}(k)\mathbf{x}(k) + \mathbf{u}'(k)\mathbf{R}(k)\mathbf{u}(k)]$, and the boundary conditions as $\mathbf{x}(k=k_0) = \mathbf{x}(k_0);\quad \mathbf{x}(k_f) = \mathbf{x}(k_f)$, find the optimal control.

B. Solution of the Problem	
Step 1	Form the Pontryagin $\mathcal{H}$ function $\mathcal{H} = \frac{1}{2}\mathbf{x}'(k)\mathbf{Q}(k)\mathbf{x}(k) + \frac{1}{2}\mathbf{u}'(k)\mathbf{R}(k)\mathbf{u}(k)$ $+\boldsymbol{\lambda}'(k+1)[\mathbf{A}(k)\mathbf{x}(k) + \mathbf{B}(k)\mathbf{u}(k)]$.
Step 2	Minimize $\mathcal{H}$ w.r.t. $\mathbf{u}(k)$ $\left(\frac{\partial \mathcal{H}}{\partial \mathbf{u}(k)}\right)_* = 0$ and obtain $\mathbf{u}^*(k) = -\mathbf{R}^{-1}(k)\mathbf{B}'(k)\boldsymbol{\lambda}^*(k+1)$.
Step 3	Using the result of Step 2, find the optimal $\mathcal{H}^*$ function as $\mathcal{H}^*(\mathbf{x}^*(k), \boldsymbol{\lambda}^*(k+1))$.
Step 4	Solve the set of $2n$ difference equations $\mathbf{x}^*(k+1) = \frac{\partial \mathcal{H}^*}{\partial \boldsymbol{\lambda}^*(k+1)} = \mathbf{A}(k)\mathbf{x}^*(k) - \mathbf{E}(k)\boldsymbol{\lambda}^*(k+1)$, $\boldsymbol{\lambda}^*(k) = \frac{\partial \mathcal{H}^*}{\partial \mathbf{x}^*(k)} = \mathbf{Q}(k)\mathbf{x}^*(k) + \mathbf{A}'(k)\boldsymbol{\lambda}^*(k+1)$, with the given boundary conditions $\mathbf{x}(k_0)$ and $\mathbf{x}(k_f)$, where $\mathbf{E}(k) = \mathbf{B}(k)\mathbf{R}^{-1}(k)\mathbf{B}'(k)$.
Step 5	Substitute the solution of $\boldsymbol{\lambda}^*(k)$ from Step 4 into the expression for $\mathbf{u}^*(k)$ of Step 2 to obtain the optimal control.

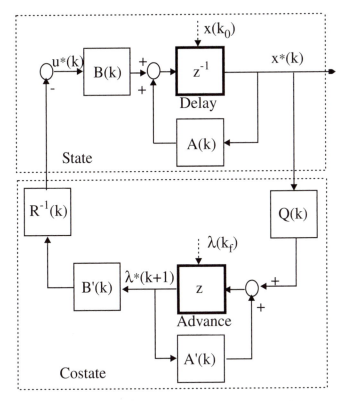

Figure 5.1   State and Costate System

Example 5.2

Consider the minimization of the performance index (PI) [120]

$$J(k_0) = \frac{1}{2} \sum_{k=k_0}^{k_f-1} u^2(k), \qquad (5.2.24)$$

subject to the boundary conditions

$$x(k_0 = 0) = 1, \qquad x(k_f = 10) = 0 \qquad (5.2.25)$$

for a simple scalar system

$$x(k+1) = x(k) + u(k). \qquad (5.2.26)$$

Solution: Let us first identify the various matrices by comparing the present state (5.2.26) and the PI (5.2.24) with the corresponding general formulation of the state (5.2.1) and the PI (5.2.3), respectively, to get

$$\mathbf{A}(k) = 1; \quad \mathbf{B}(k) = 1; \quad \mathbf{F}(k_f) = 0; \quad \mathbf{Q}(k) = 0; \quad \mathbf{R}(k) = 1.$$

$$(5.2.27)$$

Now let us use the procedure given in Table 5.1.

- **Step 1:** Form the Pontryagin $\mathcal{H}$ function as

$$\mathcal{H}(x(k), u(k), \lambda(k+1)) = \frac{1}{2}u^2(k) + \lambda(k+1)[x(k) + u(k)].$$

$$(5.2.28)$$

- **Step 2:** Minimizing $\mathcal{H}$ of (5.2.28) w.r.t. $u(k)$

$$\frac{\partial \mathcal{H}}{\partial u(k)} = 0 \longrightarrow u^*(k) + \lambda^*(k+1) = 0 \longrightarrow u^*(k) = -\lambda^*(k+1).$$

$$(5.2.29)$$

- **Step 3:** Using the control relation (5.2.29) and the Hamiltonian (5.2.28), form the optimal $\mathcal{H}^*$ function

$$\mathcal{H}^*(x^*(k), \lambda^*(k+1)) = x^*(k)\lambda^*(k+1) - \frac{1}{2}\lambda^{*2}(k+1). \quad (5.2.30)$$

- **Step 4:** Obtain the set of 2 state and costate difference equations

$$x^*(k+1) = \frac{\partial \mathcal{H}^*}{\partial \lambda^*(k+1)} \longrightarrow x^*(k+1) = x^*(k) - \lambda^*(k+1)$$

$$(5.2.31)$$

and

$$\lambda^*(k) = \frac{\partial \mathcal{H}^*}{\partial x^*(k)} \longrightarrow \lambda^*(k) = \lambda^*(k+1). \quad (5.2.32)$$

Solving these 2 equations (5.2.31) and (5.2.32) (by first eliminating $\lambda(k)$ and solving for $x(k)$) along with the boundary conditions (5.2.25), we get the optimal solutions as

$$x^*(k) = 1 - 0.1k; \quad \lambda^*(k+1) = 0.1. \quad (5.2.33)$$

- **Step 5:** Using the previous state and costate solutions, the optimal control $u^*(k)$ is obtained from (5.2.29) as

$$u^*(k) = -0.1. \quad (5.2.34)$$

5.2.2   Free-Final State and Open-Loop Optimal Control

Let us, first of all, note that for a *free-final state* system, it is usual to obtain *closed-loop* optimal control configuration. However, we reserve this to the next section. Let us consider the *free-final state* condition as

$$\mathbf{x}(k_f) \text{ is free, and } k_f \text{ is fixed.} \tag{5.2.35}$$

Then, the final condition (5.2.9) along with the Lagrangian (5.2.5) becomes

$$\left[-\boldsymbol{\lambda}^*(k) + \frac{\partial S(\mathbf{x}^*(k), k)}{\partial \mathbf{x}^*(k)}\right]' \bigg|_{k_f} \delta\mathbf{x}(k_f) = 0. \tag{5.2.36}$$

Now, for this *free-final point system* with k_f fixed, and $\mathbf{x}(k_f)$ being free, $\delta x(k_f)$ becomes *arbitrary* and its coefficient in (5.2.36) should be zero. Thus, the boundary condition (5.2.36) along with the performance index (5.2.3) becomes

$$\boldsymbol{\lambda}(k_f) = \frac{\partial S(\mathbf{x}(k_f), k_f)}{\partial \mathbf{x}(k_f)} = \frac{\partial}{\partial \mathbf{x}(k_f)}\left[\frac{1}{2}\mathbf{x}'(k_f)\mathbf{F}(k_f)\mathbf{x}(k_f)\right] \tag{5.2.37}$$

which gives

$$\boxed{\boldsymbol{\lambda}(k_f) = \mathbf{F}(k_f)\mathbf{x}(k_f).} \tag{5.2.38}$$

The state and costate system (5.2.22) along with the initial condition (5.2.2) and the final condition (5.2.38) constitute a TPBVP. The solution of this TPBVP, which is difficult because of the coupled nature of the solutions (i.e., the state $\mathbf{x}^*(k)$ has to be solved forward starting from its initial condition $\mathbf{x}(k_0)$ and the costate $\boldsymbol{\lambda}^*(k)$ has to be solved *backward* starting from its final condition $\boldsymbol{\lambda}(k_f)$) leads us to *open-loop optimal control*. The entire procedure is now summarized in Table 5.2.

5.3   Discrete-Time Linear State Regulator System

In this section, we discuss the state regulator system, and obtain closed-loop optimal control configuration for discrete-time systems. This leads us to matrix difference Riccati equation (DRE). Now, we restate the

Table 5.2 Procedure Summary for Discrete-Time Optimal Control System: Free-Final Point Condition

A. Statement of the Problem	
Given the plant as $\mathbf{x}(k+1) = \mathbf{A}(k)\mathbf{x}(k) + \mathbf{B}(k)\mathbf{u}(k)$ the performance index as $J(k_0) = \frac{1}{2}\mathbf{x}'(k_f)\mathbf{F}(k_f)\mathbf{x}(k_f)$ $+\frac{1}{2}\sum_{k=k_0}^{k_f-1}[\mathbf{x}'(k)\mathbf{Q}(k)\mathbf{x}(k) + \mathbf{u}'(k)\mathbf{R}(k)\mathbf{u}(k)]$ and the boundary conditions as $\mathbf{x}(k = k_0) = \mathbf{x}(k_0); \quad \mathbf{x}(k_f)$ is free, and k_f is fixed, find the optimal control.	
B. Solution of the Problem	
Step 1	Form the Pontryagin $\mathcal{H}$ function $\mathcal{H} = \frac{1}{2}\mathbf{x}'(k)\mathbf{Q}(k)\mathbf{x}(k) + \frac{1}{2}\mathbf{u}'(k)\mathbf{R}(k)\mathbf{u}(k)$ $+\boldsymbol{\lambda}'(k+1)[\mathbf{A}(k)\mathbf{x}(k) + \mathbf{B}(k)\mathbf{u}(k)]$.
Step 2	Minimize $\mathcal{H}$ w.r.t. $\mathbf{u}(k)$ as $\left(\frac{\partial\mathcal{H}}{\partial\mathbf{u}(k)}\right)_* = 0$ and obtain $\mathbf{u}^*(k) = -\mathbf{R}^{-1}(k)\mathbf{B}'(k)\boldsymbol{\lambda}^*(k+1)$.
Step 3	Using the result of Step 2 in Step 1, find the optimal $\mathcal{H}^*$ as $\mathcal{H}^*(\mathbf{x}^*(k), \boldsymbol{\lambda}^*(k+1))$.
Step 4	Solve the set of $2n$ difference equations $\mathbf{x}^*(k+1) = \frac{\partial\mathcal{H}^*}{\partial\boldsymbol{\lambda}^*(k+1)} = \mathbf{A}(k)\mathbf{x}^*(k) - \mathbf{E}(k)\boldsymbol{\lambda}^*(k+1)$, $\boldsymbol{\lambda}^*(k) = \frac{\partial\mathcal{H}^*}{\partial\mathbf{x}^*(k)} = \mathbf{Q}(k)\mathbf{x}^*(k) + \mathbf{A}'(k)\boldsymbol{\lambda}^*(k+1)$, with the given initial condition and the final condition $\boldsymbol{\lambda}(k_f) = \mathbf{F}(k_f)\mathbf{x}(k_f)$, where, $\mathbf{E}(k) = \mathbf{B}(k)\mathbf{R}^{-1}(k)\mathbf{B}'(k)$.
Step 5	Substitute the solution of $\boldsymbol{\lambda}^*(k)$ from Step 4 into the expression for $\mathbf{u}^*(k)$ of Step 2, to obtain the optimal control.

problem of linear state regulator and summarize the results derived in Section 5.2.

Consider the linear, time-varying discrete-time control system described by the plant (5.2.1) and the performance index (5.2.3). We are given the initial and final conditions as

$$\mathbf{x}(k = k_0) = \mathbf{x}(k_0); \quad \mathbf{x}(k_f) \text{ is free, and } k_f \text{ is fixed.} \quad (5.3.1)$$

Then the optimal control (5.2.20) and the state and costate equations (5.2.22) are reproduced, respectively here for convenience as

$$\mathbf{u}^*(k) = -\mathbf{R}^{-1}(k)\mathbf{B}'(k)\boldsymbol{\lambda}^*(k+1) \quad (5.3.2)$$

and

$$\mathbf{x}^*(k+1) = \mathbf{A}(k)\mathbf{x}^*(k) - \mathbf{E}(k)\boldsymbol{\lambda}^*(k+1), \quad (5.3.3)$$
$$\boldsymbol{\lambda}^*(k) = \mathbf{Q}(k)\mathbf{x}^*(k) + \mathbf{A}'(k)\boldsymbol{\lambda}^*(k+1), \quad (5.3.4)$$

where, $\mathbf{E}(k) = \mathbf{B}(k)\mathbf{R}^{-1}(k)\mathbf{B}'(k)$, and the final costate relation (5.2.38) is given by

$$\boldsymbol{\lambda}(k_f) = \mathbf{F}(k_f)\mathbf{x}(k_f). \quad (5.3.5)$$

5.3.1   Closed-Loop Optimal Control: Matrix Difference Riccati Equation

In order to obtain *closed-loop* optimal configuration, we need to try to express the costate function $\boldsymbol{\lambda}^*(k+1)$ in the optimal control (5.3.2) in terms of the state function $\mathbf{x}^*(k)$. The final condition (5.3.5) prompts us to express

$$\boldsymbol{\lambda}^*(k) = \mathbf{P}(k)\mathbf{x}^*(k) \quad (5.3.6)$$

where, $\mathbf{P}(k)$ is yet to be determined. This linear transformation is called the *Riccati transformation*, and is of fundamental importance in the solution of the problem. Using the transformation (5.3.6) in the state and costate equations (5.3.3) and (5.3.4), we have

$$\mathbf{P}(k)\mathbf{x}^*(k) = \mathbf{Q}(k)\mathbf{x}^*(k) + \mathbf{A}'(k)\mathbf{P}(k+1)\mathbf{x}^*(k+1) \quad (5.3.7)$$

and

$$\mathbf{x}^*(k+1) = \mathbf{A}(k)\mathbf{x}^*(k) - \mathbf{E}(k)\mathbf{P}(k+1)\mathbf{x}^*(k+1). \quad (5.3.8)$$

Solving for $\mathbf{x}^*(k+1)$ from (5.3.8)

$$\mathbf{x}^*(k+1) = [\mathbf{I} + \mathbf{E}(k)\mathbf{P}(k+1)]^{-1}\mathbf{A}(k)\mathbf{x}^*(k). \qquad (5.3.9)$$

Substituting (5.3.9) in (5.3.7) yields

$$\mathbf{P}(k)\mathbf{x}^*(k) = \mathbf{Q}(k)\mathbf{x}^*(k) + \mathbf{A}'(k)\mathbf{P}(k+1)\left[\mathbf{I} + \mathbf{E}(k)\mathbf{P}(k+1)\right]^{-1}\mathbf{A}(k)\mathbf{x}^*(k). \tag{5.3.10}$$

Since, this relation (5.3.10) must hold for all values of $\mathbf{x}^*(k)$, we have

$$\boxed{\mathbf{P}(k) = \mathbf{A}'(k)\mathbf{P}(k+1)\left[\mathbf{I} + \mathbf{E}(k)\mathbf{P}(k+1)\right]^{-1}\mathbf{A}(k) + \mathbf{Q}(k).}$$
$$(5.3.11)$$

This relation (5.3.11) is called the matrix *difference Riccati equation* (DRE). Alternatively, we can express (5.3.11) as

$$\boxed{\mathbf{P}(k) = \mathbf{A}'(k)\left[\mathbf{P}^{-1}(k+1) + \mathbf{E}(k)\right]^{-1}\mathbf{A}(k) + \mathbf{Q}(k)} \quad (5.3.12)$$

where, we assume that the inversion of $\mathbf{P}(k)$ exists for all $k \neq k_f$. The *final* condition for solving the matrix DRE (5.3.11) or (5.3.12) is obtained from (5.3.5) and (5.3.6) as

$$\boldsymbol{\lambda}(k_f) = \mathbf{F}(k_f)\mathbf{x}(k_f) = \mathbf{P}(k_f)\mathbf{x}(k_f), \qquad (5.3.13)$$

which gives

$$\boxed{\mathbf{P}(k_f) = \mathbf{F}(k_f).} \qquad (5.3.14)$$

In the matrix DRE (5.3.11), the term $\mathbf{P}(k)$ is on the left hand side and $\mathbf{P}(k+1)$ is on the right hand side and hence it needs to be solved *backwards* starting from the *final* condition (5.3.14). Since $\mathbf{Q}(k)$ and $\mathbf{F}(k_f)$ are assumed to be *positive semidefinite*, we can show that the Riccati matrix $\mathbf{P}(k)$ is positive definite. Now to obtain the closed-loop optimal control, we eliminate $\boldsymbol{\lambda}^*(k+1)$ from the control relation (5.3.2) and the state relation (5.3.4) and use the transformation (5.3.6). Thus, we get the relation for closed-loop, optimal control as

$$\boxed{\mathbf{u}^*(k) = -\mathbf{R}^{-1}(k)\mathbf{B}'(k)\mathbf{A}^{-T}(k)\left[\mathbf{P}(k) - \mathbf{Q}(k)\right]\mathbf{x}^*(k).} \quad (5.3.15)$$

Here, $\mathbf{A}^{-T}$ is the inverse of $\mathbf{A}'$ and we assume that the inverse of $\mathbf{A}(k)$ exists. This relation (5.3.15) is the desired version for the closed-loop optimal control in terms of the state. We may write the closed-loop, optimal control relation (5.3.15) in a simplified form as

$$\boxed{\mathbf{u}^*(k) = -\mathbf{L}(k)\mathbf{x}^*(k)} \qquad (5.3.16)$$

where,

$$\boxed{\mathbf{L}(k) = \mathbf{R}^{-1}(k)\mathbf{B}'(k)\mathbf{A}^{-T}(k)\left[\mathbf{P}(k) - \mathbf{Q}(k)\right].} \qquad (5.3.17)$$

This is the required relation for the *optimal feedback control law* and the feedback gain $\mathbf{L}(k)$ is called the "Kalman gain." The optimal state $\mathbf{x}^*(k)$ is obtained by substituting the optimal control $\mathbf{u}^*(k)$ given by (5.3.16) in the original state equation (5.2.1) as

$$\boxed{\mathbf{x}^*(k+1) = (\mathbf{A}(k) - \mathbf{B}(k)\mathbf{L}(k))\,\mathbf{x}^*(k).} \qquad (5.3.18)$$

Alternate Forms for the DRE

Alternate forms which do not require the inversion of the matrix $\mathbf{A}(k)$ for the matrix DRE (5.3.11) and the optimal control (5.3.16) are obtained as follows. Using the well-known matrix inversion lemma

$$\left[\mathbf{A}_1^{-1} + \mathbf{A}_2\mathbf{A}_4\mathbf{A}_3\right]^{-1} = \mathbf{A}_1 - \mathbf{A}_1\mathbf{A}_2\left[\mathbf{A}_3\mathbf{A}_1\mathbf{A}_2 + \mathbf{A}_4^{-1}\right]^{-1}\mathbf{A}_3\mathbf{A}_1$$
$$(5.3.19)$$

in (5.3.12), and manipulating, we have the matrix DRE as

$$\mathbf{P}(k) = \mathbf{A}'(k)\left\{\mathbf{P}(k+1) - \mathbf{P}(k+1)\mathbf{B}(k).\right.$$
$$\left[\mathbf{B}'(k)\mathbf{P}(k+1)\mathbf{B}(k) + \mathbf{R}(k)\right]^{-1}\mathbf{B}'(k)\mathbf{P}(k+1)\right\}\mathbf{A}(k)$$
$$+\mathbf{Q}(k). \qquad (5.3.20)$$

Next, consider the optimal control (5.3.2) and the transformation (5.3.6), to get

$$\mathbf{u}^*(k) = -\mathbf{R}^{-1}(k)\mathbf{B}'(k)\mathbf{P}(k+1)\mathbf{x}^*(k+1) \qquad (5.3.21)$$

which upon using the state equation (5.2.1) becomes

$$\mathbf{u}^*(k) = -\mathbf{R}^{-1}(k)\mathbf{B}'(k)\mathbf{P}(k+1)\left[\mathbf{A}(k)\mathbf{x}^*(k) + \mathbf{B}(k)\mathbf{u}^*(k)\right]. \quad (5.3.22)$$

Rearranging, we have

$$\left[\mathbf{I} + \mathbf{R}^{-1}(k)\mathbf{B}'(k)\mathbf{P}(k+1)\mathbf{B}(k)\right]\mathbf{u}^*(k) =$$
$$-\mathbf{R}^{-1}(k)\mathbf{B}'(k)\mathbf{P}(k+1)\mathbf{A}(k)\mathbf{x}^*(k).$$

(5.3.23)

Premultiplying by $\mathbf{R}(k)$ and solving for $\mathbf{u}^*(k)$,

$$\boxed{\mathbf{u}^*(k) = -\mathbf{L}_a(k)\mathbf{x}^*(k)}$$

(5.3.24)

where, $\mathbf{L}_a(k)$, called the *Kalman gain matrix* is

$$\boxed{\mathbf{L}_a(k) = [\mathbf{B}'(k)\mathbf{P}(k+1)\mathbf{B}(k) + \mathbf{R}(k)]^{-1}\mathbf{B}'(k)\mathbf{P}(k+1)\mathbf{A}(k).}$$

(5.3.25)

Let us note from the optimal feedback control law (5.3.24) that the Kalman gains are dependent on the solution of the matrix DRE (5.3.20) involving the system matrices and performance index matrices. Finally, the closed-loop, optimal control (5.3.24) with the state (5.2.1) gives us the optimal system

$$\mathbf{x}^*(k+1) = [\mathbf{A}(k) - \mathbf{B}(k)\mathbf{L}_a(k)]\,\mathbf{x}^*(k).$$

(5.3.26)

Using the gain relation (5.3.25), an alternate form for the matrix DRE (5.3.20) becomes

$$\boxed{\mathbf{P}(k) = \mathbf{A}'(k)\mathbf{P}(k+1)\left[\mathbf{A}(k) - \mathbf{B}(k)\mathbf{L}_a(k)\right] + \mathbf{Q}(k).}$$

(5.3.27)

Let us now make some notes:

1. There is essentially more than one form of the matrix DRE given by (5.3.11) or (5.3.12), (5.3.20), and (5.3.27).

2. However, the Kalman feedback gain matrix has only two forms given by the first form (5.3.17) which goes with the DRE (5.3.11) or (5.3.12) and the second form (5.3.25) which corresponds to the DRE (5.3.20) or (5.3.27).

3. It is a simple matter to see that the matrix DRE (5.3.11) and the associated Kalman feedback gain matrix (5.3.17) involve the inversion of the matrix $\mathbf{I} + \mathbf{E}(k)\mathbf{P}(k+1)$ *once* only, whereas the

matrix DRE (5.3.20) and the associated Kalman feedback gain matrix (5.3.25) together involve *two* matrix inversions. The number of matrix inversions directly affects the overall computation time, especially if one is looking for *on-line* implementation of closed-loop optimal control strategy.

5.3.2   *Optimal Cost Function*

For finding the optimal cost function $J^*(k_0)$, we can follow the same procedure as the one used for the continuous-time systems in Chapter 3 to get

$$J^* = \frac{1}{2}\mathbf{x}^{*\prime}(k_0)\mathbf{P}(k_0)\mathbf{x}(k_0).$$ (5.3.28)

Let us note that the Riccati function $\mathbf{P}(k)$ is generated *off-line* before we obtain the optimal control $\mathbf{u}^*(k)$ to be applied to the system. Thus, in general for any initial state k, we have the optimal cost as

$$J^*(k) = \frac{1}{2}\mathbf{x}^{*\prime}(k)\mathbf{P}(k)\mathbf{x}^*(k).$$ (5.3.29)

The entire procedure is now summarized in Table 5.3. The actual implementation of this control law is shown in Figure 5.2. We now illustrate the previous procedure by considering a second order system with a general cost function.

Example 5.3

Consider the minimization of a functional [33]

$$J = \left[x_1^2(k_f) + 2x_2^2(k_f)\right]$$

$$+ \sum_{k=k_0}^{k_f-1} \left[0.5x_1^2(k) + 0.5x_2^2(k) + 0.5u^2(k)\right]$$ (5.3.30)

for the second order system

$$x_1(k+1) = 0.8x_1(k) + x_2(k) + u(k)$$
$$x_2(k+1) = 0.6x_2(k) + 0.5u(k)$$ (5.3.31)

subject to the initial conditions

$$x_1(k_0 = 0) = 5, \quad x_2(k_0 = 0) = 3; \quad k_f = 10, \text{ and } \mathbf{x}(k_f) \text{ is free.}$$ (5.3.32)

Table 5.3 Procedure Summary of Discrete-Time, Linear
Quadratic Regulator System

A. Statement of the Problem	
Given the plant as $\mathbf{x}(k+1) = \mathbf{A}(k)\mathbf{x}(k) + \mathbf{B}(k)\mathbf{u}(k)$ the performance index as $J(k_0) = \frac{1}{2}\mathbf{x}'(k_f)\mathbf{F}(k_f)\mathbf{x}(k_f)$ $+ \frac{1}{2}\sum_{k=k_0}^{k_f-1} [\mathbf{x}'(k)\mathbf{Q}(k)\mathbf{x}(k) + \mathbf{u}'(k)\mathbf{R}(k)\mathbf{u}(k)]$ and the boundary conditions as $\mathbf{x}(k = k_0) = \mathbf{x}(k_0);\quad \mathbf{x}(k_f)$ is free, and k_f is free, find the closed-loop optimal control, state and performance index.	
B. Solution of the Problem	
Step 1	Solve the matrix difference Riccati equation (DRE) $\mathbf{P}(k) = \mathbf{A}'(k)\mathbf{P}(k+1)\left[\mathbf{I} + \mathbf{E}(k)\mathbf{P}(k+1)\right]^{-1}\mathbf{A}(k) + \mathbf{Q}(k)$ with final condition $\mathbf{P}(k = k_f) = \mathbf{F}(k_f)$, where $\mathbf{E}(k) = \mathbf{B}(k)\mathbf{R}^{-1}(k)\mathbf{B}'(k)$.
Step 2	Solve the optimal state $\mathbf{x}^*(k)$ from $\mathbf{x}^*(k+1) = [\mathbf{A}(k) - \mathbf{B}(k)\mathbf{L}(k)]\,\mathbf{x}^*(k)$ with initial condition $\mathbf{x}(k_0) = \mathbf{x}_0$, where $\mathbf{L}(k) = \mathbf{R}^{-1}(k)\mathbf{B}'(k)\mathbf{A}^{-T}(k)\left[\mathbf{P}(k) - \mathbf{Q}(k)\right].$
Step 3	Obtain the optimal control $\mathbf{u}^*(k)$ from $\mathbf{u}^*(k) = -\mathbf{L}(k)\mathbf{x}^*(k)$, where $\mathbf{L}(k)$ is the Kalman gain.
Step 4	Obtain the optimal performance index from $J^* = \frac{1}{2}\mathbf{x}^{*\prime}(k)\mathbf{P}(k)\mathbf{x}^*(k).$

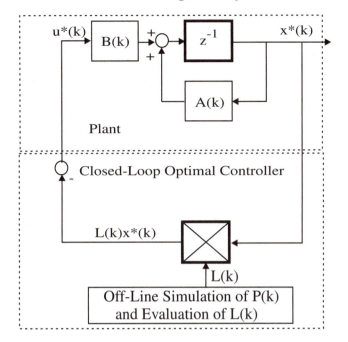

Figure 5.2   Closed-Loop Optimal Controller for Linear
Discrete-Time Regulator

Solution: Let us first identify the various matrices by comparing
the system (5.3.31) and the PI (5.3.30) of the system with the
system (5.2.1) and the PI (5.2.3) of the general formulation as

$$\mathbf{A}(k) = \begin{bmatrix} 0.8 & 1.0 \\ 0.0 & 0.6 \end{bmatrix} ; \quad \mathbf{B}(k) = \begin{bmatrix} 1.0 \\ 0.5 \end{bmatrix} ;$$

$$\mathbf{F}(k_f) = \begin{bmatrix} 2.0 & 0 \\ 0 & 4.0 \end{bmatrix} ; \quad \mathbf{Q}(k) = \begin{bmatrix} 1 & 0 \\ 0 & 1 \end{bmatrix} ; \quad \mathbf{R} = 1. \qquad (5.3.33)$$

Now let us use the procedure given in Table 5.3.

- **Step 1:** Solve the matrix difference Riccati equation (5.3.11)

$$\begin{bmatrix} p_{11}(k) & p_{12}(k) \\ p_{12}(k) & p_{22}(k) \end{bmatrix} = \begin{bmatrix} 1 & 0 \\ 0 & 1 \end{bmatrix} + \begin{bmatrix} 0.8 & 0.0 \\ 1.0 & 0.6 \end{bmatrix} \begin{bmatrix} p_{11}(k+1) & p_{12}(k+1) \\ p_{12}(k+1) & p_{22}(k+1) \end{bmatrix}.$$

$$\left[\begin{bmatrix} 1 & 0 \\ 0 & 1 \end{bmatrix} + \begin{bmatrix} 1.0 \\ 0.5 \end{bmatrix} [1]^{-1} [1.0 \ 0.5] \begin{bmatrix} p_{11}(k+1) & p_{12}(k+1) \\ p_{12}(k+1) & p_{22}(k+1) \end{bmatrix} \right]^{-1} \text{x}$$

$$\begin{bmatrix} 0.8 & 1.0 \\ 0.0 & 0.6 \end{bmatrix}$$

(5.3.34)

backwards in time starting with the final condition (5.3.14) as

$$\begin{bmatrix} p_{11}(10) & p_{12}(10) \\ p_{12}(10) & p_{22}(10) \end{bmatrix} = \mathbf{F}(k_f) = \begin{bmatrix} 2.0 & 0 \\ 0 & 4.0 \end{bmatrix}. \qquad (5.3.35)$$

- **Step 2:** The optimal control $u^*(k)$ is obtained from (5.3.16) as

$$u^*(k) = - \begin{bmatrix} l_1(k) & l_2(k) \end{bmatrix} \begin{bmatrix} x_1(k) \\ x_2(k) \end{bmatrix} \qquad (5.3.36)$$

where $\mathbf{l'} = [l_1, l_2]$ is given by (5.3.17).

- **Step 3:** Using the optimal control (5.3.36) the optimal states are computed by solving the state equation (5.3.18) *forward* in time. This is an iterative process in the backward direction. Evaluation of these solutions require the use of standard software such as MATLAB© as shown below.

```
*********************************************************
% Solution Using Control System Toolbox (STB)
% MATLAB Version 6
%
A=[0.8 1;0,0.6]; %% system matrix A
B=[1;0.5]; %% system matrix B
Q=[1 0;0 1]; %% performance index state weighting matrix Q
R=[1]; %% performance index control weighting matrix R
F=[2,0;0,4]; %% performance index weighting matrix F
%
x1(1)=5; %% initial condition on state x1
x2(1)=3; %% initial condition on state x2
xk=[x1(1);x2(1)];
% note that if kf = 10 then k = [k0,kf] = [0 1 2,...,10],
% then we have 11 points and an array x1 should have subscript
% x1(N) with N=1 to 11. This is because x(o) is illegal in array
```

```
% definition in MATLAB. Let us use N = kf+1
k0=0; % the initial instant k_0
kf=10; % the final instant k_f
N=kf+1; %
[n,n]=size(A); % fixing the order of the system matrix A
I=eye(n); % identity matrix I
E=B*inv(R)*B'; % the matrix E = BR^{-1}B'
%
% solve matrix difference Riccati equation backwards
% starting from kf to k0
% use the form P(k) = A'P(k+1)[I + EP(k+1)]^{-1}A + Q
% first fix the final condition P(k_f) = F
% note that P, Q, R are all symmatric ij = ji
Pkplus1=F;
p11(N)=F(1);
p12(N)=F(2);
p21(N)=F(3);
p22(N)=F(4);
%
for k=N-1:-1:1,
    Pk = A'*Pkplus1*inv(I+E*Pkplus1)*A+Q;
    p11(k) = Pk(1);
    p12(k) = Pk(2);
    p21(k) = Pk(3);
    p22(k) = Pk(4);
    Pkplus1 = Pk;
end
%
% calcuate the feedback coefficient L
% L = R^{-1}B'A^{-T}[P(k) - Q]
%
for k = N:-1:1,
   Pk=[p11(k),p12(k);p21(k),p22(k)];
   Lk = inv(R)*B'*inv(A')*(Pk-Q);
   l1(k) = Lk(1);
   l2(k) = Lk(2);
end
%
% solve the optimal states
% x(k+1) = [A-B*L]x(k) given x(0)
%
for k=1:N-1,
   Lk = [l1(k),l2(k)];
   xk = [x1(k);x2(k)];
   xkplus1 = (A-B*Lk)*xk;
```

```
    x1(k+1) = xkplus1(1);
    x2(k+1) = xkplus1(2);
end
%
% solve for optimal control u(k) = - L(k)x(k)
%
for k=1:N,
    Lk = [l1(k),l2(k)];
    xk = [x1(k);x2(k)];
    u(k) = - Lk*xk;
end
%
% plot various values: P(k), x(k), u(k)
% let us first reorder the values of k = 0 to 10
figure(1)
plot(k,p11,'k:o',k,p12,'k:+',k,p22,'k:*')
xlabel('k')
ylabel('Riccati Coefficients')
gtext('p_{11}(k)')
gtext('p_{12}(k)=p_{21}(k)')
gtext('p_{22}(k)')
%
figure(2)
plot(k,x1,'k:o',k,x2,'k:+')
xlabel('k')
ylabel('Optimal States')
gtext('x_1(k)')
gtext('x_2(k)')
%
figure(3)
plot(k,u,'k:*')
xlabel('k')
ylabel('Optimal Control')
gtext('u(k)')
% end of the program
%
*************************************************************
```

The Riccati coefficients of the matrix $\mathbf{P}(k)$ obtained using MATLAB©
are shown in Figure 5.3. The optimal states are plotted in Figure 5.4 and the optimal control is shown in Figure 5.5.

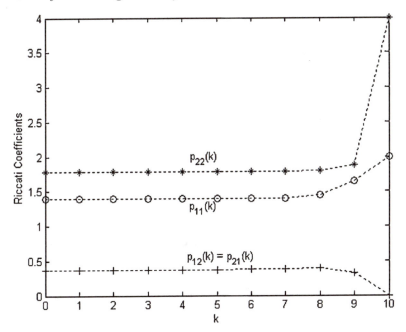

Figure 5.3   Riccati Coefficients for Example 5.3

5.4   *Steady-State Regulator System*

Here, we let k_f tend to ∞ and this necessitates that we assume the time-invariant case. Thus, the linear time-invariant plant becomes

$$x(k+1) = \mathbf{A}x(k) + \mathbf{B}u(k) \qquad (5.4.1)$$

and the performance index becomes

$$J = \frac{1}{2}\sum_{k=k_0}^{\infty}\left[\mathbf{x}^{*\prime}(k)\mathbf{Q}\mathbf{x}(k) + \mathbf{u}^{*\prime}(k)\mathbf{R}\mathbf{u}^{*}(k)\right]. \qquad (5.4.2)$$

As the final time k_f tends to ∞, we have the Riccati matrix $\mathbf{P}(k)$ attaining a steady-state value $\bar{\mathbf{P}}$ in (5.3.11). That is,

$$\mathbf{P}(k) = \mathbf{P}(k+1) = \bar{\mathbf{P}} \qquad (5.4.3)$$

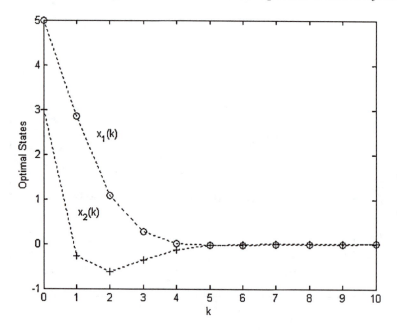

Figure 5.4   Optimal Control and States for Example 5.3

resulting in the *algebraic Riccati equation* (ARE) as

$$\boxed{\bar{\mathbf{P}} = \mathbf{A}'\bar{\mathbf{P}}\left[\mathbf{I} + \mathbf{B}\mathbf{R}^{-1}\mathbf{B}'\bar{\mathbf{P}}\right]^{-1}\mathbf{A} + \mathbf{Q}} \qquad (5.4.4)$$

or (5.3.12) as

$$\bar{\mathbf{P}} = \mathbf{A}'\left[\bar{\mathbf{P}}^{-1} + \mathbf{E}\right]^{-1}\mathbf{A} + \mathbf{Q}$$

where, $\mathbf{E} = \mathbf{B}\mathbf{R}^{-1}\mathbf{B}'$. The feedback optimal control (5.3.15) becomes

$$\boxed{\mathbf{u}^*(k) = -\mathbf{R}^{-1}\mathbf{B}'\mathbf{A}^{-T}\left[\bar{\mathbf{P}} - \mathbf{Q}\right]\mathbf{x}^*(k) = -\bar{\mathbf{L}}\mathbf{x}^*(k)} \qquad (5.4.5)$$

where, the Kalman gain (5.3.17) becomes

$$\boxed{\bar{\mathbf{L}} = \mathbf{R}^{-1}\mathbf{B}'\mathbf{A}^{-T}\left[\bar{\mathbf{P}} - \mathbf{Q}\right]} \qquad (5.4.6)$$

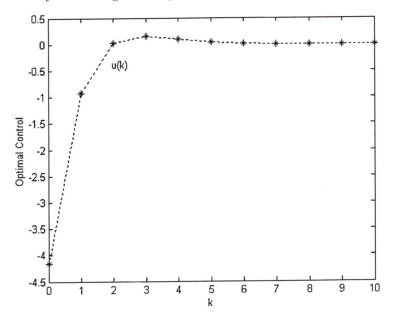

Figure 5.5   Optimal Control and States for Example 5.3

and $\mathbf{A}^{-T}$ is the inverse of $\mathbf{A}'$.

Alternate Form: An alternate form for the ARE (5.4.4) is obtained by considering steady-state form of the DRE (5.3.20) as

$$\bar{\mathbf{P}} = \mathbf{A}'\{\bar{\mathbf{P}} - \bar{\mathbf{P}}\mathbf{B}\left[\mathbf{B}'\bar{\mathbf{P}}\mathbf{B} + \mathbf{R}\right]^{-1}\mathbf{B}'\bar{\mathbf{P}}\}\mathbf{A} + \mathbf{Q}. \qquad (5.4.7)$$

The optimal feedback control (5.3.24) becomes

$$\mathbf{u}^*(k) = -\bar{\mathbf{L}}_a\mathbf{x}^*(k) \qquad (5.4.8)$$

where, the alternate optimal-feedback gain matrix (5.3.25) becomes

$$\bar{\mathbf{L}}_a = \left[\mathbf{B}'\bar{\mathbf{P}}\mathbf{B} + \mathbf{R}\right]^{-1}\mathbf{B}'\bar{\mathbf{P}}\mathbf{A}. \qquad (5.4.9)$$

The optimal control (5.4.9) with the state (5.4.1) gives us the optimal system

$$\mathbf{x}^*(k+1) = \left[\mathbf{A} - \mathbf{B}\bar{\mathbf{L}}_a\right]\mathbf{x}^*(k). \qquad (5.4.10)$$

Table 5.4   Procedure Summary of Discrete-Time, Linear Quadratic Regulator System: Steady-State Condition

A. Statement of the Problem	
Given the plant as $$\mathbf{x}(k+1) = \mathbf{A}\mathbf{x}(k) + \mathbf{B}\mathbf{u}(k)$$ the performance index as $$J(k_0) = \tfrac{1}{2}\sum_{k=k_0}^{\infty} [\mathbf{x}'(k)\mathbf{Q}\mathbf{x}(k) + \mathbf{u}'(k)\mathbf{R}\mathbf{u}(k)]$$ where, $k_f = \infty$, find the optimal control, state and the performance index.	
B. Solution of the Problem	
Step 1	Solve the matrix algebraic Riccati equation $$\bar{\mathbf{P}} = \mathbf{A}'\bar{\mathbf{P}}\left[\mathbf{I} + \mathbf{B}\mathbf{R}^{-1}\mathbf{B}'\bar{\mathbf{P}}\right]^{-1}\mathbf{A} + \mathbf{Q},\ \text{or}$$ $$\bar{\mathbf{P}} = \mathbf{A}'\{\bar{\mathbf{P}} - \bar{\mathbf{P}}\mathbf{B}\left[\mathbf{B}'\bar{\mathbf{P}}\mathbf{B} + \mathbf{R}\right]^{-1}\mathbf{B}'\bar{\mathbf{P}}\}\mathbf{A} + \mathbf{Q}.$$
Step 2	Solve the optimal state $\mathbf{x}^*(k)$ from $$\mathbf{x}^*(k+1) = \left[\mathbf{A} - \mathbf{B}\bar{\mathbf{L}}\right]\mathbf{x}^*(k)\ \text{or}$$ $$\mathbf{x}^*(k+1) = \left[\mathbf{A} - \mathbf{B}\bar{\mathbf{L}}_a\right]\mathbf{x}^*(k).$$ with initial condition $\mathbf{x}(k_0) = \mathbf{x}_0$, where $$\bar{\mathbf{L}} = \mathbf{R}^{-1}\mathbf{B}'\mathbf{A}^{-T}\left[\bar{\mathbf{P}} - \mathbf{Q}\right]\ \text{and}$$ $$\bar{\mathbf{L}}_a = \left[\mathbf{B}'\bar{\mathbf{P}}\mathbf{B} + \mathbf{R}\right]^{-1}\mathbf{B}'\bar{\mathbf{P}}\mathbf{A}.$$
Step 3	Obtain the optimal control $\mathbf{u}^*(k)$ from $$\mathbf{u}^*(k) = -\bar{\mathbf{L}}\mathbf{x}^*(k),\ \text{or}$$ $$\mathbf{u}^*(k) = -\bar{\mathbf{L}}_a\mathbf{x}^*(k).$$
Step 4	Obtain the optimal performance index from $$J^*(k_0) = \tfrac{1}{2}\mathbf{x}^{*'}(k)\bar{\mathbf{P}}\mathbf{x}^*(k).$$

The optimal cost function (5.3.29) becomes

$$\boxed{J^*(k) = \mathbf{x}^{*'}(k)\bar{\mathbf{P}}\mathbf{x}^*(k).} \tag{5.4.11}$$

The entire procedure is now summarized in Table 5.4. The implementation of this closed-loop optimal control for steady-state ($k_f \to \infty$) case is shown in Figure 5.6. We now illustrate the previous procedure by considering the same system of Example 5.3.

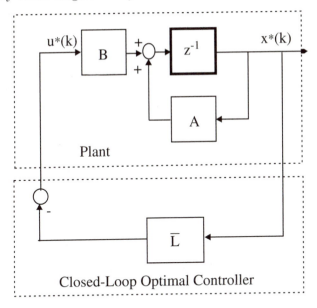

Figure 5.6   Closed-Loop Optimal Control for Discrete-Time
Steady-State Regulator System

Example 5.4

Consider the minimization of the performance index (5.3.30) without the terminal cost function and the plant (5.3.31) with $k_f = \infty$. Let us find the closed-loop optimal control for this system.

Solution: We have already identified the various matrices of the state (5.2.1) and the performance index (5.2.3) as given by (5.3.33), except that we now have $\mathbf{F}(k_f) = \mathbf{0}$.
Then, the solution of the algebraic Riccati equation (5.4.7), the closed-loop, optimal control (5.4.8) and the optimal states (5.4.10) are best solved using MATLAB© as shown below.

```
************************************************************
% Solution Using Control System Toolbox (STB) in
% the MATLAB, Version 6
%
A=[0.8 1;0,0.6]; %% system matrix A
B=[1;0.5]; %% system matrix B
Q=[1 0;0 1]; %% performance index state weighting matrix Q0.5
R=[1]; %% performance index control weighting matrix R
%
x1(1)=5; %% initial condition on state x1
```

```
x2(1)=3; %% initial condition on state x2
xk=[x1(1);x2(1)];
% note that if kf = 10 then k = [k0,kf] = [0 1 2,...,10],
% then we have 11 points and an array x1 should have subscript
% x1(N) with N=1 to 11. This is because x(o) is illegal in array
% definition in MATLAB. Let us use N = kf+1
k0=0; % the initial instant k_0
kf=10; % the final instant k_f
N=kf+1; %
[n,n]=size(A); % fixing the order of the system matrix A
I=eye(n); % identity matrix I
E=B*inv(R)*B'; % the matrix E = BR^{-1}B'
%
% solve matrix algebraic Riccati equation
% use the form P = A'PA - A'PB[B'PB+R]^{-1}B'PA + Q
% note that P, Q, R are all symmatric ij = ji
% calcuate the feedback coefficient L
% L = [B'PB+R]^{-1}B'PA
%
[P,EIGVAL,L,RR] = dare(A,B,Q,R)
%
P =

    1.3944    0.3738
    0.3738    1.7803


EIGVAL =

    0.3211 + 0.2151i
    0.3211 - 0.2151i


L =

    0.3937    0.7281
%
% solve the optimal states
% x(k+1) = (A-B*L)x(k) given x(0)
%
for k=1:N-1,
    xk = [x1(k);x2(k)];
    xkplus1 = (A-B*L)*xk;
    x1(k+1) = xkplus1(1);
    x2(k+1) = xkplus1(2);
end
%
% solve for optimal control u(k) = - Lx(k)
```

```
%
for k=1:N,
    xk = [x1(k);x2(k)];
    u(k) = -L*xk;
  end
%
k=0:10;
plot(k,x1,'k:o',k,x2,'k:+')
xlabel('k')
ylabel('Optimal States')
gtext('x_1(k)')
gtext('x_2(k)')
plot(k,u,'k:*')
xlabel('k')
ylabel('Optimal Control')
gtext('u(k)')
%
% end of the program
%
*****************************************************
```

Note that the value obtained for **P** previously is the same as the steady-state value for the Example 5.3 (see Figure 5.3). The optimal states are shown in Figure 5.7 and the optimal control is shown in Figure 5.8.

5.4.1   *Analytical Solution to the Riccati Equation*

This subsection is based on [89, 138]. The solution of the matrix difference Riccati equation is critical for linear quadratic regulator system. Thus, traditionally, the solution to the DRE (5.3.20) is obtained by *iteration in a recursive* manner using the final condition (5.3.14). Alternatively, one can obtain *analytical solution* to the DRE.

Let us rewrite the Hamiltonian system (5.2.22) arising in the time-invariant LQR system as (omitting the optimal notation (*) for clarity)

$$\begin{bmatrix} \mathbf{x}(k) \\ \boldsymbol{\lambda}(k) \end{bmatrix} = \mathbf{H} \begin{bmatrix} \mathbf{x}(k+1) \\ \boldsymbol{\lambda}(k+1) \end{bmatrix}. \tag{5.4.12}$$

Here, the Hamiltonian matrix **H** is

$$\mathbf{H} = \begin{bmatrix} \mathbf{H}_{11} & \mathbf{H}_{12} \\ \mathbf{H}_{21} & \mathbf{H}_{22} \end{bmatrix} = \begin{bmatrix} \mathbf{A}^{-1} & \mathbf{A}^{-1}\mathbf{E} \\ \mathbf{Q}\mathbf{A}^{-1} & \mathbf{A}' + \mathbf{Q}\mathbf{A}^{-1}\mathbf{E} \end{bmatrix} \tag{5.4.13}$$

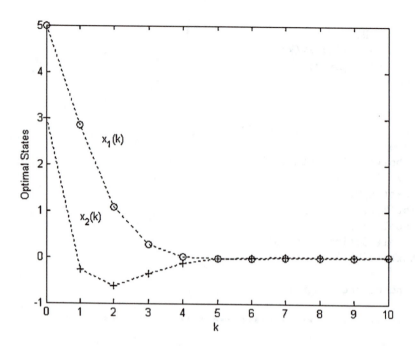

Figure 5.7 Implementation of Optimal Control for Example 5.4

where,

$$E = BR^{-1}B'.$$

Let us note that

$$H_{11}H_{22} - H_{12}H_{21} = I. \tag{5.4.14}$$

The boundary conditions for (5.4.12) are reproduced here from (5.3.1) and (5.3.5) as

$$\mathbf{x}(k = k_0) = \mathbf{x}_0; \quad \lambda(k = k_f) = \mathbf{P}(k = k_f)\mathbf{x}(k_f). \tag{5.4.15}$$

Also, in trying to obtain the difference Riccati equation, we assumed a transformation (5.3.6) between the *state* and *costate* as

$$\lambda(k) = \mathbf{P}(k)\mathbf{x}(k); \quad \forall \, k \leq k_f. \tag{5.4.16}$$

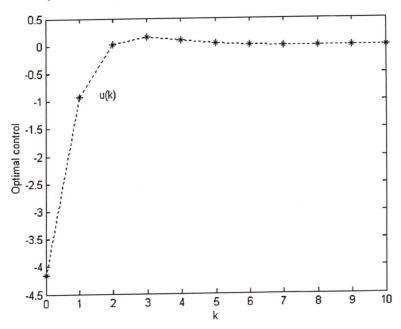

Figure 5.8 Implementation of Optimal Control for Example 5.4

Now, we will show that the solution $\mathbf{P}(k)$ to the Riccati equation (5.3.20) can be obtained in terms of the eigenvalues and eigenvectors of the Hamiltonian matrix $\mathbf{H}$.

Let us first define

$$\mathbf{J} = \begin{bmatrix} 0 & \mathbf{I} \\ -\mathbf{I} & 0 \end{bmatrix} \quad \text{such that} \quad \mathbf{J}^{-1} = -\mathbf{J}. \tag{5.4.17}$$

Now, using (5.4.13), (5.4.14) and (5.4.17), it is a simple thing to show that

$$\mathbf{H'JH} = \mathbf{J}. \tag{5.4.18}$$

By premultiplying and postmultiplying (5.4.18) with $\mathbf{H}^{-1}$ and $\mathbf{J}^{-1}$, respectively, we have

$$\mathbf{H'J} = \mathbf{JH}^{-1}, \quad \mathbf{J}^{-1}\mathbf{H'J} = \mathbf{H}^{-1} \tag{5.4.19}$$

and using (5.4.17), we get

$$\mathbf{H}^{-1} = -\mathbf{J}\mathbf{H}'\mathbf{J}. \tag{5.4.20}$$

Now, substituting for the quantities on the right hand side of the previous equation, we have

$$\mathbf{H}^{-1} = \begin{bmatrix} \mathbf{A} + \mathbf{E}\mathbf{A}^{-T}\mathbf{Q} & -\mathbf{E}\mathbf{A}^{-T} \\ -\mathbf{A}^{-T}\mathbf{Q} & \mathbf{A}^{-T} \end{bmatrix} \tag{5.4.21}$$

where, $\mathbf{A}^{-T}$ is the transpose of $\mathbf{A}^{-1}$. Now, let us show that if μ is an eigenvalue of $\mathbf{H}$, then $1/\mu$ is also the eigenvalue of $\mathbf{H}$. First, if μ is an eigenvalue with a corresponding eigenvector $[f, g]'$, then

$$\begin{bmatrix} \mathbf{A}^{-1} & \mathbf{A}^{-1}\mathbf{E} \\ \mathbf{Q}\mathbf{A}^{-T} & \mathbf{A}' + \mathbf{Q}\mathbf{A}^{-1}\mathbf{E} \end{bmatrix} \begin{bmatrix} f \\ g \end{bmatrix} = \mu \begin{bmatrix} f \\ g \end{bmatrix} \tag{5.4.22}$$

and rearranging, we have

$$\begin{bmatrix} \mathbf{A}' + \mathbf{Q}\mathbf{A}^{-1}\mathbf{E} & -\mathbf{Q}\mathbf{A}^{-1} \\ -\mathbf{A}^{-1}\mathbf{E} & \mathbf{A}^{-1} \end{bmatrix} \begin{bmatrix} g \\ -f \end{bmatrix} = \mu \begin{bmatrix} g \\ -f \end{bmatrix}.$$

Using (5.4.21), we have

$$\begin{bmatrix} \mathbf{H}^{-T} \end{bmatrix} \begin{bmatrix} g \\ -f \end{bmatrix} = \mu \begin{bmatrix} g \\ -f \end{bmatrix}. \tag{5.4.23}$$

This shows that μ is also an eigenvalue of $\mathbf{H}^{-T}$, and hence of $\mathbf{H}^{-1}$. We know that from elementary matrix algebra that if α is an eigenvalue of a matrix $\mathbf{A}$, then $1/\alpha$ is also an eigenvalue of matrix $\mathbf{A}^{-1}$. Therefore, $1/\mu$ is an eigenvalue of $\mathbf{H}$, and hence the result. This means that the eigenvalues of $\mathbf{H}$ can be arranged as

$$\mathbf{D} = \begin{bmatrix} \mathbf{M} & 0 \\ 0 & \mathbf{M}^{-1} \end{bmatrix} \tag{5.4.24}$$

where, $\mathbf{M}$ is a diagonal (Jordon) matrix containing n eigenvalues outside the unit circle and $\mathbf{M}^{-1}$ is a diagonal matrix containing n eigenvalues inside the unit circle which means that $\mathbf{M}^{-1}$ is stable. Now, we note that $\mathbf{D}$ can be written in terms of a nonsingular matrix $\mathbf{W}$ whose columns are the eigenvectors of $\mathbf{H}$ as

$$\mathbf{W}^{-1}\mathbf{H}\mathbf{W} = \mathbf{D}. \tag{5.4.25}$$

Let us now introduce a transformation as

$$\begin{bmatrix} \mathbf{x}(k) \\ \boldsymbol{\lambda}(k) \end{bmatrix} = \mathbf{W} \begin{bmatrix} \mathbf{v}(k) \\ \mathbf{z}(k) \end{bmatrix} = \begin{bmatrix} \mathbf{W}_{11} & \mathbf{W}_{12} \\ \mathbf{W}_{21} & \mathbf{W}_{22} \end{bmatrix} \begin{bmatrix} \mathbf{v}(k) \\ \mathbf{z}(k) \end{bmatrix}. \qquad (5.4.26)$$

Then, using (5.4.26), (5.4.24) and the Hamiltonian system (5.4.12), we get

$$\begin{bmatrix} \mathbf{v}(k) \\ \mathbf{z}(k) \end{bmatrix} = \mathbf{W}^{-1} \begin{bmatrix} \mathbf{x}^*(k) \\ \boldsymbol{\lambda}^*(k) \end{bmatrix} = \mathbf{W}^{-1}\mathbf{H} \begin{bmatrix} \mathbf{x}^*(k+1) \\ \boldsymbol{\lambda}^*(k+1) \end{bmatrix}$$

$$= \mathbf{W}^{-1}\mathbf{H}\mathbf{W} \begin{bmatrix} \mathbf{v}(k+1) \\ \mathbf{z}(k+1) \end{bmatrix} = \mathbf{D} \begin{bmatrix} \mathbf{v}(k+1) \\ \mathbf{z}(k+1) \end{bmatrix}$$

$$= \begin{bmatrix} \mathbf{M} & 0 \\ 0 & \mathbf{M}^{-1} \end{bmatrix} \begin{bmatrix} \mathbf{v}(k+1) \\ \mathbf{z}(k+1) \end{bmatrix}. \qquad (5.4.27)$$

The solution of (5.4.27) in terms of the final conditions is found by first writing it as

$$\begin{bmatrix} \mathbf{v}(k+1) \\ \mathbf{z}(k+1) \end{bmatrix} = \mathbf{D}^{-1} \begin{bmatrix} \mathbf{v}(k) \\ \mathbf{z}(k) \end{bmatrix}.$$

Then solving this for the given final conditions, we get

$$\begin{bmatrix} \mathbf{v}(k) \\ \mathbf{z}(k) \end{bmatrix} = \begin{bmatrix} \mathbf{M}^{(k_f-k)} & 0 \\ 0 & \mathbf{M}^{-(k_f-k)} \end{bmatrix} \begin{bmatrix} \mathbf{v}(k_f) \\ \mathbf{z}(k_f) \end{bmatrix}. \qquad (5.4.28)$$

Here, since $\mathbf{M}$ is unstable (i.e., it does not go to zero as $(k_f - k) \to \infty$), we rewrite (5.4.28) as

$$\begin{bmatrix} \mathbf{v}(k_f) \\ \mathbf{z}(k) \end{bmatrix} = \begin{bmatrix} \mathbf{M}^{-(k_f-k)} & 0 \\ 0 & \mathbf{M}^{-(k_f-k)} \end{bmatrix} \begin{bmatrix} \mathbf{v}(k) \\ \mathbf{z}(k_f) \end{bmatrix}. \qquad (5.4.29)$$

Next, at the final time k_f, using (5.4.16) and (5.4.26) we have

$$\boldsymbol{\lambda}(k_f) = \mathbf{W}_{21}\mathbf{v}(k_f) + \mathbf{W}_{22}\mathbf{z}(k_f) = \mathbf{P}(k_f)\mathbf{x}(k_f) = \mathbf{F}(k_f)\mathbf{x}(k_f),$$
$$= \mathbf{F}(k_f)\left[\mathbf{W}_{11}\mathbf{v}(k_f) + \mathbf{W}_{12}\mathbf{z}(k_f)\right] \qquad (5.4.30)$$

the solution of which becomes

$$\mathbf{z}(k_f) = \mathbf{T}(k_f)\mathbf{v}(k_f) \qquad (5.4.31)$$

where,

$$\mathbf{T}(k_f) = -\left[\mathbf{W}_{22} - \mathbf{F}(k_f)\mathbf{W}_{12}\right]^{-1}\left[\mathbf{W}_{21} - \mathbf{F}(k_f)\mathbf{W}_{11}\right]. \qquad (5.4.32)$$

Now, using (5.4.29) and (5.4.31)

$$\begin{aligned}
\mathbf{z}(k) &= \mathbf{M}^{-(k_f-k)}\mathbf{z}(k_f) \\
&= \mathbf{M}^{-(k_f-k)}\mathbf{T}(k_f)\mathbf{v}(k_f) \\
&= \mathbf{M}^{-(k_f-k)}\mathbf{T}(k_f)\mathbf{M}^{-(k_f-k)}\mathbf{v}(k).
\end{aligned} \tag{5.4.33}$$

This means that at each value of k,

$$\mathbf{z}(k) = \mathbf{T}(k)\mathbf{v}(k) \tag{5.4.34}$$

where,

$$\mathbf{T}(k) = \mathbf{M}^{-(k_f-k)}\mathbf{T}(k_f)\mathbf{M}^{-(k_f-k)}. \tag{5.4.35}$$

Now we relate $\mathbf{P}(k)$ in (5.4.16) with $\mathbf{T}(k)$ in (5.4.34) by first using (5.4.26) and (5.4.16) to get

$$\begin{aligned}
\boldsymbol{\lambda}^*(k) &= \mathbf{W}_{21}\mathbf{v}(k) + \mathbf{W}_{22}\mathbf{z}(k) \\
&= \mathbf{P}(k)\mathbf{x}^*(k) = \mathbf{P}(k)\left[\mathbf{W}_{11}\mathbf{v}(k) + \mathbf{W}_{12}\mathbf{z}(k)\right]
\end{aligned} \tag{5.4.36}$$

and then using (5.4.34)

$$\left[\mathbf{W}_{21} + \mathbf{W}_{22}\mathbf{T}(k)\right]\mathbf{v}(k) = \mathbf{P}(k)\left[\mathbf{W}_{11} + \mathbf{W}_{12}\mathbf{T}(k)\right]\mathbf{v}(k). \tag{5.4.37}$$

Since this must hold good for all $\mathbf{x}(0)$ and hence for all $\mathbf{v}(k)$, leading to

$$\boxed{\mathbf{P}(k) = \left[\mathbf{W}_{21} + \mathbf{W}_{22}\mathbf{T}(k)\right]\left[\mathbf{W}_{11} + \mathbf{W}_{12}\mathbf{T}(k)\right]^{-1}.} \tag{5.4.38}$$

Finally, we have the *nonrecursive analytical* solution (5.4.38) to the matrix difference Riccati equation (5.3.20). Let us note that the solution (5.4.38) requires the relations (5.4.35) and (5.4.32) and the eigenvalues (5.4.24), eigenvectors (5.4.25), and the given terminal cost matrix $\mathbf{F}(k_f)$.

Steady-State Condition

As the terminal time $k_f \to \infty$, the matrix difference Riccati equation (5.3.20) becomes matrix algebraic Riccati equation (5.4.7). Now, let us find the analytical expression for the solution $\overline{\mathbf{P}}$ of this ARE. As $(k_f - k) \to \infty$, $\mathbf{M}^{-(k_f-k)} \to \infty$, since $\mathbf{M}^{-1}$ is stable. This means that in (5.4.35) $\mathbf{T}(k) \to 0$. Then the steady-state solution of (5.4.38) gives

$$\boxed{\overline{\mathbf{P}} = \mathbf{W}_{21}\mathbf{W}_{11}^{-1}.} \tag{5.4.39}$$

Let us note that the previous steady-state solution (5.4.39) requires the *unstable* eigenvalues (5.4.24) and eigenvectors (5.4.25). Thus, we have the analytical solution (5.4.39) of the ARE (5.4.7).

Example 5.5

Consider the same Example 5.3 to use analytical solution of matrix Riccati difference equation based on [138]. The results are obtained using Control System Toolbox of MATLAB©, Version 6 as shown below. The solution of matrix DRE is not readily available with MATLAB© and hence a program was developed based on the analytical solution of the matrix DRE [138]. The following MATLAB© m file for Example 5.5 requires two additional MATLAB© files *lqrdnss.m* and *lqrdnssf.m* given in Appendix C. The solutions are shown in Figure 5.9. Using these Riccati gains, the optimal states $x_1^*(k)$ and $x_2^*(k)$ are shown in Figure 5.10 and the optimal control $u^*(k)$ is shown in Figure 5.11.

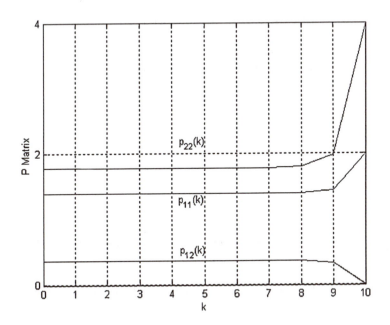

Figure 5.9 Riccati Coefficients for Example 5.5

```
***************************************************
%% Solution using Control System Toolbox and
```

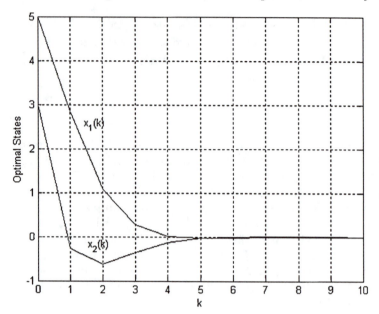

Figure 5.10   Optimal States for Example 5.5

```
%% the MATLAB. Version 6
%% The following file example.m requires
%% two other files lqrnss.m and lqrnssf.m
%% which are given in Appendix C
clear all
A=[.8,1;0,.6];
B=[1;.5];
F=[2,0;0,4];
Q=[1,0;0,1];
R=1;
kspan=[0 10];
x0(:,1)=[5.;3.];
[x,u]=lqrdnss_dsn(A,B,F,Q,R,x0,kspan);
********************************************************
```

5.5   *Discrete-Time Linear Quadratic Tracking System*

In this section, we address linear quadratic tracking (LQT) problem for a discrete-time system and are interested in obtaining a closed-loop control scheme that enables a given system track (or follow) a *desired* trajectory over the given interval of time. We essentially deal

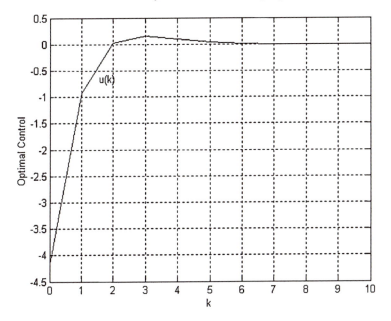

Figure 5.11   Optimal Control for Example 5.5

with linear, time-invariant systems in order to get some elegant results although the method can as well be applied to nonlinear, time-varying case [89].

Let us consider a linear, time-invariant system described by the state equation

$$\mathbf{x}(k+1) = \mathbf{A}\mathbf{x}(k) + \mathbf{B}\mathbf{u}(k) \tag{5.5.1}$$

and the output relation

$$\mathbf{y}(k) = \mathbf{C}\mathbf{x}(k). \tag{5.5.2}$$

The performance index to be minimized is

$$J = \frac{1}{2}\left[\mathbf{C}\mathbf{x}(k_f) - \mathbf{z}(k_f)\right]'\mathbf{F}\left[\mathbf{C}\mathbf{x}(k_f) - \mathbf{z}(k_f)\right]$$

$$+ \frac{1}{2}\sum_{k=k_0}^{k_f-1}\left\{\left[\mathbf{C}\mathbf{x}(k) - \mathbf{z}(k)\right]'\mathbf{Q}\left[\mathbf{C}\mathbf{x}(k) - \mathbf{z}(k)\right] + \mathbf{u}'(k)\mathbf{R}\mathbf{u}(k)\right\}$$

$$\tag{5.5.3}$$

where, $\mathbf{x}(k)$, $\mathbf{u}(k)$, and $\mathbf{y}(k)$ are n, r, and n order *state*, *control*, and *output* vectors, respectively. Also, we assume that $\mathbf{F}$ and $\mathbf{Q}$ are each

nxn dimensional *positive semidefinite* symmetric matrices, and $\mathbf{R}$ is an rxr *positive definite* symmetric matrix. The initial condition is given as $\mathbf{x}(k_0)$ and the final condition $\mathbf{x}(k_f)$ is *free* with k_f fixed. We want the error $\mathbf{e}(k) = \mathbf{y}(k) - \mathbf{z}(k)$ as small as possible with minimum control effort, where $\mathbf{z}(k)$ is n dimensional *reference* vector. The methodology to obtain the solution for the optimal tracking system is carried out using the following steps.

- **Step 1:** *Hamiltonian*

- **Step 2:** *State and Costate System*

- **Step 3:** *Open-Loop Optimal Control*

- **Step 4:** *Riccati and Vector Equations*

- **Step 5:** *Closed-Loop Optimal Control*

Now the details follow.

- **Step 1:** *Hamiltonian:* We first formulate the Hamiltonian as

$$\mathcal{H}(\mathbf{x}(k), \mathbf{u}(k), \lambda(k+1)) = \frac{1}{2} \sum_{k=k_0}^{k_f-1} \{[\mathbf{C}\mathbf{x}(k) - \mathbf{z}(k)]' \, \mathbf{Q} \, [\mathbf{C}\mathbf{x}(k) - \mathbf{z}(k)]$$
$$+ \, \mathbf{u}'(k)\mathbf{R}\mathbf{u}(k)\} + \lambda'(k+1) \, [\mathbf{A}\mathbf{x}(k) + \mathbf{B}\mathbf{u}(k)]$$
$$(5.5.4)$$

and follow the identical approach of the state regulator system described in the previous section. For the sake of simplicity, let us define

$$\mathbf{E} = \mathbf{B}\mathbf{R}^{-1}\mathbf{B}', \quad \mathbf{V} = \mathbf{C}'\mathbf{Q}\mathbf{C} \text{ and } \mathbf{W} = \mathbf{C}'\mathbf{Q}. \qquad (5.5.5)$$

- **Step 2:** *State and Costate System:* Using (5.2.15), (5.2.13) and (5.2.14) for the state, costate, and control, respectively, we obtain the state equation as

$$\frac{\partial \mathcal{H}}{\partial \lambda^*(k+1)} = \mathbf{x}^*(k+1) \longrightarrow \mathbf{x}^*(k+1) = \mathbf{A}\mathbf{x}^*(k) + \mathbf{B}\mathbf{u}^*(k),$$
$$(5.5.6)$$

the costate equation as

$$\frac{\partial \mathcal{H}}{\partial \mathbf{x}^*(k)} = \lambda^*(k) \longrightarrow \lambda^*(k) = \mathbf{A}'\lambda^*(k+1) + \mathbf{V}\mathbf{x}^*(k) - \mathbf{W}\mathbf{z}(k),$$
$$(5.5.7)$$

and the control equation as

$$\frac{\partial \mathcal{H}}{\partial \mathbf{u}^*(k)} = 0 \longrightarrow 0 = \mathbf{B}'\boldsymbol{\lambda}^*(k+1) + \mathbf{R}\mathbf{u}^*(k).$$

$$(5.5.8)$$

The final condition (5.2.37) becomes

$$\boldsymbol{\lambda}(k_f) = \mathbf{C}'\mathbf{F}\mathbf{C}\mathbf{x}(k_f) - \mathbf{C}'\mathbf{F}\mathbf{z}(k_f). \qquad (5.5.9)$$

- **Step 3:** *Open-Loop Optimal Control:* The relation (5.5.8) yields the *open-loop* optimal control as

$$\mathbf{u}^*(k) = -\mathbf{R}^{-1}\mathbf{B}'\boldsymbol{\lambda}^*(k+1) \qquad (5.5.10)$$

and using this in the state (5.5.6) and costate (5.5.7) system (also called Hamiltonian system), we have the Hamiltonian (canonical) system as

$$\begin{bmatrix} \mathbf{x}^*(k+1) \\ \boldsymbol{\lambda}^*(k) \end{bmatrix} = \begin{bmatrix} \mathbf{A} & -\mathbf{E} \\ \mathbf{V} & \mathbf{A}' \end{bmatrix} \begin{bmatrix} \mathbf{x}^*(k) \\ \boldsymbol{\lambda}^*(k+1) \end{bmatrix} + \begin{bmatrix} \mathbf{0} \\ -\mathbf{W} \end{bmatrix} \mathbf{z}(k). \quad (5.5.11)$$

Thus, we see that the Hamiltonian system is similar to that obtained for state regulator system in the previous section, except for the *nonhomogeneous* nature due to the forcing term $\mathbf{z}(k)$.

- **Step 4:** *Riccati and Vector Equations:* Now to obtain *closed-loop* configuration for the optimal control (5.5.10), we may assume from the nature of the boundary condition (5.5.9) a transformation

$$\boxed{\boldsymbol{\lambda}^*(k) = \mathbf{P}(k)\mathbf{x}^*(k) - \mathbf{g}(k)} \qquad (5.5.12)$$

where, the matrix $\mathbf{P}(k)$ and the vector $\mathbf{g}(k)$ are yet to be determined. In order to do so we essentially eliminate the costate $\boldsymbol{\lambda}^*(k)$ from the canonical system (5.5.11) using the transformation (5.5.12). Thus,

$$\mathbf{x}^*(k+1) = \mathbf{A}\mathbf{x}^*(k) - \mathbf{E}\mathbf{P}(k+1)\mathbf{x}^*(k+1) + \mathbf{E}\mathbf{g}(k+1) \quad (5.5.13)$$

which is solved for $\mathbf{x}^*(k+1)$ to yield

$$\mathbf{x}^*(k+1) = [\mathbf{I} + \mathbf{E}\mathbf{P}(k+1)]^{-1}[\mathbf{A}\mathbf{x}^*(k) + \mathbf{E}\mathbf{g}(k+1)]. \quad (5.5.14)$$

Now using (5.5.14) and (5.5.12) in the costate relation in (5.5.11), we have

$$\left[-\mathbf{P}(k) + \mathbf{A}'\mathbf{P}(k+1) \left[\mathbf{I} + \mathbf{E}\mathbf{P}(k+1) \right]^{-1} \mathbf{A} + \mathbf{V} \right] \mathbf{x}(k) +$$

$$\left[\mathbf{g}(k) + \mathbf{A}'\mathbf{P}(k+1) \left[\mathbf{I} + \mathbf{E}\mathbf{P}(k+1) \right]^{-1} \mathbf{E}\mathbf{g}(k+1) - \right.$$

$$\mathbf{A}'\mathbf{g}(k+1) - \mathbf{W}\mathbf{z}(k) \right] = 0$$

$$(5.5.15)$$

This equation must hold for all values of the state $\mathbf{x}^*(k)$ which in turn leads to the fact that the coefficient of $\mathbf{x}(k)$ and the rest of the terms in (5.5.15) must individually vanish. That is

$$\boxed{\mathbf{P}(k) = \mathbf{A}'\mathbf{P}(k+1) \left[\mathbf{I} + \mathbf{E}\mathbf{P}(k+1) \right]^{-1} \mathbf{A} + \mathbf{V}} \quad \text{or}$$

$$\boxed{\mathbf{P}(k) = \mathbf{A}' \left[\mathbf{P}^{-1}(k+1) + \mathbf{E} \right]^{-1} \mathbf{A} + \mathbf{V}}$$

$$(5.5.16)$$

and

$$\boxed{\mathbf{g}(k) = \mathbf{A}' \left\{ \mathbf{I} - \left[\mathbf{P}^{-1}(k+1) + \mathbf{E} \right]^{-1} \mathbf{E} \right\} \mathbf{g}(k+1) + \mathbf{W}\mathbf{z}(k)} \quad \text{or}$$

$$\boxed{\mathbf{g}(k) = \left\{ \mathbf{A}' - \mathbf{A}'\mathbf{P}(k+1) \left[\mathbf{I} + \mathbf{E}\mathbf{P}(k+1) \right]^{-1} \mathbf{E} \right\} \mathbf{g}(k+1) + \mathbf{W}\mathbf{z}(k).}$$

$$(5.5.17)$$

To obtain the boundary conditions for (5.5.16) and (5.5.17), let us compare (5.5.9) and (5.5.12) to yield

$$\boxed{\mathbf{P}(k_f) = \mathbf{C}'\mathbf{F}\mathbf{C}} \tag{5.5.18}$$

$$\boxed{\mathbf{g}(k_f) = \mathbf{C}'\mathbf{F}\mathbf{z}(k_f).} \tag{5.5.19}$$

Let us note that (5.5.16) is the *nonlinear, matrix difference Riccati equation* (DRE) to be solved *backwards* using the final condition (5.5.18), and the *linear, vector difference equation* (5.5.17) is solved *backwards* using the final condition (5.5.19).

- **Step 5:** *Closed-Loop Optimal Control:* Once we obtain these solutions *off-line*, we are ready to use the transformation (5.5.12) in the control relation (5.5.10) to get the closed-loop optimal control as

$$\mathbf{u}^*(k) = -\mathbf{R}^{-1}\mathbf{B}' \left[\mathbf{P}(k+1)\mathbf{x}(k+1) - \mathbf{g}(k+1) \right] \quad (5.5.20)$$

and substituting for the state from (5.5.6) in (5.5.20),

$$\mathbf{u}^*(k) = -\mathbf{R}^{-1}\mathbf{B}'\mathbf{P}(k+1)\left[\mathbf{A}\mathbf{x}^*(k) + \mathbf{B}\mathbf{u}^*(k)\right] + \mathbf{R}^{-1}\mathbf{B}'\mathbf{g}(k+1).$$
(5.5.21)

Now premultiplying by $\mathbf{R}$ and solving for the optimal control $\mathbf{u}^*(k)$ we have

$$\boxed{\mathbf{u}^*(k) = -\mathbf{L}(k)\mathbf{x}^*(k) + \mathbf{L}_g(k)\mathbf{g}(k+1)}$$
(5.5.22)

where, the *feedback gain* $\mathbf{L}(k)$ and the *feed forward gain* $\mathbf{L}_g(k)$ are given by

$$\mathbf{L}(k) = \left[\mathbf{R} + \mathbf{B}'\mathbf{P}(k+1)\mathbf{B}\right]^{-1}\mathbf{B}'\mathbf{P}(k+1)\mathbf{A} \quad (5.5.23)$$

$$\mathbf{L}_g(k) = \left[\mathbf{R} + \mathbf{B}'\mathbf{P}(k+1)\mathbf{B}\right]^{-1}\mathbf{B}' \quad (5.5.24)$$

The optimal state trajectory is now given from (5.5.6) and (5.5.22) as

$$\mathbf{x}^*(k+1) = \left[\mathbf{A} - \mathbf{B}\mathbf{L}(k)\right]\mathbf{x}(k) + \mathbf{B}\mathbf{L}_g(k)\mathbf{g}(k+1). \quad (5.5.25)$$

The implementation of the discrete-time optimal tracker is shown in Figure 5.12. The complete procedure for the linear quadratic tracking system is summarized in Table 5.5.

Example 5.6

We now illustrate the previous procedure by considering the same system of the Example 5.3. Let us say that we are interested in tracking $x_1(k)$ with respect to the desired trajectory $z_1(k) = 2$ and we do not have any condition on the second state $x_2(k)$. Then the various matrices are

$$\mathbf{A}(k) = \begin{bmatrix} 0.8 & 1.0 \\ 0.0 & 0.6 \end{bmatrix}; \quad \mathbf{B}(k) = \begin{bmatrix} 1.0 \\ 0.5 \end{bmatrix}; \quad \mathbf{C} = \begin{bmatrix} 1 & 0 \\ 0 & 1 \end{bmatrix}$$

$$\mathbf{F}(k_f) = \begin{bmatrix} 1 & 0 \\ 0 & 0 \end{bmatrix}; \quad \mathbf{Q}(k) = \begin{bmatrix} 1 & 0 \\ 0 & 0 \end{bmatrix}; \quad \mathbf{R} = 0.01. \quad (5.5.26)$$

Now let us use the procedure given in Table 5.5. Note that one has to try various values of the matrix $\mathbf{R}$ in order to get a better tracking of the states.

The various solutions obtained using MATLAB© Version 6. Figure 5.13 shows Riccati functions $p_{11}(k), p_{12}(k)$, and $p_{22}(k)$; Figure 5.14 shows vector coefficients $g_1(k)$ and $g_2(k)$; Figure 5.6 gives the optimal states and Figure 5.6 gives optimal control.

The MATLAB© program used is given in Appendix C.

Table 5.5 Procedure Summary of Discrete-Time Linear Quadratic Tracking System

A. Statement of the Problem
Given the plant as
$\mathbf{x}(k+1) = \mathbf{Ax}(k) + \mathbf{Bu}(k)$,
the output relation as
$\mathbf{y}(k) = \mathbf{Cx}(k)$,
the performance index as
$J(k_0) = \frac{1}{2}\left[\mathbf{Cx}(k_f) - \mathbf{z}(k_f)\right]' \mathbf{F}\left[\mathbf{Cx}(k_f) - \mathbf{z}(k_f)\right]$
$+\frac{1}{2}\sum_{k=k_0}^{k_f-1}\left\{\left[\mathbf{Cx}(k) - \mathbf{z}(k)\right]' \mathbf{Q}\left[\mathbf{Cx}(k) - \mathbf{z}(k)\right] + \mathbf{u}'(k)\mathbf{Ru}(k)\right\}$
and the boundary conditions as
$\mathbf{x}(k_0) = \mathbf{x}_0,\quad \mathbf{x}(k_f)$ is free, and k is fixed,
find the optimal control and state.

	B. Solution of the Problem
Step 1	Solve the matrix difference Riccati equation $\mathbf{P}(k) = \mathbf{A}'\mathbf{P}(k+1)\left[\mathbf{I} + \mathbf{EP}(k+1)\right]^{-1}\mathbf{A} + \mathbf{V}$ with $\mathbf{P}(k_f) = \mathbf{C}'\mathbf{FC}$, where $\mathbf{V} = \mathbf{C}'\mathbf{QC}$ and $\mathbf{E} = \mathbf{BR}^{-1}\mathbf{B}'$.
Step 2	Solve the vector difference equation $\mathbf{g}(k) = \mathbf{A}'\left\{\mathbf{I} - \left[\mathbf{P}^{-1}(k+1) + \mathbf{E}\right]^{-1}\mathbf{E}\right\}\mathbf{g}(k+1) + \mathbf{Wz}(k)$ with $\mathbf{g}(k_f) = \mathbf{C}'\mathbf{Fz}(k_f)$, where, $\mathbf{W} = \mathbf{C}'\mathbf{Q}$.
Step 3	Solve for the optimal state $\mathbf{x}^*(k)$ as $\mathbf{x}^*(k+1) = \left[\mathbf{A} - \mathbf{BL}(k)\right]\mathbf{x}^*(k) + \mathbf{BL}_g(k)\mathbf{g}(k+1)$ where, $\mathbf{L}(k) = \left[\mathbf{R} + \mathbf{B}'\mathbf{P}(k+1)\mathbf{B}\right]^{-1}\mathbf{B}'\mathbf{P}(k+1)\mathbf{A}$, $\mathbf{L}_g(k) = \left[\mathbf{R} + \mathbf{B}'\mathbf{P}(k+1)\mathbf{B}\right]^{-1}\mathbf{B}'$.
Step 4	Obtain the optimal control as $\mathbf{u}^*(k) = -\mathbf{L}(k)\mathbf{x}^*(k) + \mathbf{L}_g(k)\mathbf{g}(k+1)$.

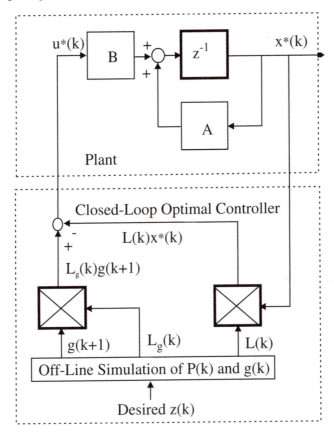

Figure 5.12   Implementation of Discrete-Time Optimal Tracker

5.6   *Frequency-Domain Interpretation*

This section is based on [89]. In this section, we use frequency domain to derive some results from the classical control point of view for a linear, time-invariant, discrete-time, optimal control system with infinite-time case, described earlier in Section 5.4. For this, we know that the optimal control involves the solution of matrix algebraic Riccati equation. For ready reference, we reproduce here results of the time-invariant case described earlier in this chapter. For the plant

$$\mathbf{x}(k+1) = \mathbf{A}\mathbf{x}(k) + \mathbf{B}\mathbf{u}(k), \qquad (5.6.1)$$

the optimal feedback control is

$$\mathbf{u}^*(k) = -\bar{\mathbf{L}}\mathbf{x}^*(k) \qquad (5.6.2)$$

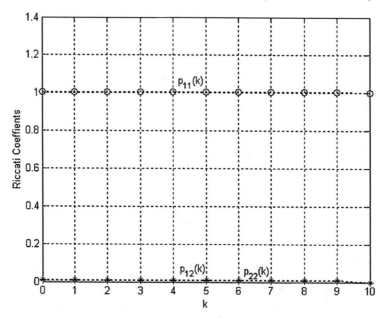

Figure 5.13   Riccati Coefficients for Example 5.6

where,

$$\bar{\mathbf{L}} = \left[\bar{\mathbf{B}}'\bar{\mathbf{P}}\mathbf{B} + \mathbf{R}\right]^{-1}\mathbf{B}'\bar{\mathbf{P}}\mathbf{A}. \qquad (5.6.3)$$

Rewriting the above by postmultiplying by $\left[\mathbf{B}'\bar{\mathbf{P}}\mathbf{B} + \mathbf{R}\right]$

$$\left[\mathbf{B}'\bar{\mathbf{P}}\mathbf{B} + \mathbf{R}\right]\bar{\mathbf{L}} = \mathbf{B}'\bar{\mathbf{P}}\mathbf{A}. \qquad (5.6.4)$$

Here, $\bar{\mathbf{P}}$ is the solution of the ARE

$$\bar{\mathbf{P}} = \mathbf{A}'\left[\bar{\mathbf{P}} - \bar{\mathbf{P}}\mathbf{B}\left[\mathbf{B}'\bar{\mathbf{P}}\mathbf{B} + \mathbf{R}\right]^{-1}\mathbf{B}'\bar{\mathbf{P}}\right]\mathbf{A} + \mathbf{Q} \qquad (5.6.5)$$

where, we assume that $[\mathbf{A}, \mathbf{B}]$ is stabilizable and $[\mathbf{A}, \sqrt{\mathbf{Q}}]$ is observable. With this optimal control (5.6.2), the optimal system becomes

$$\mathbf{x}^*(k+1) = \left[\mathbf{A} - \mathbf{B}\bar{\mathbf{L}}\right]\mathbf{x}^*(k) \qquad (5.6.6)$$

and is asymptotically stable. Here, the open-loop characteristic polynomial of the system is

$$\boldsymbol{\Delta}_o(z) = |z\mathbf{I} - \mathbf{A}| \qquad (5.6.7)$$

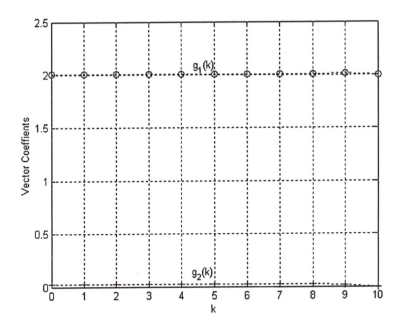

Figure 5.14   Coefficients $g_1(t)$ and $g_2(t)$ for Example 5.6

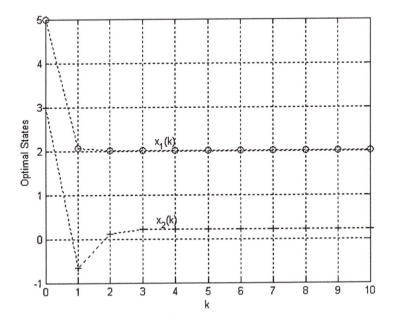

Figure 5.15   Optimal States for Example 5.6

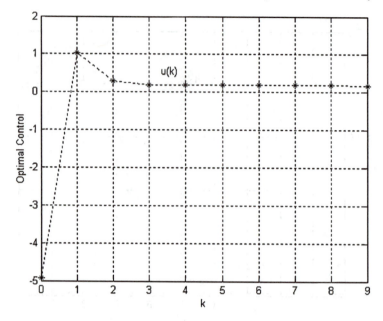

Figure 5.16   Optimal Control for Example 5.6

whereas, the optimal closed-loop characteristic polynomial is [89]

$$\boldsymbol{\Delta}_c(z) = |z\mathbf{I} - \mathbf{A} + \mathbf{B}\bar{\mathbf{L}}|,$$
$$= |\mathbf{I} + \mathbf{B}\bar{\mathbf{L}}[z\mathbf{I} - \mathbf{A}]^{-1}|.|z\mathbf{I} - \mathbf{A}|,$$
$$= |\mathbf{I} + \bar{\mathbf{L}}[z\mathbf{I} - \mathbf{A}]^{-1}\mathbf{B}|.\boldsymbol{\Delta}_o(z). \tag{5.6.8}$$

From Figure 5.17, we note that

1. $-\bar{\mathbf{L}}[z\mathbf{I} - \mathbf{A}]^{-1}\mathbf{B}$ is called the *loop gain matrix*, and

2. $\mathbf{I} + \bar{\mathbf{L}}[z\mathbf{I} - \mathbf{A}]^{-1}\mathbf{B}$ is termed *return difference matrix*.

First of all, let us note that

$$\bar{\mathbf{P}} - \mathbf{A}'\bar{\mathbf{P}}\mathbf{A} = [z^{-1}\mathbf{I} - \mathbf{A}]'\bar{\mathbf{P}}[z\mathbf{I} - \mathbf{A}] + [z^{-1}\mathbf{I} - \mathbf{A}]'\bar{\mathbf{P}}\mathbf{A} + \mathbf{A}'\bar{\mathbf{P}}[z\mathbf{I} - \mathbf{A}].$$
$$\tag{5.6.9}$$

Using the ARE (5.6.5) to replace the left-hand side of the previous equation, we have

$$[z^{-1}\mathbf{I} - \mathbf{A}]'\bar{\mathbf{P}}[z\mathbf{I} - \mathbf{A}] + [z^{-1}\mathbf{I} - \mathbf{A}]'\bar{\mathbf{P}}\mathbf{A} + \mathbf{A}'\bar{\mathbf{P}}[z\mathbf{I} - \mathbf{A}]$$
$$+ \mathbf{A}'\bar{\mathbf{P}}\mathbf{B}\left[\mathbf{B}'\bar{\mathbf{P}}\mathbf{B} + \mathbf{R}\right]^{-1}\mathbf{B}'\bar{\mathbf{P}}\mathbf{A} = \mathbf{Q}. \tag{5.6.10}$$

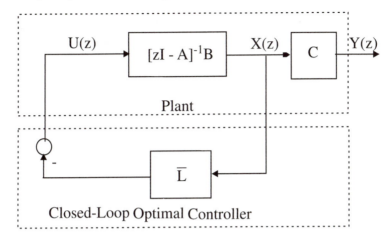

Figure 5.17   Closed-Loop Discrete-Time Optimal Control System

Premultiplying by $\mathbf{B}'[z^{-1}\mathbf{I} - \mathbf{A}]^{-T}$ (where for example, we define $\mathbf{M}^{-T}$ as the transpose of $\mathbf{M}^{-1}$) and postmultiplying by $[z\mathbf{I} - \mathbf{A}]^{-1}\mathbf{B}$, the previous relation becomes

$$\mathbf{B}'\bar{\mathbf{P}}\mathbf{B} + \mathbf{B}\bar{\mathbf{P}}\mathbf{A}[z\mathbf{I} - \mathbf{A}]^{-1}\mathbf{B} + \mathbf{B}'[z^{-1}\mathbf{I} - \mathbf{A}]^{-T}\mathbf{A}'\bar{\mathbf{P}}\mathbf{B}$$
$$+ \mathbf{B}'[z^{-1}\mathbf{I} - \mathbf{A}]^{-T}\mathbf{A}'\bar{\mathbf{P}}\mathbf{B}\left[\mathbf{B}'\bar{\mathbf{P}}\mathbf{B} + \mathbf{R}\right]^{-1}\mathbf{B}'\bar{\mathbf{P}}\mathbf{A}[z\mathbf{I} - \mathbf{A}]\mathbf{B},$$
$$= \mathbf{B}'[z^{-1}\mathbf{I} - \mathbf{A}]^{-T}\mathbf{Q}[z\mathbf{I} - \mathbf{A}]^{-1}\mathbf{B}. \qquad (5.6.11)$$

Using (5.6.4) in the above, we get

$$\mathbf{B}'\bar{\mathbf{P}}\mathbf{B} + [\mathbf{B}'\bar{\mathbf{P}}\mathbf{B} + \mathbf{R}]\bar{\mathbf{L}}[z\mathbf{I} - \mathbf{A}]^{-1}\mathbf{B} + \mathbf{B}'[z^{-1}\mathbf{I} - \mathbf{A}]^{-T}\bar{\mathbf{P}}'[\mathbf{B}'\bar{\mathbf{P}}\mathbf{B} + \mathbf{R}]$$
$$+ \mathbf{B}'[z^{-1}\mathbf{I} - \mathbf{A}]^{-T}\bar{\mathbf{L}}^{-T}[\mathbf{B}'\bar{\mathbf{P}}\mathbf{B} + \mathbf{R}]\bar{\mathbf{L}}[z^{-1}\mathbf{I} - \mathbf{A}]^{-1}\mathbf{B},$$
$$= \mathbf{B}'[z^{-1}\mathbf{I} - \mathbf{A}]^{-T}\mathbf{Q}[z\mathbf{I} - \mathbf{A}]^{-1}\mathbf{B}. \qquad (5.6.12)$$

Now, adding the positive definite matrix $\mathbf{R}$ to both sides of the previous equation and factoring we get

$$\mathbf{B}'[z^{-1}\mathbf{I} - \mathbf{A}']^{-1}\mathbf{Q}[z\mathbf{I} - \mathbf{A}]^{-1}\mathbf{B} + \mathbf{R}$$
$$= \left[\mathbf{I} + \bar{\mathbf{L}}[z^{-1}\mathbf{I} - \mathbf{A}]^{-1}\mathbf{B}\right]'[\mathbf{B}'\bar{\mathbf{P}}\mathbf{B} + \mathbf{R}]\left[\mathbf{I} + \bar{\mathbf{L}}[z\mathbf{I} - \mathbf{A}]^{-1}\mathbf{B}\right]. \qquad (5.6.13)$$

With fictitious output equation as

$$\mathbf{y}(k) = \mathbf{C}\mathbf{x}(k) + \mathbf{D}\mathbf{u}(k) \qquad (5.6.14)$$

where, $\mathbf{Q} = \mathbf{C}'\mathbf{C}$ and $\mathbf{R} = \mathbf{D}'\mathbf{D}$, and using (5.6.1), the transfer function relating the output $\mathbf{y}(k)$ and the control $\mathbf{u}^*(k)$ becomes

$$\mathbf{Y}(z) = \begin{bmatrix} \mathbf{C}[z\mathbf{I} - \mathbf{A}]^{-1}\mathbf{B} \\ \mathbf{D} \end{bmatrix} \mathbf{U}(z). \qquad (5.6.15)$$

Also, we see that

$$\begin{bmatrix} \mathbf{C}[z^{-1}\mathbf{I} - \mathbf{A}]^{-1}\mathbf{B} \\ \mathbf{D} \end{bmatrix}' \begin{bmatrix} \mathbf{C}[z\mathbf{I} - \mathbf{A}]^{-1}\mathbf{B} \\ \mathbf{D} \end{bmatrix}$$

$$= \mathbf{B}[z^{-1}\mathbf{I} - \mathbf{A}]^{-T}\mathbf{Q}[z\mathbf{I} - \mathbf{A}]^{-1}\mathbf{B} + \mathbf{R}. \qquad (5.6.16)$$

Then, we can easily see from (5.6.13) that the previous transfer function product can be expressed in terms of the return difference matrix. Hence, the relation (5.6.15) implies that Figure 5.17 shows a return difference product.

5.7 Problems

1. Make reasonable assumptions wherever necessary.

2. Use MATLAB© wherever possible to solve the problems and plot all the optimal controls and states for all problems. Provide the relevant MATLAB© m files.

Problem 5.1 Show that the coefficient matrix $\mathbf{P}(k)$ in the matrix difference Riccati equation (5.3.11)

$$\mathbf{P}(k) = \mathbf{A}'(k)\mathbf{P}(k+1)\left[\mathbf{I} + \mathbf{E}(k)\mathbf{P}(k+1)\right]^{-1}\mathbf{A}(k) + \mathbf{Q}(k)$$

is positive definite.

Problem 5.2 A second order discrete-time system is given by

$$x_1(k+1) = 2x_1(k) + 0.2x_2(k), \qquad x_1(0) = 2$$
$$x_2(k+1) = 2x_2(k) + 0.2u(k), \qquad x_2(0) = 0.$$

The performance index to be minimized is

$$J = \frac{1}{20} \sum_{k=k_0}^{k_f-1} \left[x_1(k)^2 + u(k)^2\right]$$

where, $k_0 = 0$, and $k_f = 10$. In order to drive the above system to the final states $x_1(10) = x_2(10) = 0$,
(a) find the open-loop optimal control, and
(b) find the closed-loop optimal control.

Problem 5.3 Find the open-loop optimal control sequence $u(0), u(1), u(2)$ for the second order discrete-time system

$$\mathbf{x}(k+1) = \begin{bmatrix} 0 & 1 \\ -1 & 1 \end{bmatrix} \mathbf{x}(k) + \begin{bmatrix} 0 \\ 1 \end{bmatrix} u(k), \qquad \mathbf{x}(0) = \begin{bmatrix} 1 \\ 2 \end{bmatrix}$$

and the performance index

$$J = \sum_{k=k_0}^{k_f-1} \left[x_1^2(k) + u^2(k)\right]$$

where, $k_f = 5$, $x_1(5)$ is unspecified, and $x_2(5) = 0$.

Problem 5.4 Derive the relation for the optimal cost function given by (5.3.28) as

$$J^* = \frac{1}{2}\mathbf{x}^{*\prime}(k_0)\mathbf{P}(k_0)\mathbf{x}(k_0).$$

Problem 5.5 Given the plant as

$$\mathbf{x}(k+1) = \mathbf{A}(k)\mathbf{x}(k) + \mathbf{B}(k)\mathbf{u}(k)$$

the performance index as

$$J(k_0) = \frac{1}{2}\mathbf{x}'(k_f)\mathbf{F}(k_f)\mathbf{x}(k_f)$$

$$+ \frac{1}{2}\sum_{k=k_0}^{k_f-1}[\mathbf{x}'(k)\mathbf{Q}(k)\mathbf{x}(k) + \mathbf{u}'(k)\mathbf{R}(k)\mathbf{u}(k)]$$

and the fixed-end boundary conditions as $\mathbf{x}(k = k_0) = \mathbf{x}(k_0)$, $\mathbf{x}(k = k_f) = \mathbf{x}(k_f)$ and k_f is fixed, obtain the closed-loop, optimal control based on similar results for continuous-time, optimal control system described in Chapter 4.

Problem 5.6 For Problem 5.3, find the closed-loop optimal control for both (a) $k_f = 3$ and (b) $k_f = \infty$.

Problem 5.7 For Problem 5.3, add a capability to track a ramp reference signal $z_1(k) = 0.5k$, $z_2(k) = 0$.

Problem 5.8 For Problem 5.3, add a capability to track a low frequency sinusoidal function.

Problem 5.9 Obtain the Hamilton-Jacobi-Bellman equation for discrete-time optimal control problem.

Problem 5.10 For the D.C. motor speed control system described in Problem 1.1, obtain

1. the discrete-time model based on zero-order hold,

2. the closed-loop optimal control to keep the speed constant at a particular value, and

3. the closed-loop optimal control to track the speed along a ramp function $0.5t$.

Problem 5.11 For the liquid-level control system described in Problem 1.2, obtain

1. the discrete-time model based on zero-order hold,

2. the closed-loop optimal control to keep the liquid level constant at a particular value, and

3. the closed-loop optimal control to track the liquid level along a ramp function $0.25t$.

Problem 5.12 For the mechanical control system described in Problem 1.4, find the discrete-time model based on zero-order hold and then find the closed-loop optimal control to track the system along (i) a constant value and (ii) a ramp function.

Problem 5.13 [105] The discretized model using zero-order hold of the longitudinal motion of an aircraft is given by

$$\dot{\mathbf{x}}(t) = \mathbf{A}\mathbf{x}(k) + \mathbf{B}\mathbf{u}(t)$$

where,

$$\mathbf{A} = \begin{bmatrix} 0.9237 & -0.3081 & 0.0 & 0.0530 & -0.0904 \\ 0.0397 & 0.9955 & 0.0 & -0.1075 & 0.5889 \\ 0.0871 & 1.900 & 1.0 & -0.6353 & 0.3940 \\ -0.0356 & 0.0101 & 0.0 & 0.0078 & 0.1374 \\ 0.0696 & -0.0127 & 0.0 & -0.0971 & 0.2874 \end{bmatrix}$$

$$\mathbf{B} = \begin{bmatrix} 0.0428 & -0.0004 & -0.1540 \\ -0.4846 & -0.5154 & -0.0023 \\ -0.1615 & -0.0675 & -0.0053 \\ -0.2020 & -0.2893 & 0.0051 \\ -0.8068 & -0.8522 & -0.0064 \end{bmatrix}$$

$$x_1(k) = \text{velocity, ft/sec}$$
$$x_2(k) = \text{pitch angle, deg}$$
$$x_3(k) = \text{altitude, ft}$$
$$x_4(k) = \text{angle of attack, deg}$$
$$x_5(k) = \text{pitch angular velocity, deg/sec}$$
$$u_1(k) = \text{elevator deflection, deg}$$
$$u_2(k) = \text{flap deflection, deg}$$
$$u_3(k) = \text{throttle position, deg}$$

Formulate a performance index to minimize the errors in states and to minimize the control effort. Obtain optimal controls and states for the system.

@@@@@@@

Chapter 6

Pontryagin Minimum Principle

In previous chapters, we introduced Pontryagin principle in getting optimal condition of a plant or system along with a performance index and boundary conditions. In this chapter, we present Pontryagin Minimum Principle (PMP) and the related topics of dynamic programming and Hamilton-Jacobi-Bellman (HJB) equation. This chapter also bridges the previous chapters dealing with *unconstrained* optimal control with the next chapter focusing on *constrained* optimal control. This chapter is motivated by [79, 89][1].

6.1 *Constrained System*

In this section, we consider a practical limitation on controls and states [79]. Let us reconsider the optimal control system (see Chapter 2, Table 2.1)

$$\dot{\mathbf{x}}(t) = \mathbf{f}(\mathbf{x}(t), \mathbf{u}(t), t) \qquad (6.1.1)$$

where, $\mathbf{x}(t)$ and $\mathbf{u}(t)$ are the $n-$ and $r-$ dimensional state and control *unconstrained* variables respectively, and the performance index

$$J = S(\mathbf{x}(t_f), t_f) + \int_{t_0}^{t_f} V(\mathbf{x}(t), \mathbf{u}(t), t) dt \qquad (6.1.2)$$

---

[1]The permissions given by John Wiley for F. L. Lewis, *Optimal Control*, John Wiley & Sons, Inc., New York, NY, 1986, and Prentice Hall for D. E. Kirk, *Optimal Control Theory: An Introduction*, Prentice Hall, Englewood Cliffs, NJ, 1970, are hereby acknowledged.

with given boundary conditions

$$\mathbf{x}(t=0) = \mathbf{x}_0, \quad \mathbf{x}(t=t_f) = \mathbf{x}_f \text{ is free}, t_f \text{ is free.} \qquad (6.1.3)$$

The important stages in obtaining optimal control for the previous system are

1. the formulation of the Hamiltonian

$$\mathcal{H}(\mathbf{x}(t), \mathbf{u}(t), \boldsymbol{\lambda}(t), t) = V(\mathbf{x}(t), \mathbf{u}(t), t) + \boldsymbol{\lambda}'(t)\mathbf{f}(\mathbf{x}(t), \mathbf{u}(t), t),$$
$$(6.1.4)$$

where, $\boldsymbol{\lambda}(t)$ is the costate variable, and

2. the three relations for control, state and costate as

$$0 = + \left(\frac{\partial \mathcal{H}}{\partial \mathbf{u}}\right)_* \qquad control \text{ relation,} \qquad (6.1.5)$$

$$\dot{\mathbf{x}}^*(t) = + \left(\frac{\partial \mathcal{H}}{\partial \boldsymbol{\lambda}}\right)_* \qquad state \text{ relation, and} \qquad (6.1.6)$$

$$\dot{\boldsymbol{\lambda}}^*(t) = - \left(\frac{\partial \mathcal{H}}{\partial \mathbf{x}}\right)_* \qquad costate \text{ relation} \qquad (6.1.7)$$

to solve for the optimal values $\mathbf{x}^*(t), \mathbf{u}^*(t)$, and $\boldsymbol{\lambda}^*(t)$, respectively, along with the general boundary condition

$$\left[\frac{\partial S}{\partial \mathbf{x}} - \boldsymbol{\lambda}(t)\right]_{*t_f} \delta \mathbf{x}_f + \left[\mathcal{H} + \frac{\partial S}{\partial t}\right]_{*t_f} \delta t_f = 0. \qquad (6.1.8)$$

In the previous problem formulation, we assumed that the control $\mathbf{u}(t)$ and the state $\mathbf{x}(t)$ are *unconstrained*, that is, there are no limitations (restrictions or bounds) on the magnitudes of the control and state variables. But, in reality, the physical systems to be controlled in an optimum manner have some *constraints* on their inputs (controls), internal variables (states) and/or outputs due to considerations mainly regarding safety, cost and other inherent limitations. For example, consider the following cases.

1. In a D.C. motor used in a typical positional control system, the input voltage to field or armature circuit are *limited* to certain standard values, say, 110 or 220 volts. Also, the magnetic flux in the field circuit *saturates* after a certain value of field current.

2. Thrust of a rocket engine used in a space shuttle launch control system *cannot exceed* a certain designed value.

3. Speed of an electric motor used in a typical speed control system *cannot exceed* a certain value without damaging some of the mechanical components such as bearings and shaft.

This optimal control problem for a system with *constraints* was addressed by Pontryagin et al., and the results were enunciated in the celebrated *Pontryagin Minimum Principle* [109]. In their original works, the previous optimization problem was addressed to *maximize* the Hamiltonian

$$\mathcal{H}(\mathbf{x}(t), \mathbf{u}(t), \boldsymbol{\lambda}(t), t) = -V(\mathbf{x}(t), \mathbf{u}(t), t) + \boldsymbol{\lambda}'(t)\mathbf{f}(\mathbf{x}(t), \mathbf{u}(t), t) \quad (6.1.9)$$

which is equivalent to *minimization* of the Hamiltonian defined in (6.1.4). However, here and throughout this book, we define the Hamiltonian as in (6.1.4) and use *Pontryagin Minimum Principle* (PMP) or simply Pontryagin Principle. Before we consider the optimal control system with *control constraints*, a historical perspective is in order regarding Pontryagin Minimum Principle [52, 53].

> *Lev Semyonovich Pontryagin (born September 3, 1908, Moscow, Russia and died May 3, 1988, Moscow, Russia) lost his eyesight when he was about 14 years old due to an explosion. His mother, although does not know mathematics herself, became his tutor by just reading and describing the various mathematical symbols as they appeared to her.*
>
> *L. S. Pontryagin (also known as L. S. by his associates) entered Moscow State University in 1925 and during the 1930s and 1940s, he made significant contributions to topology leading to the publication of Topological Groups which was translated into several languages. Later, as head of the Steklov Mathematical Institute, he devoted to engineering problems of mathematics and soon focused on two major problems of general theory of singularly perturbed systems of ordinary differential equations and the maximum principle in optimal control theory. In particular, in 1955, he formulated a general time-optimal control problem for a fifth-order dynamical system describing optimal maneuvers of an aircraft with bounded control functions. In trying to*

*"invent a new calculus of variations," he spent "three con-
secutive sleepless nights," and came up with the idea of the
Hamiltonian formulation for the problem and the adjoint
differential equations, "thanks to his wonderful geometric
insight."*

*Pontryagin suggested to his former students and close asso-
ciates V. Boltyanski and R. V. Gamkrelidze to join him in
continuing investigations into the optimal control systems.
Pontryagin and Boltyanski focused on controllability and
Gamkrelidze investigated the second variation of the optimal control system which in fact led to the maximum principle. It took nearly a year to complete the full proof of the
maximum principle and published as a short note in [25].
E. F. Mishchenko closely collaborated with L. S. during his
later years and contributed to the development of the theory of singularly perturbed ordinary differential equations
and differential games. The maximum principle was first
presented at the International Congress of Mathematicians
held in Edigburgh, UK, in 1958. The summary of the works
on optimal control was published in English translation in
1962 by this group of mathematicians [109]. Pontryagin
and his associates, for their works, were awarded the Lenin
Prize in 1961.*

6.2   *Pontryagin Minimum Principle*

In Chapter 2, for finding the optimal control $\mathbf{u}^*(t)$ for the problem
described by the plant (6.1.1), performance index (6.1.2), and boundary
conditions (6.1.3), we used *arbitrary variations* in control $\mathbf{u}(t) = \mathbf{u}^*(t) + \delta\mathbf{u}(t)$ to define the *increment* ΔJ and the (first) *variation* δJ in J as

$$\Delta J(\mathbf{u}^*(t), \delta\mathbf{u}(t)) = J(\mathbf{u}(t)) - J(\mathbf{u}^*(t)) \geq 0 \quad \text{for minimum}$$
$$= \delta J(\mathbf{u}^*(t), \delta\mathbf{u}(t)) + \text{higher-order terms} \quad (6.2.1)$$

where, the first variation

$$\delta J = \frac{\partial J}{\partial \mathbf{u}} \delta\mathbf{u}(t). \tag{6.2.2}$$

Also, in Chapter 2, in order to obtain optimal control of *unconstrained*
systems, we applied the *fundamental theorem of calculus of variations*

(Theorem 2.1), i.e., the necessary condition of minimization is that the first *variation* δJ must be zero for an *arbitrary* variation $\delta \mathbf{u}(t)$. But now we place *restrictions* on the control $\mathbf{u}(t)$ such as

$$||\mathbf{u}(t)|| \leq \mathbf{U} \tag{6.2.3}$$

or component wise,

$$|u_j(t)| \leq U_j \longrightarrow U_j^- \leq u_j(t) \leq U_j^+ \tag{6.2.4}$$

where, U_j^- and U_j^+ are the lower and upper bounds or limits on the control function $u_j(t)$. Then, we can *no longer* assume that the control *variation* $\delta \mathbf{u}(t)$ is *arbitrary* for all $t \in [t_0, t_f]$. In other words, the variation $\delta \mathbf{u}(t)$ is not arbitrary if the extremal control $\mathbf{u}^*(t)$ lies on the boundary condition or reaches a limit. If, for example, an extremal control $\mathbf{u}^*(t)$ lies on the boundary during some interval $[t_a, t_b]$ of the entire interval $[t_0, t_f]$, as shown in Figure 6.1(a), then the negative admissible control variation $-\delta \mathbf{u}(t)$ is not allowable as shown in Figure 6.1(b) [79]. The reason for taking $+\delta \mathbf{u}(t)$ and $-\delta \mathbf{u}(t)$ the way it is shown will be apparent later. Then, assuming that all the admissible variations $||\delta \mathbf{u}(t)||$ is small enough that the sign of the increment ΔJ is determined by the sign of the variation δJ, the necessary condition for $\mathbf{u}^*(t)$ to minimize J is that the first variation

$$\delta J(\mathbf{u}^*(t), \delta \mathbf{u}(t)) \geq 0. \tag{6.2.5}$$

Summarizing, the relation for the first variation (6.2.5) is valid if $\mathbf{u}^*(t)$ lies on the boundary (or has a constraint) during any portion of the time interval $[t_0, t_f]$ and the first variation $\delta J = 0$ if $\mathbf{u}^*(t)$ lies within the boundary (or has no constraint) during the entire time interval $[t_0, t_f]$. Next, let us see how the constraint affects the necessary conditions (6.1.5) to (6.1.6) which were derived by using the assumption that the admissible control values $\mathbf{u}(t)$ are *unconstrained*. Using the results of Chapter 2, we have the first variation as

$$\begin{aligned}
\delta J(\mathbf{u}^*(t), \delta \mathbf{u}(t)) = \int_{t_0}^{t_f} & \left\{ \left[\frac{\partial \mathcal{H}}{\partial \mathbf{x}} + \dot{\boldsymbol{\lambda}}(t) \right]_* \delta \mathbf{x}(t) \right. \\
& + \left[\frac{\partial \mathcal{H}}{\partial \mathbf{u}} \right]_*' \delta \mathbf{u}(t) + \left[\frac{\partial \mathcal{H}}{\partial \boldsymbol{\lambda}} - \dot{\mathbf{x}}(t) \right]_*' \delta \boldsymbol{\lambda}(t) \right\} dt \\
& + \left[\frac{\partial S}{\partial \mathbf{x}} - \boldsymbol{\lambda}(t) \right]_{*t_f}' \delta \mathbf{x}_f + \left[\mathcal{H} + \frac{\partial S}{\partial t} \right]_{*t_f} \delta t_f.
\end{aligned}$$
$$\tag{6.2.6}$$

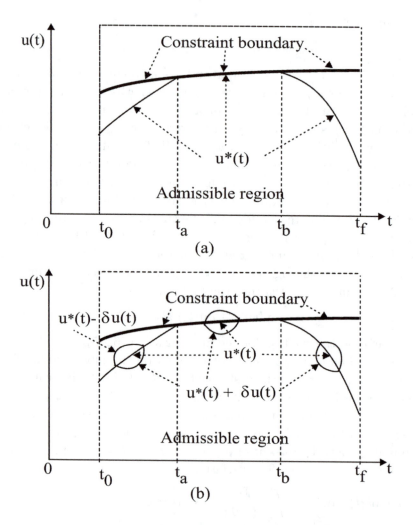

Figure 6.1   (a) An Optimal Control Function Constrained by a Boundary (b) A Control Variation for Which $-\delta u(t)$ Is Not Admissible [79]

In the above,

1. if the optimal state $\mathbf{x}^*(t)$ equations are satisfied, it results in the state relation (6.1.6),

2. if the costate $\boldsymbol{\lambda}^*(t)$ is selected so that the coefficient of the dependent variation $\delta\mathbf{x}(t)$ in the integrand is identically zero, it results in the costate condition (6.1.7), and

3. the boundary condition is selected such that it results in the auxiliary boundary condition (6.1.8).

When the previous items are satisfied, then the first variation (6.2.6) becomes

$$\delta J(\mathbf{u}^*(t), \delta\mathbf{u}(t)) = \int_{t_0}^{t_f} \left[\frac{\partial\mathcal{H}}{\partial\mathbf{u}}\right]' \delta\mathbf{u}(t)dt. \qquad (6.2.7)$$

The integrand in the previous relation is the *first order* approximation to *change* in the Hamiltonian $\mathcal{H}$ due to a *change* in $\mathbf{u}(t)$ alone. This means that by definition

$$\left[\frac{\partial\mathcal{H}}{\partial\mathbf{u}}(\mathbf{x}^*(t), \mathbf{u}^*(t), \boldsymbol{\lambda}^*(t), t)\right]' \delta\mathbf{u}(t) \equiv$$
$$\mathcal{H}(\mathbf{x}^*(t), \mathbf{u}^*(t) + \delta\mathbf{u}(t), \boldsymbol{\lambda}^*(t), t) - \mathcal{H}(\mathbf{x}^*(t), \mathbf{u}^*(t), \boldsymbol{\lambda}^*(t), t). \quad (6.2.8)$$

Then, using (6.2.8) in the first variation (6.2.7), we have

$$\delta J(\mathbf{u}^*(t), \delta\mathbf{u}(t)) = \int_{t_0}^{t_f} [\mathcal{H}(\mathbf{x}^*(t), \mathbf{u}^*(t) + \delta\mathbf{u}(t), \boldsymbol{\lambda}^*(t), t)$$
$$- \mathcal{H}(\mathbf{x}^*(t), \mathbf{u}^*(t), \boldsymbol{\lambda}^*(t), t)] dt. \qquad (6.2.9)$$

Now, using the above, the necessary condition (6.2.5) becomes

$$\int_{t_0}^{t_f} [\mathcal{H}(\mathbf{x}^*(t), \mathbf{u}^*(t) + \delta\mathbf{u}(t), \boldsymbol{\lambda}^*(t), t) - \mathcal{H}(\mathbf{x}^*(t), \mathbf{u}^*(t), \boldsymbol{\lambda}^*(t), t)] dt \geq 0$$
$$(6.2.10)$$

for all admissible $\delta\mathbf{u}(t)$ less than a small value. The relation (6.2.10) becomes

$$\mathcal{H}(\mathbf{x}^*(t), \mathbf{u}^*(t) + \delta\mathbf{u}(t), \boldsymbol{\lambda}^*(t), t) \geq \mathcal{H}(\mathbf{x}^*(t), \mathbf{u}^*(t), \boldsymbol{\lambda}^*(t), t). \quad (6.2.11)$$

Replacing $\mathbf{u}^*(t) + \delta\mathbf{u}(t)$ by $\mathbf{u}(t)$, the *necessary* condition (6.2.10) becomes

$$\mathcal{H}(\mathbf{x}^*(t), \mathbf{u}^*(t), \boldsymbol{\lambda}^*(t), t) \leq \mathcal{H}(\mathbf{x}^*(t), \mathbf{u}(t), \boldsymbol{\lambda}^*(t), t) \qquad (6.2.12)$$

or, in other words,

$$\min_{|\mathbf{u}(t)| \leq \mathbf{U}} \{\mathcal{H}(\mathbf{x}^*(t), \mathbf{u}(t), \boldsymbol{\lambda}^*(t), t)\} = \mathcal{H}(\mathbf{x}^*(t), \mathbf{u}^*(t), \boldsymbol{\lambda}^*(t), t). \qquad (6.2.13)$$

The previous relation, which means that the *necessary condition for the constrained optimal control system is that the optimal control should minimize the Hamiltonian*, is the main contribution of the Pontryagin Minimum Principle. We note that this is only the *necessary condition* and is not in general *sufficient* for optimality.

6.2.1 Summary of Pontryagin Principle

The Pontryagin Principle is now summarized below. Given the plant as

$$\dot{\mathbf{x}}(t) = \mathbf{f}(\mathbf{x}(t), \mathbf{u}(t), t), \qquad (6.2.14)$$

the performance index as

$$J = S(\mathbf{x}(t_f), t_f) + \int_{t_0}^{t_f} V(\mathbf{x}(t), \mathbf{u}(t), t) dt, \qquad (6.2.15)$$

and the boundary conditions as

$$\mathbf{x}(t_0) = \mathbf{x}_0 \text{ and } t_f, \quad \mathbf{x}(t_f) = \mathbf{x}_f \text{ are free}, \qquad (6.2.16)$$

to find the optimal control, form the Pontryagin $\mathcal{H}$ function

$$\mathcal{H}(\mathbf{x}(t), \mathbf{u}(t), \boldsymbol{\lambda}(t), t) = V(\mathbf{x}(t), \mathbf{u}(t), t) + \boldsymbol{\lambda}'(t)\mathbf{f}(\mathbf{x}(t), \mathbf{u}(t), t), \qquad (6.2.17)$$

minimize $\mathcal{H}$ w.r.t. $\mathbf{u}(t) (\leq \mathbf{U})$ as

$$\mathcal{H}(\mathbf{x}^*(t), \mathbf{u}^*(t), \boldsymbol{\lambda}^*(t), t) \leq \mathcal{H}(\mathbf{x}^*(t), \mathbf{u}(t), \boldsymbol{\lambda}^*(t), t), \qquad (6.2.18)$$

and solve the set of 2n state and costate equations

$$\dot{\mathbf{x}}^*(t) = \left(\frac{\partial \mathcal{H}}{\partial \boldsymbol{\lambda}}\right)_* \text{ and } \dot{\boldsymbol{\lambda}}^*(t) = -\left(\frac{\partial \mathcal{H}}{\partial \mathbf{x}}\right)_* \qquad (6.2.19)$$

Table 6.1   Summary of Pontryagin Minimum Principle

A. Statement of the Problem
Given the plant as $\dot{\mathbf{x}}(t) = \mathbf{f}(\mathbf{x}(t), \mathbf{u}(t), t)$, the performance index as $J = S(\mathbf{x}(t_f), t_f) + \int_{t_0}^{t_f} V(\mathbf{x}(t), \mathbf{u}(t), t)dt$, and the boundary conditions as $\mathbf{x}(t_0) = \mathbf{x}_0$ and t_f and $\mathbf{x}(t_f) = \mathbf{x}_f$ are free, find the optimal control.

B. Solution of the Problem	
Step 1	Form the Pontryagin $\mathcal{H}$ function $\mathcal{H}(\mathbf{x}(t), \mathbf{u}(t), \boldsymbol{\lambda}(t), t) = V(\mathbf{x}(t), \mathbf{u}(t), t) + \boldsymbol{\lambda}'(t)\mathbf{f}(\mathbf{x}(t), \mathbf{u}(t), t)$
Step 2	Minimize $\mathcal{H}$ w.r.t. $\mathbf{u}(t)(\le \mathbf{U})$ $\mathcal{H}(\mathbf{x}^*(t), \mathbf{u}^*(t), \boldsymbol{\lambda}^*(t), t) \le \mathcal{H}(\mathbf{x}^*(t), \mathbf{u}(t), \boldsymbol{\lambda}^*(t), t)$
Step 3	Solve the set of 2n state and costate equations $\dot{\mathbf{x}}^*(t) = \left(\frac{\partial \mathcal{H}}{\partial \boldsymbol{\lambda}}\right)_*$ and $\dot{\boldsymbol{\lambda}}^*(t) = -\left(\frac{\partial \mathcal{H}}{\partial \mathbf{x}}\right)_*$ with boundary conditions $\mathbf{x}_0$ and $\left[\mathcal{H} + \frac{\partial S}{\partial t}\right]_{*t_f} \delta t_f + \left[\frac{\partial S}{\partial \mathbf{x}} - \boldsymbol{\lambda}\right]'_{*t_f} \delta\mathbf{x}_f = 0$.

with the boundary conditions $\mathbf{x}_0$ and

$$\left[\mathcal{H} + \frac{\partial S}{\partial t}\right]_{*t_f} \delta t_f + \left[\frac{\partial S}{\partial \mathbf{x}} - \boldsymbol{\lambda}\right]_{*t_f}' \delta \mathbf{x}_f = 0 \qquad (6.2.20)$$

The entire procedure is now summarized in Table 6.1. Note that in Figure 6.1, the variations $+\delta\mathbf{u}(t)$ and $-\delta\mathbf{u}(t)$ are taken in such a way that the negative variation $-\delta\mathbf{u}(t)$ is not admissible and thus we get the condition (6.2.10). On the other hand, by taking the variations $+\delta\mathbf{u}(t)$ and $-\delta\mathbf{u}(t)$ in such a way that the positive variation $+\delta\mathbf{u}(t)$ is not admissible, we get the corresponding condition as

$$\int_{t_0}^{t_f} [\mathcal{H}(\mathbf{x}^*(t), \mathbf{u}^*(t) - \delta\mathbf{u}(t), \boldsymbol{\lambda}^*(t), t) - \mathcal{H}(\mathbf{x}^*(t), \mathbf{u}^*(t), \boldsymbol{\lambda}^*(t), t)] \, dt \geq 0$$

$$(6.2.21)$$

which can again be written as (6.2.11) or (6.2.12). It should be noted that

1. the optimality condition (6.2.12) is valid for both *constrained* and *unconstrained* control systems, whereas the control relation (6.1.5) is valid for *unconstrained* systems only,

2. the results given in the Table 6.1 provide the *necessary* conditions only, and

3. the *sufficient* condition for *unconstrained* control systems is that the second derivative of the Hamiltonian

$$\frac{\partial^2 \mathcal{H}}{\partial \mathbf{u}^2}(\mathbf{x}^*(t), \mathbf{u}^*(t), \boldsymbol{\lambda}^*(t), t) = \left(\frac{\partial \mathcal{H}}{\partial \mathbf{u}^2}\right)_* \qquad (6.2.22)$$

must be *positive definite*.

Let us illustrate the previous principle by a simple example in *static* optimization which is described by *algebraic* equations unlike the *dynamic* optimization described by *differential* equations.

Example 6.1

We are interested in minimizing a scalar function

$$H = u^2 - 6u + 7 \qquad (6.2.23)$$

subject to the constraint relation

$$|u| \leq 2, \longrightarrow -2 \leq u \leq +2. \tag{6.2.24}$$

Solution: First let us use a relation similar to (6.1.5) for *unconstrained* control as

$$\frac{\partial H}{\partial u} = 0 \longrightarrow 2u^* - 6 = 0 \longrightarrow u^* = 3 \tag{6.2.25}$$

and the corresponding optimal H^* from (6.2.23) becomes

$$H^* = 3^2 - 6 \text{x} 3 + 7 = -2. \tag{6.2.26}$$

This value of $u^* = 3$ is certainly outside the constraint (admissible) region specified by (6.2.24). But, using the relation (6.2.18) for the *constrained* control, we have

$$H(u^*) \leq H(u),$$
$$H(u^{*^2} - 6u^* + 7) \leq H(u^2 - 6u + 7). \tag{6.2.27}$$

The complete situation is depicted in Figure 6.2 which shows that the admissible optimal value is $u^* = +2$ and the corresponding optimal H^* is

$$H^* = 2^2 - 6 \text{x} 2 + 7 = -1. \tag{6.2.28}$$

However, let us note if our constraint relation (6.2.24) had been

$$|u| \leq 3, \longrightarrow -3 \leq u \leq +3 \tag{6.2.29}$$

then, we are lucky to use the relation similar to (6.1.5) or (6.2.25) and obtain the optimal value as $u^* = 3$. But, in general this is not true.

6.2.2 *Additional Necessary Conditions*

In their celebrated works [109], Pontryagin and his co-workers also obtained *additional necessary conditions* for constrained optimal control systems. These are stated below without proof [109].

1. If the final time t_f is *fixed* and the Hamiltonian $\mathcal{H}$ does not depend on time t explicitly, then the Hamiltonian $\mathcal{H}$ must be *constant* when evaluated along the optimal trajectory; that is

$$\mathcal{H}(\mathbf{x}^*(t), \mathbf{u}^*(t), \boldsymbol{\lambda}^*(t)) = \text{ constant } = C_1 \ \forall t \in [t_0, t_f]. \tag{6.2.30}$$

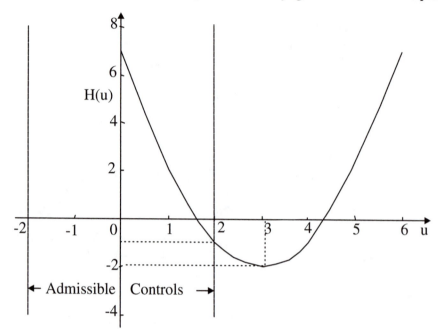

Figure 6.2 Illustration of Constrained (Admissible) Controls

2. If the final time t_f is *free* or *not specified priori* and the Hamiltonian does not depend explicitly on time t, then the Hamiltonian must be identically zero when evaluated along the optimal trajectory; that is,

$$\mathcal{H}(\mathbf{x}^*(t), \mathbf{u}^*(t), \boldsymbol{\lambda}^*(t)) = 0 \quad \forall\, t \in [t_0, t_f] \qquad (6.2.31)$$

Further treatment of *constrained* optimal control systems is carried out in Chapter 7. According to Gregory and Lin [61] the credit for formulating the optimal control problem for the first time in 1950 is given to M. R. Hestenes [64], the detailed proof of the problem was given by a group of Russian mathematicians led by Pontryagin and hence called the Pontryagin Minimum Principle (PMP) [109]. The PMP is the heart of the optimal control theory. However, the original proof given by Pontryagin *et al.* is highly rigorous and lengthy. There are several books devoting lengthy proof of PMP such as Athans and Falb [6], Lee and Markus [86] and Machki and Strauss [97]. Also see recent books (Pinch [108] and Hocking [66]) for a simplified treatment of the proof.

6.3   Dynamic Programming

Given a dynamical process or plant and the corresponding performance
index, there are basically two ways of solving for the optimal control of
the problem, one is the Pontryagin maximum principle [109] and the
other is Bellman's dynamic programming [12, 14, 15]. Here we concen-
trate on the latter, the dynamic programming (DP). The technique is
called dynamic programming because it is a technique based on computer "programming" and suitable for "dynamic" systems. The basic
idea of DP is a discrete, multistage optimization problem in the sense
that at each of the finite set of times, a decision is chosen from a finite
number of decisions based on some optimization criterion. The central
theme of DP is based on a simple intuitive concept called *principle of
optimality*.

6.3.1   Principle of Optimality

Consider a simple multistage decision optimization process shown in
Figure 6.3. Here, let the optimizing cost function for the segment AC

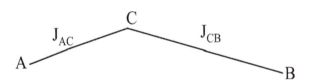

Figure 6.3   Optimal Path from A to B

be J_{AC} and for the segment CB be J_{CB}. Then the optimizing cost for
the entire segment AB is

$$J_{AB} = J_{AC} + J_{CB}. \qquad (6.3.1)$$

That is, if J_{AC} is the optimal cost of the segment AC of the entire optimal path AB, then J_{CB} is the optimal cost of the remaining segment
CB. In other words, one can break the total optimal path into smaller
segments which are themselves optimal. Conversely, if one finds the
optimal values for these smaller segments, then one can obtain the optimal value for the entire path. This obvious looking property is called
the *principle of optimality* (PO) and stated as follows [79]:

> *An optimal policy has the property that whatever the pre-
> vious state and decision (i.e., control), the remaining deci-*

sions must constitute an optimal policy with regard to the
state resulting from the previous decision.

Backward Solution

It looks natural to start working *backward* from the final stage or point,
although one can also work *forward* from the initial stage or point. To illustrate the principle of optimality, let us consider a multistage decision
process as shown in Figure 6.4. This may represent an aircraft routing

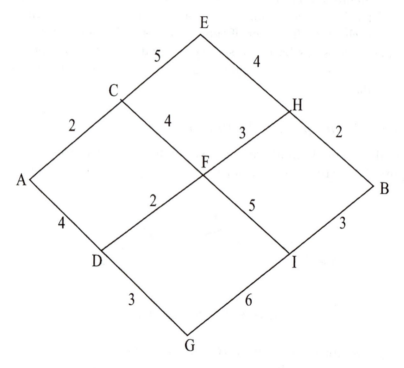

Figure 6.4   A Multistage Decision Process

network or a simple message (telephone) network system. In an aircraft
routing system, both the initial point A and the final point B represent
the two cities to be connected and the other nodes C, D, E, F, G, H, I
represent the intermediate cities. The numbers (called units) over each
segment indicate the cost (or performance index) of flying between the
two cities. Now we are interested in finding the most *economical* route
to fly from city A to city B. We have 5 stages starting from $k = 0$ to
$k = N = 4$. Also, we can associate the current *state* as the junction
or the node. The decision is made at each state. Let the decision or

control be $u = \pm 1$, where $u = +1$ indicates *move up or left* and $u = -1$ indicates *move down or right* looking from each junction towards right.

Now, our working of the dynamic programming algorithm is shown in Figure 6.5.

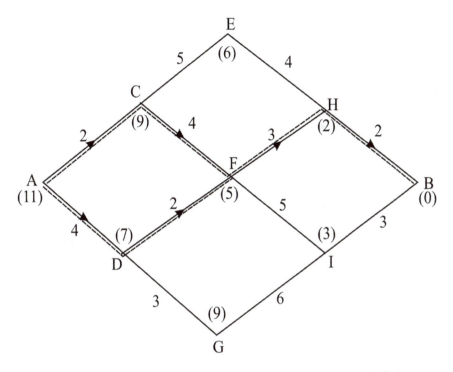

Figure 6.5   A Multistage Decision Process: Backward Solution

Stage 5: $k = k_f = N = 4$

This is just the starting point, there is only one city B and hence there is no cost involved.

Stage 4: $k = 3$

There are two cities H and I at this stage and we need to find the most economical route from this stage to stage 5. Working *backward*, we begin with B which can be reached by H or I. It takes 2 units to fly from H to B by using control or decision $u = -1$ (downward or right) and hence let us place the number 2 within parenthesis under

H. Similarly, it takes 3 units to fly from I to B by using control or decision $u = +1$ (upward or left) and hence place the number 3 just near to I. Let us also place an arrow head to the corresponding paths or routes. Note there is no other way of flying from H to B and I to B except as shown by the arrows.

Stage 3: $k = 2$

Here, there are three cities E, F, G and from these nodes we can fly to H and I. Consider first E. The total cost to fly from E to B will be $2 + 4 = 6$ by using control or decision $u = -1$ (downward or right) and let us place units 6 in parenthesis at the node E. Secondly, from F, we can take two routes F, H, B and F, I, B, by using decisions $u = +1$ (upward or left) and $u = -1$ (downward or right) and the corresponding costs are $2 + 3 = 5$ and $3 + 5 = 8$, respectively. Note, that we placed 5 instead of 8 at the node F and an arrow head on the segment FH to indicate the *optimal* cost to fly the route F, H, B instead of the costlier route F, I, B. Finally, consider G. There is only one route which is G, I, B to go to B starting from G. The cost is the cost to fly from G to I and the cost to fly from I to B.

Stage 2: $k = 1$

By the same procedure as explained above, we see that the node C has minimum cost 9 and the node D has minimum cost 7.

Stage 1: $k = 0$

Here, note that from A, the two segments AC and AD have the same minimum cost indicating either route is economical.

Optimal Solution

This is easy to find, we just follow the route of the arrow heads from A to B. Note that there are two routes to go from stage 0 to stage 1. Thus, the most economical (optimal) route is either A, C, F, H, B or A, D, F, H, B. The total minimum cost is 11 units.

Forward Solution

One can solve the previous system using *forward* solution, starting from
A at stage 0 and working forward to stages 1, 2, 3 and finally to stage
5 to reach *B*. We do get the identical result as in *backward* solution as
shown in Figure 6.6.

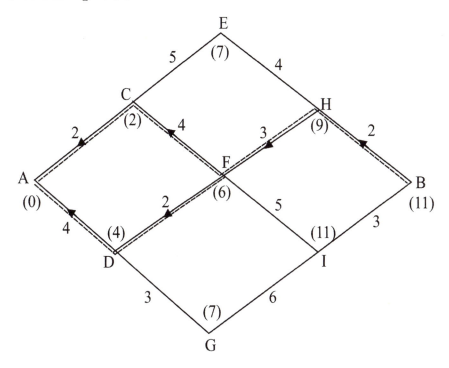

Figure 6.6 A Multistage Decision Process: Forward Solution

Thus, as shown in both the previous cases, we

1. divide the entire route into several stages,

2. find the optimal (economical) route for each stage, and

3. finally, using the principle of optimality, we are able to combine
   the different optimal segments into one single optimal route (or
   trajectory).

In the previous cases, we have fixed both the initial and final points
and thus we have a *fixed-end-point* system. We can similarly address
the *variable end point* system.

Next, we explore how the principle of optimality in the dynamic programming can be used to optimal control systems. We notice that the dynamic programming approach is naturally a discrete-time system. Also, it can be easily applied to either linear or nonlinear systems, whereas the optimal control of a nonlinear system using Pontryagin principle leads to nonlinear two-point boundary value problem (TP-BVP) which is usually very difficult to solve for optimal solutions.

6.3.2 Optimal Control Using Dynamic Programming

Let us first consider the optimal control of a discrete-time system. Or even if there is a continuous-time system, one can easily discretize it to obtain the discrete-time system by using one of the several approaches [82]. Let the plant be described by

$$\mathbf{x}(k + 1) = \mathbf{f}(\mathbf{x}(k), \mathbf{u}(k), k) \tag{6.3.2}$$

and the cost function be

$$J_i(\mathbf{x}(k_i)) = J = S(\mathbf{x}(k_f), k_f) + \sum_{k=i}^{k_f-1} V(\mathbf{x}(k), \mathbf{u}(k)) \tag{6.3.3}$$

where, $\mathbf{x}(k), \mathbf{u}(k)$ are the n and r state and control vectors, respectively. Note, we showed the dependence of J on the initial time (k) and state $(x(k))$.

We are interested in using the principle of optimality to find the optimal control $\mathbf{u}^*(k)$ which applied to the plant (6.3.2) gives optimal state $\mathbf{x}^*(k)$. Let us assume that we evaluated the *optimal* control, state and cost for all values starting from $k + 1$ to k_f. Then, at any time or stage k, we use the principle of optimality to write as

$$J_k^*(\mathbf{x}(k)) = \min_{\mathbf{u}(k)} \left[V[\mathbf{x}(k), \mathbf{u}(k)] + J_{k+1}^*(\mathbf{x}^*(k + 1)) \right]. \tag{6.3.4}$$

The previous relation is the mathematical form of the principle of optimality as applied to optimal control system. It is also called *functional equation of dynamic programming*. Thus, it means that if one had found the optimal control, state and cost from any stage $k + 1$ to the final stage k_f, then one can find the optimal values for a single stage from k to $k + 1$.

Example 6.2

Consider a simple scalar example to illustrate the procedure underlying the dynamic programming method [79, 89].

$$x(k+1) = x(k) + u(k) \tag{6.3.5}$$

and the performance criterion to be optimized as

$$J = \frac{1}{2}x^2(k_f) + \frac{1}{2}\sum_{k=k_0}^{k_f-1}\left[x^2(k) + u^2(k)\right] \tag{6.3.6}$$

where, for simplicity of calculations, we take $k_f = 2$. Let the constraints and the quantization values on the control be

$$-1.0 \le u(k) \le +1.0, \quad k = 0, 1, 2 \quad \text{or}$$
$$u(k) = -1.0, \quad -0.5, \quad 0, \quad +0.5, \quad +1.0 \tag{6.3.7}$$

and on the state be

$$0 \le x(k) \le +2.0, \quad k = 0, 1 \quad \text{or}$$
$$x(k) = 0, \quad 0.5, \quad 1.0 \quad 1.5 \quad 2.0. \tag{6.3.8}$$

Find the optimal control sequence $u^*(k)$ and the state $x^*(k)$ which minimize the performance criterion (6.3.6).

Solution: To use the principle of optimality, to solve the previous system, we first set up a grid between $x(k)$ and k, omitting all the arrows, arrow heads, etc. We divide the stages into two sets: one for $k = 2$ and the other for $k = 1, 0$. We start with $k = 2$ and first find the optimal values and work backward for $k = 1, 0$ using the state (6.3.5), the cost function (6.3.6) and the optimal control (6.3.7).

Stage: $k = 2$

First calculate the state $x(2)$ using the state relation (6.3.5)for all admissible values of $x(k)$ and $u(k)$ given by (6.3.7) and (6.3.8). Thus, for example, for the admissible value of $x(1) = 2.0$ and $u(1) = -1, -0.5, 0, 0.5, 1$, we have

$$x(2) = x(1) + u(1)$$
$$x(2) = 2.0 + (-1) = 1.0$$
$$x(2) = 2.0 + (-0.5) = 1.5$$
$$x(2) = 2.0 + 0 = 2.0$$
$$x(2) = 2.0 + 0.5 = \cancel{2.5}$$
$$x(2) = 1.5 + 1 = \cancel{3.0} \,. \tag{6.3.9}$$

Note the values 2.5 and 3.0 (shown by a striking arrow) of state $x(2)$ are not allowed due to exceeding the state constraint (6.3.8).

Also, corresponding to the functional equation (6.3.4), we have for this example,

$$J_k(x(k)) = \min_{u(k)} \left[\frac{1}{2} u^2(k) + x^2(k) + J^*_{k+1} \right] \qquad (6.3.10)$$

from which we have for the optimal cost at $k = 2$

$$J^*_{k_f} = \frac{1}{2} x^2(2) \qquad (6.3.11)$$

which is evaluated for all admissible values of $x(2)$ as

$$
\begin{aligned}
J^*_{k_f} &= 2.000 & \text{for} & \quad x(2) = 2.0 \\
&= 1.125 & \text{for} & \quad x(2) = 1.5 \\
&= 0.500 & \text{for} & \quad x(2) = 1.0 \\
&= 0.125 & \text{for} & \quad x(2) = 0.5 \\
&= 0.000 & \text{for} & \quad x(2) = 0. \qquad (6.3.12)
\end{aligned}
$$

The entire computations are shown in Table 6.2 for $k = 2$ and in Table 6.3 for $k = 1, 0$.

The data from Tables 6.2 and 6.3 corresponding to optimal conditions is represented in the dynamic programming context in Figure 6.7. Here in this figure, $u^*_0 = u^*(x(0), 0)$ and $u^*_1 = u^*(x(1), 1)$ and the quantities within parenthesis are the optimal cost values at that stage and state. For example, at stage $k = 1$ and state $x(k) = 1.0$, the value $J^*_{12} = 0.75$ indicates that the cost of transfer the state from $x(1)$ to $x(2)$ is 0.75. Thus, in Figure 6.7, for finding the optimal trajectories for any initial state, we simply follow the arrows. For example, to transfer the state $x(0) = 1$ to state $x(2) = 0$, we need to apply first $u^*_0 = -1$ to transfer it to $x(1) = 0$ and then $u^*_1 = 0$.

Note: In the previous example, it so happened that for the given control and state quantization and constraint values (6.3.7) and (6.3.8), respectively, the calculated values using $x(k+1) = x(k) + u(k)$ either exactly coincide with the quantized values or outside the range. In some cases, it may happen that for the given control and state quantization and constraint values, the corresponding values of states may not exactly coincide with the quantized values, in which case, we need to perform some kind of *interpolation* on the values. For example, let us say, the state constraint and quantization is

$$-1 \le x(k) \le +2, \ k = 0, 1 \quad \text{or}$$
$$x(k) = -1.0, \ 0, \ 0.5, \ 1.0 \ 2.0. \qquad (6.3.13)$$

Table 6.2   Computation of Cost during the Last Stage $k = 2$

Current State $x(1)$	Current Control $u(1)$	Next State $x(2)$	Cost J_{12}	Optimal Cost $J_{12}^*(x(1))$	Optimal Control $u^*(x(1), 1)$
	-1.0	1.0	3.0		
	-0.5	1.5	2.25	$J_{12}^*(2.0)=2.25$	$u^*(1.5,1) = -0.5$
2.0	0	2.0	4.0		
	0.5	~~2.5~~			
	1.0	~~3.0~~			
	-1.0	0.5	1.75	$J_{12}^*(1.5)=1.75$	$u^*(1.5,1) = -1.0$
	-0.5	1.0	1.75	$J_{12}^*(1.5)=1.75$	$u^*(1.5,1) = -0.5$
1.5	0	1.5	2.25		
	0.5	2.0	3.25		
	1.0	~~2.5~~			
	-1.0	0	1.0		
	-0.5	0.5	0.75	$J_{12}^*(1.0) = 0.75$	$u^*(1,1) = -0.5$
1.0	0	1.0	1.0		
	0.5	1.5	1.75		
	1.0	2.0	3.0		
	-1.0	~~0.5~~			
	-0.5	0	0.25	$J_{12}^*(0.5)=0.25$	$u^*(0.5,1)=-0.5$
0.5	0	0.5	0.25	$J_{12}^*(0.5)=0.25$	$u^*(0.5,1)=0$
	0.5	1.0	0.75		
	1.0	1.5	1.75		
	-1.0	~~1.0~~			
	-0.5	~~0.5~~			
0	0	0	0	$J_{12}^*(0)=0$	$u^*(0,1)=0$
	0.5	0.5	0.25		
	1.0	1.0	1.0		
Use these to calculate the above: $x(2) = x(1) + u(1)$; $J_{12} = 0.5x^2(2) + 0.5u^2(1) + 0.5x^2(1)$ A strikeout ($\longrightarrow$) indicates the value is not admissible.					

Table 6.3 Computation of Cost during the Stage $k = 1,0$

Current State $x(0)$	Current Control $u(0)$	Next State $x(1)$	Cost J_{02}	Optimal Cost $J_{02}^*(x(0))$	Optimal Control $u^*(x(0),0)$
	-1.0	1.0	3.25	$J_{02}^*(2.0) = 3.25$	$u^*(2.0,0) = -1.0$
	-0.5	1.5	3.875		
2.0	0	2.0	4.25		
	0.5	~~2.5~~			
	1.0	~~3.0~~			
	-1.0	0.5	1.875	$J_{02}^*(1.5) = 1.875$	$u^*(1.5,0) = -1.0$
	-0.5	1.0	2.0		
1.5	0	1.5	2.875		
	0.5	2.0	3.25		
	1.0	~~2.5~~			
	-1.0	0	1.0	$J_{02}^*(1)=1$	$u^*(1,0)=-1.0$
	-0.5	0.5	0.875		
1.0	0	1.0	1.25		
	0.5	1.5	2.375		
	1.0	2.0	3.0		
	-1.0	~~0.5~~			
	-0.5	0	0.25	$J_{02}^*(0.5)=0.25$	$u^*(1,0)=-0.5$
0.5	0	0.5	0.375		
	0.5	1.0	1.0		
	1.0	1.5	2.375		
	-1.0	~~1.0~~			
	-0.5	~~0.5~~			
0	0	0	0	$J_{02}^*(0)=0$	$u^*(0,0) = 0$
	0.5	0.5	0.375		
	1.0	1.0	1.25		
Use these to calculate the above: $x(1) = x(0) + u(0)$; $J_{02} = 0.5u^2(0) + 0.5x^2(0) + J_{12}^*(x(1))$ A strikeout ($\longrightarrow$) indicates the value is not admissible.					

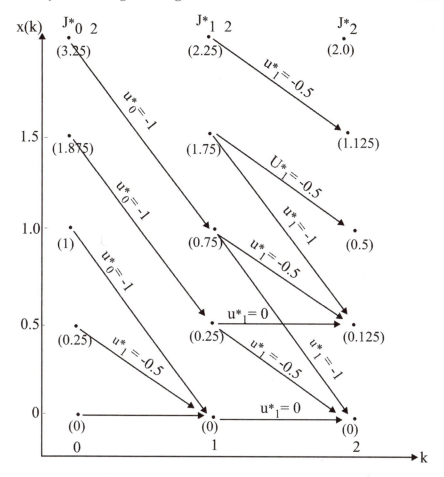

Figure 6.7   Dynamic Programming Framework of Optimal State
Feedback Control

Then, for $x(1) = 2.0$ and $u(1) = -0.5$, when we try to use the state
equation (6.3.5) to find the $x(2) = x(1) + u(1)$, we get $x(2) = 1.5$ which,
although is not an allowable quantized value, is within the constraint
(limit). Hence, we cannot simply calculate the quantity $J_2 = 0.5x^2(2)$
as $J_2 = 0.5(1.5)^2 = 1.125$, instead using the interpolation we calculate
it as

$$J_2 = 0.5[x(2) = 1.5]^2$$
$$= 0.5[x(2) = 1]^2 + \frac{0.5[x(2) = 2]^2 - 0.5[x(2) = 1]^2}{2}$$
$$= 0.5 + \frac{2 - 0.5}{2} = 1.25. \tag{6.3.14}$$

We notice that the dynamic programming technique is a computation-ally intensive method especially with increase in the order and the number of stages of the system. However, with the tremendous ad-vances in high-speed computational tools since Bellman [15] branded this increased computational burden inherent in dynamic programming as "curse of dimensionality," the "curse" may be a "boon" due to the special advantages of dynamic programming in treating both *linear* and *nonlinear* systems with ease and in handling the *constraints* on states and/or controls.

6.3.3   Optimal Control of Discrete-Time Systems

Here, we try to derive the optimal feedback control of a discrete-time system using the principle of optimality of dynamic programming [79, 89]. Consider a linear, time-invariant, discrete-time plant,

$$\mathbf{x}(k+1) = \mathbf{A}\mathbf{x}(k) + \mathbf{B}\mathbf{u}(k) \tag{6.3.15}$$

and the associated performance index

$$J_i = \frac{1}{2}\mathbf{x}'(k_f)\mathbf{F}\mathbf{x}(k_f)$$

$$+ \frac{1}{2}\sum_i^{k_f-1} [\mathbf{x}'(k)\mathbf{Q}\mathbf{x}(k) + \mathbf{u}'(k)\mathbf{R}\mathbf{u}(k)] \tag{6.3.16}$$

where, $\mathbf{x}(k)$ and $\mathbf{u}(k)$ are n and r dimensional state and control vec-tors, and $\mathbf{A}(k)$ and $\mathbf{B}(k)$ are matrices of nxn and nxr dimensions, respectively. Further, $\mathbf{F}$ and $\mathbf{Q}$ are each nxn order symmetric, positive *semidefinite* matrices, and $\mathbf{R}$ is rxr symmetric, positive *definite* matrix. For our present discussion, let us assume that there are no constraints on the state or control.

The problem is to find the optimal control $\mathbf{u}^*(k)$ for $i \leq k \leq k_f$ that minimizes the performance index J_k using the principle of optimality. Let us assume further that the initial state $\mathbf{x}(k_0)$ is *fixed* and the final state $\mathbf{x}(k_f)$ is *free*. In using dynamic programming, we start with the final stage $\mathbf{x}(k_f)$ and work backwards. At each stage, we find the opti-mal control and state. Let us start with last stage $k = k_f$.

Last Stage: $k = k_f$

Let us first note that at $i = k_f$,

$$J_{k_f} = \frac{1}{2}\mathbf{x}'(k_f)\mathbf{F}\mathbf{x}(k_f). \tag{6.3.17}$$

Previous to Last Stage: $k = k_f - 1$

At $i = k_f - 1$, the cost function (6.3.16) becomes

$$J_{k_f-1} = \frac{1}{2}\mathbf{x}'(k_f - 1)\mathbf{Q}\mathbf{x}(k_f - 1) + \frac{1}{2}\mathbf{u}'(k_f - 1)\mathbf{R}\mathbf{u}(k_f - 1)$$
$$+ \frac{1}{2}\mathbf{x}'(k_f)\mathbf{F}\mathbf{x}(k_f). \tag{6.3.18}$$

According to the functional equation of the principle of optimality (6.3.4), we need to find the optimal control $\mathbf{u}^*(k_f - 1)$ to minimize the cost function (6.3.18). Before that, let us rewrite the relation (6.3.18) to make all the terms in (6.3.18) to belong to stage $k_f - 1$. For this, using (6.3.15) in (6.3.18), we have

$$J_{k_f-1} = \frac{1}{2}\mathbf{x}'(k_f - 1)\mathbf{Q}\mathbf{x}(k_f - 1) + \frac{1}{2}\mathbf{u}'(k_f - 1)\mathbf{R}\mathbf{u}(k_f - 1)$$
$$+ \frac{1}{2}\left[\mathbf{A}\mathbf{x}(k_f - 1) + \mathbf{B}\mathbf{u}(k_f - 1)\right]' \mathbf{F}\left[\mathbf{A}\mathbf{x}(k_f - 1) + \mathbf{B}\mathbf{u}(k_f - 1)\right]. \tag{6.3.19}$$

Since there are no constraints on states or controls, we can easily find the minimum value of (6.3.19) w.r.t. $\mathbf{u}(k_f - 1)$ by simply making

$$\frac{\partial J_{k_f-1}}{\partial \mathbf{u}(k_f - 1)} = \mathbf{R}\mathbf{u}^*(k_f - 1) + \mathbf{B}'\mathbf{F}\left[\mathbf{A}\mathbf{x}(k_f - 1) + \mathbf{B}\mathbf{u}^*(k_f - 1)\right] = 0. \tag{6.3.20}$$

Solving for $\mathbf{u}^*(k_f - 1)$, we have

$$\mathbf{u}^*(k_f - 1) = -\left[\mathbf{R} + \mathbf{B}'\mathbf{F}\mathbf{B}\right]^{-1} \mathbf{B}'\mathbf{F}\mathbf{A}\mathbf{x}(k_f - 1)$$
$$= -\mathbf{L}(k_f - 1)\mathbf{x}(k_f - 1) \tag{6.3.21}$$

where,

$$\mathbf{L}(k_f - 1) = \left[\mathbf{R} + \mathbf{B}'\mathbf{F}\mathbf{B}\right]^{-1} \mathbf{B}'\mathbf{F}\mathbf{A} \tag{6.3.22}$$

is also called the *Kalman* gain. Now the optimal cost $J^*_{k_f-1}$ for this stage $k_f - 1$ is found by substituting the optimal control $\mathbf{u}^*(k_f - 1)$ from (6.3.21) into the cost function (6.3.19) to get

$$J^*_{k_f-1} = \frac{1}{2}\mathbf{x}'(k_f - 1)\left[\{\mathbf{A} - \mathbf{BL}(k_f - 1)\}'\,\mathbf{F}\,\{\mathbf{A} - \mathbf{BL}(k_f - 1)\}\right.$$
$$\left. + \mathbf{L}'(k_f - 1)\mathbf{RL}(k_f - 1) + \mathbf{Q}\right]\mathbf{x}(k_f - 1)$$
$$= \frac{1}{2}\mathbf{x}'(k_f - 1)\mathbf{P}(k_f - 1)\mathbf{x}(k_f - 1) \tag{6.3.23}$$

where,

$$\mathbf{P}(k_f - 1) = \{\mathbf{A} - \mathbf{BL}(k_f - 1)\}'\,\mathbf{F}\,\{\mathbf{A} - \mathbf{BL}(k_f - 1)\}$$
$$+ \mathbf{L}'(k_f - 1)\mathbf{RL}(k_f - 1) + \mathbf{Q} \tag{6.3.24}$$

Stage: $k_f - 2$

Using $i = k_f - 2$ in the cost function (6.3.16), we have

$$J_{k_f-2} = \frac{1}{2}\mathbf{x}'(k_f)\mathbf{Fx}(k_f) + \frac{1}{2}\mathbf{x}'(k_f - 2)\mathbf{Qx}(k_f - 2)$$
$$+ \frac{1}{2}\mathbf{u}'(k_f - 2)\mathbf{Ru}(k_f - 2) + \frac{1}{2}\mathbf{x}'(k_f - 1)\mathbf{Qx}(k_f - 1)$$
$$+ \frac{1}{2}\mathbf{u}'(k_f - 1)\mathbf{Ru}(k_f - 1). \tag{6.3.25}$$

Now, using (6.3.15) to replace k_f, (6.3.21) to replace $u(k_f - 1)$ and (6.3.24) in (6.3.25), we get

$$J_{k_f-2} = \frac{1}{2}\mathbf{x}'(k_f - 2)\mathbf{Qx}(k_f - 2) + \frac{1}{2}\mathbf{u}'(k_f - 2)\mathbf{Ru}(k_f - 2)$$
$$+ \frac{1}{2}\mathbf{x}'(k_f - 1)\mathbf{P}(k_f - 1)\mathbf{x}(k_f - 1) \tag{6.3.26}$$

where, $\mathbf{P}(k_f - 1)$ is given by (6.3.24). At this stage, we need to express all functions at stage $k_f - 2$. Then, once again, for this stage, to determine $\mathbf{u}^*(k_f - 2)$ according to the optimality principle (6.3.4), we minimize J_{k_f-2} in (6.3.26) w.r.t. $\mathbf{u}(k_f - 2)$ and get relations similar to (6.3.21), (6.3.22), (6.3.23), and (6.3.24). For example, the optimal cost function becomes

$$J^*_{k_f-2} = \frac{1}{2}\mathbf{x}'(k_f - 2)\mathbf{P}(k_f - 2)\mathbf{x}(k_f - 2) \tag{6.3.27}$$

where, $\mathbf{P}(k_f - 2)$ is obtained similar to (6.3.24) except we replace $k_f - 1$ by $k_f - 2$. We continue this procedure for all other stages $k_f - 3, k_f - 4, ..., k_0$.

Any Stage k

Now we are in a position to generalize the previous set of relations for any k. Thus, the optimal control is given by

$$\mathbf{u}^*(k) = -\mathbf{L}(k)\mathbf{x}^*(k), \tag{6.3.28}$$

where, the Kalman gain $\mathbf{L}(k)$ is given by

$$\mathbf{L}(k) = \left[\mathbf{R} + \mathbf{B}'\mathbf{P}(k+1)\mathbf{B}\right]^{-1}\mathbf{B}'\mathbf{P}(k+1)\mathbf{A}, \tag{6.3.29}$$

the matrix $\mathbf{P}(k)$, also called the Riccati matrix, is the *backward* solution of

$$\begin{aligned}
\mathbf{P}(k) = \left[\mathbf{A} - \mathbf{B}\mathbf{L}(k)\right]' \mathbf{P}(k+1) \left[\mathbf{A} - \mathbf{B}\mathbf{L}(k)\right] \\
+ \mathbf{L}'(k)\mathbf{R}\mathbf{L}(k) + \mathbf{Q}
\end{aligned} \tag{6.3.30}$$

with the final condition $\mathbf{P}(k_f) = \mathbf{F}$, and the optimal cost function as

$$J_k^* = \frac{1}{2}\mathbf{x}^{*\prime}(k)\mathbf{P}(k)\mathbf{x}^*(k). \tag{6.3.31}$$

We notice that these are the same set of relations we obtained in Chapter 5 by using Pontryagin principle.

6.3.4 *Optimal Control of Continuous-Time Systems*

Here, we describe dynamic programming (DP) technique as applied to finding optimal control of continuous-time systems. First of all, we note that although in the previous sections, the DP is explained w.r.t. the discrete-time situation, DP can also be applied to continuous-time systems. However, one can either

1. discretize the continuous-time systems in one or other ways and use the DP as applicable to discrete-time systems, as explained in previous sections, or

2. apply directly the DP to continuous-time systems leading to the celebrated Hamilton-Jacobi-Bellman (HJB) equation, as presented in the next section.

In using the discretization of continuous-time processes, we can either employ

1. the Euler method, or

2. sampler and zero-order hold method.

Let us now briefly discuss these two approaches.

1. *Euler Method:* Let us first take up the Euler approximation of a linear time invariant (LTI) system (although it can be used for nonlinear systems as well) for which the plant is

$$\dot{\mathbf{x}}(t) = \mathbf{A}\mathbf{x}(t) + \mathbf{B}\mathbf{u}(t) \qquad (6.3.32)$$

and the cost function is

$$J(0) = \frac{1}{2}\mathbf{x}'(t_f)\mathbf{F}(t_f)\mathbf{x}(t_f)$$
$$+ \frac{1}{2}\int_0^{t_f} \left[\mathbf{x}'(t)\mathbf{Q}\mathbf{x}(t) + \mathbf{u}'(t)\mathbf{R}\mathbf{u}(t)\right] dt \qquad (6.3.33)$$

where, the state vector $\mathbf{x}(t)$ and control vector $\mathbf{u}(t)$ and the various system and weighted matrices and are defined in the usual manner. Assume some typical boundary conditions for finding the optimal control $\mathbf{u}^*(t)$.

Using the Euler approximation of the derivative in (6.3.32) as

$$\dot{\mathbf{x}}(t) = \frac{\mathbf{x}(k+1) - \mathbf{x}(k)}{T} \qquad (6.3.34)$$

where, T is the discretization (sampling) interval and $\mathbf{x}(k) = \mathbf{x}(kT)$, the discretized version of the state model (6.3.32) becomes

$$\mathbf{x}(k+1) = [\mathbf{I} + T\mathbf{A}]\,\mathbf{x}(k) + T\mathbf{B}\mathbf{u}(k). \qquad (6.3.35)$$

Also, replacing the *integration* process in continuous-time cost function (6.3.33) by the *summation* process, we get

$$J(0) = \frac{1}{2}\mathbf{x}'(k_f)\mathbf{F}\mathbf{x}(t_f)$$
$$+ \frac{1}{2}\sum_{k=k_0}^{k_f-1} \left[\mathbf{x}'(k)\mathbf{Q}_d\mathbf{x}(k) + \mathbf{u}(k)\mathbf{R}_d\mathbf{u}(k)\right] \qquad (6.3.36)$$

where, $\mathbf{Q}_d = T\mathbf{Q}$, and $\mathbf{R}_d = T\mathbf{R}$.

2. *Zero-Order Hold:* Alternatively, using sampler and zero-order hold [83], the continuous-time state model (6.3.32) becomes

$$\mathbf{x}(k+1) = \mathbf{A}_d\mathbf{x}(k) + \mathbf{B}_d\mathbf{u}(k)$$

$$\mathbf{A}_d = e^{\mathbf{A}T}, \text{ and } \mathbf{B}_d = \int_0^T e^{\mathbf{A}\tau}\mathbf{B}d\tau. \qquad (6.3.37)$$

Thus, we have the discrete-time state model (6.3.35) or (6.3.37) and the corresponding discrete-time cost function (6.3.36) for which we can now apply the DP method explained in the previous sections.

6.4   The Hamilton-Jacobi-Bellman Equation

In this section, we present an alternate method of obtaining the *closed-loop* optimal control, using the *principle of optimality* and the *Hamilton-Jacobi-Bellman* (HJB) equation. First we need to state Bellman's principle of optimality [12]. It simply states that *any portion of the optimal trajectory is optimal.* Alternatively, the optimal policy (control) has the property that no matter what the previous decisions (i.e., controls) have been, the remaining decision must constitute an optimal policy. In Chapter 2, we considered the plant as

$$\dot{\mathbf{x}}(t) = \mathbf{f}(\mathbf{x}(t), \mathbf{u}(t), t) \qquad (6.4.1)$$

and the performance index (PI) as

$$J(\mathbf{x}(t_0), t_0) = \int_{t_0}^{t_f} V(\mathbf{x}(t), \mathbf{u}(t), t)dt. \qquad (6.4.2)$$

Now, we provide the alternative approach, called Hamilton-Jacobi-Bellman approach and obtain a control law as a function of the state variables, leading to *closed-loop* optimal control. This is important from the practical point of view in implementation of the optimal control.

Let us define a scalar function $J^*(\mathbf{x}^*(t), t)$ as the *minimum* value of the performance index J for an initial state $\mathbf{x}^*(t)$ at time t, i.e.,

$$J^*(\mathbf{x}^*(t), t) = \int_t^{t_f} V(\mathbf{x}^*(\tau), \mathbf{u}^*(\tau), \tau)d\tau. \qquad (6.4.3)$$

In other words, $J^*(\mathbf{x}^*(t), t)$ is the value of *the performance index when evaluated along the optimal trajectory starting at* $\mathbf{x}(t)$. Here, we used the principle of optimality in saying that the *trajectory from t to t_f is optimal*. However, we are not interested in finding the optimal control for specific initial state $\mathbf{x}(t)$, but for any unspecified initial conditions. Thus, our interest is in $J(\mathbf{x}(t_0), t_0)$ as a function of $\mathbf{x}(t_0)$ and t_0. Now consider

$$
\begin{aligned}
\frac{dJ^*(\mathbf{x}^*(t), t)}{dt} &= \left(\frac{\partial J^*(\mathbf{x}^*(t), t)}{\partial \mathbf{x}^*}\right)' \dot{\mathbf{x}}^*(t) + \frac{\partial J^*(\mathbf{x}^*(t), t)}{\partial t}, \\
&= \left(\frac{\partial J^*(\mathbf{x}^*(t), t)}{\partial \mathbf{x}^*}\right)' \mathbf{f}(\mathbf{x}^*(t), \mathbf{u}^*(t), t) + \frac{\partial J^*(\mathbf{x}^*(t), t)}{\partial t}.
\end{aligned}
$$

$$(6.4.4)$$

From (6.4.3), we have

$$
\frac{dJ^*(\mathbf{x}^*(t), t)}{dt} = -V(\mathbf{x}^*(t), \mathbf{u}^*(t), t).
$$

$$(6.4.5)$$

Using (6.4.4) and (6.4.5), we get

$$
\frac{\partial J^*(\mathbf{x}^*(t), t)}{\partial t} + V(\mathbf{x}^*(t), \mathbf{u}^*(t), t)
$$
$$
+ \left(\frac{\partial J^*(\mathbf{x}^*(t), t)}{\partial \mathbf{x}^*}\right)' \mathbf{f}(\mathbf{x}^*(t), \mathbf{u}^*(t), t) = 0.
$$

$$(6.4.6)$$

Let us introduce the Hamiltonian as

$$
\mathcal{H} = V(\mathbf{x}(t), \mathbf{u}(t), t) + \left(\frac{\partial J^*(\mathbf{x}^*(t), t)}{\partial \mathbf{x}^*}\right)' \mathbf{f}(\mathbf{x}(t), \mathbf{u}(t), t) \quad (6.4.7)
$$

Using (6.4.7) in (6.4.6), we have

$$
\boxed{\frac{\partial J^*(\mathbf{x}^*(t), t)}{\partial t} + \mathcal{H}\left(\mathbf{x}^*(t), \frac{\partial J^*(\mathbf{x}^*(t), t)}{\partial \mathbf{x}^*}, \mathbf{u}^*(t), t\right) = 0; \forall\, t \in [t_0, t_f)}
$$

$$(6.4.8)$$

with boundary condition from (6.4.3) as

$$
J^*(\mathbf{x}^*(t_f), t_f) = 0,
$$

$$(6.4.9)$$

or

$$
J^*(\mathbf{x}^*(t_f), t_f) = S(\mathbf{x}^*(t_f), t_f)
$$

$$(6.4.10)$$

if the original PI (6.4.2) contains a terminal cost function. This equation (6.4.8) is called the *Hamilton-Jacobi equation*. Since this equation is the continuous-time analog of Bellman's recurrence equations in dynamic programming [15], it is also called the Hamilton-Jacobi-Bellman (HJB) equation. Comparing the Hamiltonian function (6.4.7) with that given in earlier chapters, we see that the costate function $\boldsymbol{\lambda}^*(t)$ is given by

$$\boldsymbol{\lambda}^*(t) = \frac{\partial J^*(\mathbf{x}^*(t), t)}{\partial \mathbf{x}^*}. \tag{6.4.11}$$

Also, we know from Chapter 2 that the state and costate are related by

$$\dot{\boldsymbol{\lambda}}^*(t) = -\left(\frac{\partial \mathcal{H}}{\partial \mathbf{x}}\right)_* \tag{6.4.12}$$

and the optimal control $\mathbf{u}^*(t)$ is obtained from

$$\left(\frac{\partial \mathcal{H}}{\partial \mathbf{u}}\right)_* = 0 \longrightarrow \mathbf{u}^*(t) = \mathbf{h}(\mathbf{x}^*(t), J_{\mathbf{x}}^*, t). \tag{6.4.13}$$

Here, comparing (6.4.11) and (6.4.12), we get

$$
\begin{aligned}
\frac{d}{dt}\left(\frac{\partial J^*(\mathbf{x}^*(t), t)}{\partial \mathbf{x}^*}\right) &= \frac{d}{dt}\left[\boldsymbol{\lambda}^*(t)\right] \\
&= -\frac{\partial \mathcal{H}\left(\mathbf{x}^*(t), \frac{\partial J^*(\mathbf{x}^*(t),t)}{\partial \mathbf{x}}, \mathbf{u}^*(t), t\right)}{\partial \mathbf{x}^*}.
\end{aligned}
\tag{6.4.14}
$$

Using

$$J_t^* = \frac{\partial J^*(\mathbf{x}^*(t), t)}{\partial t}; \quad J_{\mathbf{x}}^* = \frac{\partial J^*(\mathbf{x}^*(t), t)}{\partial \mathbf{x}^*} \tag{6.4.15}$$

The HJB equation (6.4.8) becomes

$$\boxed{J_t^* + \mathcal{H}\left(\mathbf{x}^*(t), J_{\mathbf{x}}^*, \mathbf{u}^*(t), t\right) = 0.} \tag{6.4.16}$$

This equation, in general, is a *nonlinear partial differential equation* in J^*, which can be solved for J^*. Once J^* is known, its gradient $J_{\mathbf{x}}^*$ can be calculated and the optimal control $\mathbf{u}^*(t)$ is obtained from (6.4.13). Often, the solution of HJB equation is very difficult. The entire procedure is summarized in Table 6.4. Let us now illustrate the

Table 6.4 Procedure Summary of Hamilton-Jacobi-Bellman (HJB) Approach

	A. Statement of the Problem
	Given the plant as $\dot{\mathbf{x}}(t) = \mathbf{f}(\mathbf{x}(t), \mathbf{u}(t), t),$ the performance index as $J = S(\mathbf{x}(t_f), t_f) + \int_{t_0}^{t_f} V(\mathbf{x}(t), \mathbf{u}(t), t)dt,$ and the boundary conditions as $\mathbf{x}(t_0) = \mathbf{x}_0; \quad \mathbf{x}(t_f)$ is free find the optimal control.
	B. Solution of the Problem
Step 1	Form the Pontryagin $\mathcal{H}$ function $\mathcal{H}(\mathbf{x}(t), \mathbf{u}(t), J_{\mathbf{x}}^*, t) = V(\mathbf{x}(t), \mathbf{u}(t), t) + J_{\mathbf{x}}^{*'}\mathbf{f}(\mathbf{x}(t), \mathbf{u}(t), t).$
Step 2	Minimize $\mathcal{H}$ w.r.t. $\mathbf{u}(t)$ as $\left(\frac{\partial \mathcal{H}}{\partial \mathbf{u}}\right)_* = 0$ and obtain $\mathbf{u}^*(t) = \mathbf{h}(\mathbf{x}^*(t), J_{\mathbf{x}}^*, t).$
Step 3	Using the result of Step 2, find the optimal $\mathcal{H}^*$ function $\mathcal{H}^*(\mathbf{x}^*(t), \mathbf{h}(\mathbf{x}^*(t), J_{\mathbf{x}}^*, t), J_{\mathbf{x}}^*, t) = \mathcal{H}^*(\mathbf{x}^*(t), J_{\mathbf{x}}^*, t)$ and obtain the HJB equation.
Step 4	Solve the HJB equation $J_t^* + \mathcal{H}(\mathbf{x}^*(t), J_{\mathbf{x}}^*, t) = 0.$ with boundary condition $J^*(\mathbf{x}^*(t_f), t_f) = S(\mathbf{x}(t_f), t_f).$
Step 5	Use the solution J^*, from Step 4 to evaluate $J_{\mathbf{x}}^*$ and substitute into the expression for $\mathbf{u}^*(t)$ of Step 2, to obtain the optimal control.

HJB procedure using a simple first-order system.

Example 6.3

Given a first-order system

$$\dot{x}(t) = -2x(t) + u(t) \tag{6.4.17}$$

and the performance index (PI)

$$J = \frac{1}{2}x^2(t_f) + \frac{1}{2}\int_0^{t_f} [x^2(t) + u^2(t)]dt \tag{6.4.18}$$

find the optimal control.

Solution: First of all, comparing the present plant (6.4.17) and the PI (6.4.18) with the general formulation of the plant (6.4.1) and the PI (6.4.2), respectively, we see that

$$V(\mathbf{x}(t), \mathbf{u}(t), t) = \frac{1}{2}u^2(t) + \frac{1}{2}x^2(t); \quad S(\mathbf{x}(t_f), t_f) = \frac{1}{2}x^2(t_f)$$
$$f(\mathbf{x}(t), \mathbf{u}(t), t) = -2x(t) + u(t). \tag{6.4.19}$$

Now we use the procedure summarized in Table 6.4.

- **Step 1:** The Hamiltonian (6.4.7) is

$$\mathcal{H}\left[\mathbf{x}^*(t), J_\mathbf{x}, \mathbf{u}^*(t), t\right] = V(\mathbf{x}(t), \mathbf{u}(t), t) + J_\mathbf{x}\mathbf{f}(\mathbf{x}(t), \mathbf{u}(t), t)$$
$$= \frac{1}{2}u^2(t) + \frac{1}{2}x^2(t) + J_\mathbf{x}(-2x(t) + u(t)). \tag{6.4.20}$$

- **Step 2:** For an unconstrained control, a necessary condition for optimization is

$$\frac{\partial \mathcal{H}}{\partial u} = 0 \longrightarrow u(t) + J_x = 0 \tag{6.4.21}$$

and solving

$$u^*(t) = -J_x. \tag{6.4.22}$$

- **Step 3**: Using the optimal control (6.4.22) and (6.4.20), form the optimal $\mathcal{H}$ function as

$$\mathcal{H} = \frac{1}{2}(-J_x)^2 + \frac{1}{2}x^2(t) + J_x(-2x(t) - J_x)$$
$$= -\frac{1}{2}J_x^2 + \frac{1}{2}x^2(t) - 2x(t)J_x. \tag{6.4.23}$$

Now using the previous relations, the H-J-B equation (6.4.16) becomes

$$J_t - \frac{1}{2}J_x^2 + \frac{1}{2}x^2(t) - 2x(t)J_x = 0 \qquad (6.4.24)$$

with boundary condition (6.4.10) as

$$J(x(t_f), t_f) = S(x(t_f), t_f) = \frac{1}{2}x^2(t_f). \qquad (6.4.25)$$

- **Step 4:** One way to solve the HJB equation (6.4.24) with the boundary condition (6.4.25) is to assume a solution and check if it satisfies the equation. In this simple case, since we want the optimal control (6.4.22) in terms of the states and the PI is a quadratic function of states and controls, we can guess the solution as

$$J(x(t)) = \frac{1}{2}p(t)x^2(t), \qquad (6.4.26)$$

where, $p(t)$, the unknown function to be determined, has the boundary condition as

$$J(x(t_f)) = \frac{1}{2}x^2(t_f) = \frac{1}{2}p(t_f)x^2(t_f), \qquad (6.4.27)$$

which gives us

$$p(t_f) = 1. \qquad (6.4.28)$$

Then using (6.4.26), we get

$$J_x = p(t)x(t); \qquad J_t = \frac{1}{2}\dot{p}(t)x^2(t), \qquad (6.4.29)$$

leading to the closed-loop optimal control (6.4.22), as

$$u^*(t) = -p(t)x^*(t). \qquad (6.4.30)$$

Using the optimal control (6.4.29) into the HJB equation (6.4.24), we have

$$\left(\frac{1}{2}\dot{p}(t) - \frac{1}{2}p^2(t) - 2p(t) + \frac{1}{2}\right)x^{*2}(t) = 0. \qquad (6.4.31)$$

For any $x^*(t)$, the previous relation becomes

$$\frac{1}{2}\dot{p}(t) - \frac{1}{2}p^2(t) - 2p(t) + \frac{1}{2} = 0, \qquad (6.4.32)$$

which upon solving with the boundary condition (6.4.28) becomes

$$p(t) = \frac{(\sqrt{5} - 2) + (\sqrt{5} + 2)\left[\frac{3-\sqrt{5}}{3+\sqrt{5}}\right]e^{2\sqrt{5}(t-t_f)}}{1 - \left[\frac{3-\sqrt{5}}{3+\sqrt{5}}\right]e^{2\sqrt{5}(t-t_f)}}. \qquad (6.4.33)$$

Note, the relation (6.4.32) is the *scalar* version of the matrix DRE (3.2.34) for the finite-time LQR system.

- **Step 5:** Using the relation (6.4.33), we have the closed-loop optimal control (6.4.30).

Note: Let us note that as $t_f \to \infty$, $p(t)$ in (6.4.33) becomes $p(\infty) = \bar{p} = \sqrt{5} - 2$, and the optimal control (6.4.30) is

$$u(t) = -(\sqrt{5} - 2)x(t). \qquad (6.4.34)$$

6.5   *LQR System Using H-J-B Equation*

We employ the H-J-B equation to obtain the closed-loop optimal control of linear quadratic regulator system. Consider the plant described by

$$\dot{\mathbf{x}}(t) = \mathbf{A}(t)\mathbf{x}(t) + \mathbf{B}(t)\mathbf{u}(t) \qquad (6.5.1)$$

where, $\mathbf{x}(t)$ and $\mathbf{u}(t)$ are n and r dimensional state and control vectors respectively, and the performance index to be minimized as

$$J = \frac{1}{2}\mathbf{x}'(t_f)\mathbf{F}\mathbf{x}(t_f)$$
$$+ \frac{1}{2}\int_{t_0}^{t_f} \left[\mathbf{x}'(t)\mathbf{Q}(t)\mathbf{x}(t) + \mathbf{u}'(t)\mathbf{R}(t)\mathbf{u}(t)\right] dt, \qquad (6.5.2)$$

where, as defined earlier, $\mathbf{F}$, and $\mathbf{Q}(t)$ are real, symmetric, positive *semidefinite* matrices respectively, and $\mathbf{R}(t)$ is a real, symmetric, positive *definite* matrix. We will use the procedure given in Table 6.4.

- **Step 1:** As a first step in optimization, let us form the Hamiltonian as

$$\mathcal{H}(\mathbf{x}(t), \mathbf{u}(t), J_{\mathbf{x}}^*, t) = \frac{1}{2}\mathbf{x}'(t)\mathbf{Q}(t)\mathbf{x}(t) + \frac{1}{2}\mathbf{u}'(t)\mathbf{R}(t)\mathbf{u}(t)$$
$$+ J_{\mathbf{x}}^{*'}(\mathbf{x}(t), t)[\mathbf{A}(t)\mathbf{x}(t) + \mathbf{B}(t)\mathbf{u}(t)].$$
$$(6.5.3)$$

- **Step 2:** A necessary condition for optimization of $\mathcal{H}$ w.r.t. $\mathbf{u}(t)$ is that

$$\frac{\partial \mathcal{H}}{\partial \mathbf{u}} = 0 \longrightarrow \mathbf{R}(t)\mathbf{u}(t) + \mathbf{B}'(t)J_{\mathbf{x}}^{*\prime}(\mathbf{x}(t), t) = 0, \qquad (6.5.4)$$

which leads to

$$\mathbf{u}^*(t) = -\mathbf{R}^{-1}(t)\mathbf{B}'(t)J_{\mathbf{x}}^*(\mathbf{x}(t), t). \qquad (6.5.5)$$

Let us note that for the *minimum* control, the sufficient condition that

$$\frac{\partial^2 \mathcal{H}}{\partial \mathbf{u}^2} = \mathbf{R}(t) \qquad (6.5.6)$$

is positive *definite*, is satisfied due to our assumption that $\mathbf{R}(t)$ is symmetric positive *definite*.

- **Step 3:** With optimal control (6.5.5) in the Hamiltonian (6.5.3)

$$\begin{aligned}
\mathcal{H}(\mathbf{x}(t), \mathbf{u}(t), J_{\mathbf{x}}^*, t) &= \frac{1}{2}\mathbf{x}'(t)\mathbf{Q}(t)\mathbf{x}(t) + \frac{1}{2}J_{\mathbf{x}}^{*\prime}\mathbf{B}(t)\mathbf{R}^{-1}(t)\mathbf{B}'(t)J_{\mathbf{x}}^* \\
&\quad + J_{\mathbf{x}}^{*\prime}\mathbf{A}(t)\mathbf{x}(t) - J_{\mathbf{x}}^{*\prime}\mathbf{B}(t)\mathbf{R}^{-1}(t)\mathbf{B}'(t)J_{\mathbf{x}}^* \\
&= \frac{1}{2}\mathbf{x}'(t)\mathbf{Q}(t)\mathbf{x}(t) - \frac{1}{2}J_{\mathbf{x}}^{*\prime}\mathbf{B}(t)\mathbf{R}^{-1}(t)\mathbf{B}'(t)J_{\mathbf{x}}^* \\
&\quad + J_{\mathbf{x}}^{*\prime}\mathbf{A}(t)\mathbf{x}(t). \qquad (6.5.7)
\end{aligned}$$

The HJB equation is

$$J_t^* + \mathcal{H}(\mathbf{x}^*(t), \mathbf{u}^*(t), J_{\mathbf{x}}^*, t) = 0. \qquad (6.5.8)$$

With (6.5.7), the HJB equation (6.5.8) becomes

$$J_t^* + \frac{1}{2}\mathbf{x}^{*\prime}(t)\mathbf{Q}(t)\mathbf{x}^*(t) - \frac{1}{2}J_{\mathbf{x}}^{*\prime}\mathbf{B}(t)\mathbf{R}^{-1}(t)\mathbf{B}'(t)J_{\mathbf{x}}^*$$
$$+ J_{\mathbf{x}}^{*\prime}\mathbf{A}(t)\mathbf{x}^*(t) = 0, \qquad (6.5.9)$$

with boundary condition

$$J^*(\mathbf{x}^*(t_f), t_f) = \frac{1}{2}\mathbf{x}^{*\prime}(t_f)\mathbf{F}(t_f)\mathbf{x}^*(t_f). \qquad (6.5.10)$$

- **Step 4:** Since the performance index J is a *quadratic* function of the state, it seems reasonable to assume a solution as

$$J^*(\mathbf{x}(t), t) = \frac{1}{2}\mathbf{x}'(t)\mathbf{P}(t)\mathbf{x}(t) \qquad (6.5.11)$$

where, $\mathbf{P}(t)$ is a real, symmetric, positive-definite matrix to be determined (for convenience $*$ is omitted for $\mathbf{x}(t)$). With

$$\frac{\partial J^*}{\partial t} = J_t = \frac{1}{2}\mathbf{x}(t)\dot{\mathbf{P}}(t)\mathbf{x}(t),$$

$$\frac{\partial J^*}{\partial \mathbf{x}} = J_\mathbf{x} = \mathbf{P}(t)\mathbf{x}(t) \qquad (6.5.12)$$

and using the performance index (6.5.11) in the HJB equation (6.5.9), we get

$$\frac{1}{2}\mathbf{x}'(t)\dot{\mathbf{P}}(t)\mathbf{x}(t) + \frac{1}{2}\mathbf{x}(t)\mathbf{Q}(t)\mathbf{x}(t)$$

$$- \frac{1}{2}\mathbf{x}'(t)\mathbf{P}(t)\mathbf{B}(t)\mathbf{R}^{-1}(t)\mathbf{B}'(t)\mathbf{P}(t)\mathbf{x}(t)$$

$$+ \mathbf{x}'(t)\mathbf{P}(t)\mathbf{A}(t)\mathbf{x}(t) = 0. \qquad (6.5.13)$$

Expressing $\mathbf{P}(t)\mathbf{A}(t)$ as

$$\mathbf{P}(t)\mathbf{A}(t) = \frac{1}{2}\left[\mathbf{P}(t)\mathbf{A}(t) + \{\mathbf{P}(t)\mathbf{A}(t)\}'\right]$$

$$+ \frac{1}{2}\left[\mathbf{P}(t)\mathbf{A}(t) - \{\mathbf{P}(t)\mathbf{A}(t)\}'\right], \qquad (6.5.14)$$

where, the first term on the right-hand side of the above expression is *symmetric* and the second term is *not symmetric*. Also, we can easily show that since all the terms, except the last term on the right-hand side of (6.5.13), are symmetric. Using (6.5.14) in (6.5.13), we have

$$\frac{1}{2}\mathbf{x}'(t)\dot{\mathbf{P}}(t)\mathbf{x}(t) + \frac{1}{2}\mathbf{x}(t)\mathbf{Q}(t)\mathbf{x}(t)$$

$$- \frac{1}{2}\mathbf{x}'(t)\mathbf{P}(t)\mathbf{B}(t)\mathbf{R}^{-1}(t)\mathbf{B}'(t)\mathbf{P}(t)\mathbf{x}(t)$$

$$+ \frac{1}{2}\mathbf{x}'(t)\mathbf{P}(t)\mathbf{A}(t)\mathbf{x}(t) + \frac{1}{2}\mathbf{x}'(t)\mathbf{A}'(t)\mathbf{P}(t)\mathbf{x}(t) = 0. \qquad (6.5.15)$$

This equation should be valid for *any* $\mathbf{x}(t)$, which then reduces to

$$\dot{\mathbf{P}}(t) + \mathbf{Q}(t) - \mathbf{P}(t)\mathbf{B}(t)\mathbf{R}^{-1}(t)\mathbf{B}'(t)\mathbf{P}(t)$$
$$+\mathbf{P}(t)\mathbf{A}(t) + \mathbf{A}'(t)\mathbf{P}(t) = 0.$$

(6.5.16)

Rewriting the above, we have the matrix differential Riccati equation (DRE) as

$$\boxed{\dot{\mathbf{P}}(t) = -\mathbf{P}(t)\mathbf{A}(t) - \mathbf{A}'(t)\mathbf{P}(t) + \mathbf{P}(t)\mathbf{B}(t)\mathbf{R}^{-1}(t)\mathbf{B}'(t)\mathbf{P}(t) - \mathbf{Q}(t).}$$

(6.5.17)

Using (6.5.10) and (6.5.11),

$$\frac{1}{2}\mathbf{x}'(t_f)\mathbf{P}(t_f)\mathbf{x}(t_f) = \frac{1}{2}\mathbf{x}'(t_f)\mathbf{F}(t_f)\mathbf{x}(t_f), \qquad (6.5.18)$$

we have the final condition for $\mathbf{P}(t)$ as

$$\boxed{\mathbf{P}(t_f) = \mathbf{F}(t_f).} \qquad (6.5.19)$$

- **Step 5:** Using (6.5.5) and (6.5.12), we have the closed-loop optimal control as

$$\mathbf{u}^*(t) = -\mathbf{R}^{-1}(t)\mathbf{B}'(t)\mathbf{P}(t)\mathbf{x}^*(t). \qquad (6.5.20)$$

Some noteworthy features of this result follow.

1. The HJB partial differential equation (6.5.8) reduces to a nonlinear, matrix, differential equation (6.5.17).

2. The matrix $\mathbf{P}(t)$ is determined by numerically integrating *backward* from t_f to t_0. We also note that since the nxn $\mathbf{P}(t)$ matrix is symmetric, one need to solve only $n(n+1)/2$ instead of nxn equations.

3. The reason for assuming the solution of the form (6.5.11) is that we are able to obtain a closed-loop optimal control, which is linear, and time-varying w.r.t. the state.

4. A *necessary* condition: The result that has been obtained is only the necessary condition for optimality in the sense that the minimum cost function $J^*(\mathbf{x}(t), t)$ must satisfy the HJB equation.

5. A *sufficient* condition: If there exists a cost function $J^s(\mathbf{x}(t), t)$ which satisfies the HJB equation, then $J^s(\mathbf{x}(t), t)$ is the minimum cost function, i.e.,

$$J^s(\mathbf{x}(t), t) = J^*(\mathbf{x}(t), t). \qquad (6.5.21)$$

6. Solution of the *nonlinear* HJB equation: For the *linear*, time-varying plant with quadratic performance index, we are able to *guess* the solution to the *nonlinear* HJB equation. In general, we may not be able to easily find the solution, and the nonlinear HJB equation needs to be solved by numerical techniques.

7. Applications of HJB equation: The HJB equation is useful in optimal control systems. Also, this provides a bridge between dynamic programming approach and optimal control.

We provide another example with infinite-time interval for the application of HJB approach.

Example 6.4

Find the closed-loop optimal control for the first-order system

$$\dot{x}(t) = -2x(t) + u(t) \qquad (6.5.22)$$

with the performance index

$$J = \int_0^\infty \left[x^2(t) + u^2(t) \right] dt. \qquad (6.5.23)$$

Hint: Assume that $J^* = fx^2(t)$.

Solution: First of all, let us identify the various functions as

$$V(x(t), u(t)) = x^2(t) + u^2(t),$$
$$f(x(t), u(t)) = -2x(t) + u(t). \qquad (6.5.24)$$

We now follow the step-by-step procedure given in Table 6.4.

• **Step 1:** Form the $\mathcal{H}$ function as

$$\mathcal{H}(x(t), u(t), J_x^*) = V(x(t), u(t)) + J_x^* f(x(t), u(t))$$
$$= x^2(t) + u^2(t) + 2fx(t) \left[-2x(t) + u(t) \right]$$
$$= x^2(t) + u^2(t) - 4fx^2(t) + 2fx(t)u(t) \quad (6.5.25)$$

where, we used $J^* = fx^2(t)$ and $J_x^* = 2fx(t)$. Here, we use a slightly different approach by using the value of J_x^* in the beginning itself.

- **Step 2:** Minimize $\mathcal{H}$ w.r.t. u to obtain optimal control $u^*(t)$ as

$$\frac{\partial \mathcal{H}}{\partial u} = 2u^*(t) + 2fx^*(t) = 0 \longrightarrow u^*(t) = -fx^*(t). \quad (6.5.26)$$

Step 3: Using the result of Step 2 in Step 1, find the optimal $\mathcal{H}$ as

$$\mathcal{H}^*(x^*(t), J_x^*, t) = x^{*^2}(t) - 4fx^{*^2}(t) - f^2 x^{*^2}(t). \quad (6.5.27)$$

- **Step 4:** Solve the HJB equation

$$\mathcal{H}^*(x^*(t), J_x^*) + J_t^* = 0 \longrightarrow$$
$$x^{*^2}(t) - 4fx^{*^2}(t) - f^2 x^{*^2}(t) = 0. \quad (6.5.28)$$

Note that $J_t = 0$ in the previous HJB equation. For any $x^*(t)$, the previous equation becomes

$$f^2 + 4f - 1 = 0 \rightarrow f = -2 \pm \sqrt{5}. \quad (6.5.29)$$

Taking the positive value of f in (6.5.29), we get

$$J^* = fx^{*^2}(t) = (-2 + \sqrt{5})x^{*^2}(t). \quad (6.5.30)$$

Note that (6.5.29) is the scalar version of the matrix ARE (3.5.15) for the infinite-time interval regulator system.

- **Step 5:** Using the value of f from Step 4, in Step 2, we get the optimal control as

$$u^*(t) = -fx^*(t) = -(\sqrt{5} - 2)x^*(t). \quad (6.5.31)$$

6.6   *Notes and Discussion*

In this chapter, we discussed two topics: dynamic programming and HJB equation. The dynamic programming was developed by Bellman during the 1960s as an optimization tool to be adapted with the then coming up of digital computers. An excellent account of dynamic programming and optimal control is given recently by Bertsekas [18, 19], where the two-volume textbook develops in depth dynamic programming, a central algorithmic method for optimal control, sequential decision making under uncertainty, and combinatorial optimization.

Problems

1. Make reasonable assumptions wherever necessary.

2. Use MATLAB© wherever possible to solve the problems and plot all the optimal controls and states for all problems. Provide the relevant MATLAB© m files.

Problem 6.1 Prove the Pontryagin Minimum Principle based on the works of Athans and Falb [6], Lee and Markus [86], Machki and Strauss [97] and some of the recent works Pinch [108] and Hocking [66].

Problem 6.2 For the general case of the Example 6.2, develop a MATLAB© based program.

Problem 6.3 For a traveling salesperson, find out the cheapest route from city L to city N if the total costs between the intermediate cities are shown in Figure 6.8.

Problem 6.4 Consider a scalar example

$$x(k+1) = x(k) + u(k) \tag{6.6.1}$$

and the performance criterion to be optimized as

$$J = \frac{1}{2}x^2(k_f) + \frac{1}{2}\sum_{k=k_0}^{k_f-1} u^2(k)$$
$$= \frac{1}{2}x^2(k_f) + \frac{1}{2}u^2(0) + \frac{1}{2}u^2(1)$$

where, for simplicity of calculations we take $k_f = 2$. Let the constraints on the control be

$$-1.0 \le u(k) \le +1.0, \ k = 0, 1, 2 \quad \text{or}$$
$$u(k) = -1.0, \ -0.5, \ 0, \ +0.5, \ +1.0$$

and on the state be

$$0.0 \le x(k) \le +1.0, \ k = 0, 1 \quad \text{or}$$
$$x(k) = 0.0, \ 0.5, \ 1.0, \ 1.5.$$

Find the optimal control sequence $u^*(k)$ and the state $x^*(k)$ which minimize the performance criterion.

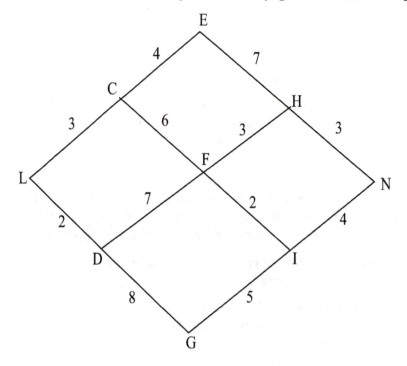

Figure 6.8   Optimal Path from A to B

Problem 6.5 Find the Hamilton-Jacobi-Bellman equation for the system

$$\dot{x}_1(t) = x_2(t)$$
$$\dot{x}_2(t) = -2x_2(t) - 3x_1^2(t) + u(t)$$

with the performance index as

$$J = \frac{1}{2} \int_0^{t_f} \left(x_1^2(t) + u^2(t) \right) dt.$$

Problem 6.6 Solve the Example 5.3 using dynamic programming approach.

Problem 6.7 For the D.C. motor speed control system described in Problem 1.1, find the HJB equation and hence the closed-loop optimal control to keep the speed at a constant value.

Problem 6.8 For the liquid-level control system described in Problem 1.2, find the HJB equation and hence the closed-loop optimal control to keep the liquid level constant at a particular value.

Problem 6.9 For the mechanical control system described in Problem 1.4, find the HJB equation and hence the closed-loop optimal control to keep the states at a constant value.

Problem 6.10 For the automobile suspension system described in Problem 1.5, find the HJB equation and hence the closed-loop control.

@@@@@@@@@@@@@@

Chapter 7

Constrained Optimal Control Systems

In the previous chapters, we considered optimization of systems without any constraints on *controls* or *state variables*. In this chapter, we present an entirely different class of systems where we impose some constraints on controls and/or states. In this way, we address the *time-optimal control* (TOC) system, where the performance measure is the minimization of the transition time from initial state to any target or desired state. Our treatment is focused on linear, time-invariant (LTI) systems. These are also called *brachistochrone* problems. Next, we address *fuel-optimal control* (FOC) system, where the performance measure is minimization of a quantity proportional to fuel consumed by the process or plant. Next, we briefly consider the energy-optimal control (EOC) system. Finally, we consider a plant with some constraints on their states. It is suggested that the student reviews the material in Appendices A and B given at the end of the book. This chapter is based on [6, 79][1].

7.1 Constrained Optimal Control

From Chapter 6 (Table 6.1) the Pontryagin Principle is now summarized below for linear, time-invariant system with a quadratic perfor-

---

[1]The permission given by McGraw-Hill for M. Athans and P. L. Falb, *Optimal Control: An Introduction to The Theory and Its Applications*, McGraw-Hill Book Company, New York, NY, 1966, is hereby acknowledged.

mance index. Given the system as

$$\dot{\mathbf{x}}(t) = \mathbf{A}(t)\mathbf{x}(t) + \mathbf{B}(t)\mathbf{u}(t) \tag{7.1.1}$$

with the control constraint as

$$U^- \leq \mathbf{u}(t) \leq U^+ \quad \longrightarrow \quad |\mathbf{u}(t)| \leq \mathbf{U} \tag{7.1.2}$$

the performance index as

$$
\begin{aligned}
J(\mathbf{x}(t_0), \mathbf{u}(t), t_0) &= J \\
&= \frac{1}{2}\mathbf{x}'(t_f)\mathbf{F}(t_f)\mathbf{x}(t_f) \\
&\quad + \frac{1}{2}\int_{t_0}^{t_f} \left[\mathbf{x}'(t)\mathbf{Q}(t)\mathbf{x}(t) + \mathbf{u}'(t)\mathbf{R}(t)\mathbf{u}(t)\right] dt
\end{aligned} \tag{7.1.3}
$$

and the boundary conditions as

$$\mathbf{x}(t_0) = \mathbf{x}_0 \text{ fixed}, \mathbf{x}(t_f) = \mathbf{x}_f \text{ is free and } t_f \text{ is free}, \tag{7.1.4}$$

to find the optimal control, form the Pontryagin $\mathcal{H}$ function

$$
\begin{aligned}
\mathcal{H}(\mathbf{x}(t), \mathbf{u}(t), \boldsymbol{\lambda}(t), t) &= \frac{1}{2}\mathbf{x}'(t)\mathbf{Q}(t)\mathbf{x}(t) + \frac{1}{2}\mathbf{u}'(t)\mathbf{R}(t)\mathbf{u}(t) \\
&\quad + \boldsymbol{\lambda}'(t)\left[\mathbf{A}(t)\mathbf{x}(t) + \mathbf{B}(t)\mathbf{u}(t)\right]
\end{aligned} \tag{7.1.5}
$$

minimize $\mathcal{H}$ w.r.t. $\mathbf{u}(t)(\leq \mathbf{U})$ as

$$\mathcal{H}(\mathbf{x}^*(t), \mathbf{u}^*(t), \boldsymbol{\lambda}^*(t), t) \leq \mathcal{H}(\mathbf{x}^*(t), \mathbf{u}(t), \boldsymbol{\lambda}^*(t), t), \tag{7.1.6}$$

and solve the set of 2n state and costate differential equations

$$
\begin{aligned}
\dot{\mathbf{x}}^*(t) &= + \left(\frac{\partial \mathcal{H}}{\partial \boldsymbol{\lambda}}\right)_*, \\
\dot{\boldsymbol{\lambda}}^*(t) &= - \left(\frac{\partial \mathcal{H}}{\partial \mathbf{x}}\right)_*
\end{aligned} \tag{7.1.7}
$$

with the boundary conditions $\mathbf{x}_0$ and

$$\left[\mathcal{H} + \frac{\partial S}{\partial t}\right]_{*t_f} \delta t_f + \left[\left(\frac{\partial S}{\partial \mathbf{x}}\right) - \lambda(t)\right]'_{*t_f} \delta \mathbf{x}_f = 0 \tag{7.1.8}$$

where,

$$S[\mathbf{x}(t_f), t_f)] = \frac{1}{2}\mathbf{x}'(t_f)\mathbf{F}(t_f)\mathbf{x}(t_f).$$

Note: Here we address the optimal control system with a *constraint* on the control $\mathbf{u}(t)$ given by (7.1.2). Thus, we cannot *in general* use the condition

$$\left(\frac{\partial \mathcal{H}}{\partial \mathbf{u}}\right)_* = 0 \qquad (7.1.9)$$

that we used earlier in Chapters 2 to 4 for continuous-time systems, where we had no constraints on the control $\mathbf{u}(t)$, because there is no guarantee that in general the optimal control $\mathbf{u}^*(t)$ obtained by using the condition (7.1.9) will satisfy the constraint on the control given by (7.1.2).

7.1.1 Time-Optimal Control of LTI System

In this section, we address the problem of minimizing the time taken for the system to go from an initial state to the desired final state of a linear, time-invariant (LTI) system. The desired final state can be conveniently taken as the origin of the state space; in this way we will be dealing with time-optimal *regulator* system.

7.1.2 Problem Formulation and Statement

Let us now present a typical time-optimal control (TOC) system. Consider a linear, time-invariant dynamical system

$$\dot{\mathbf{x}}(t) = \mathbf{A}\mathbf{x}(t) + \mathbf{B}\mathbf{u}(t) \qquad (7.1.10)$$

where, $\mathbf{x}(t)$ is nth state vector; $\mathbf{u}(t)$ is rth control vector, and the matrices $\mathbf{A}$ and $\mathbf{B}$ are constant matrices of nxn and nxr dimensions, respectively. We are also given that

1. the system (7.1.10) is *completely controllable*, that is, the matrix

$$\mathbf{G} = \begin{bmatrix} \mathbf{B} & \vdots & \mathbf{A}\mathbf{B} & \vdots & \mathbf{A}^2\mathbf{B} & \vdots & \cdots & \vdots & \mathbf{A}^{n-1}\mathbf{B} \end{bmatrix} \qquad (7.1.11)$$

   is of rank n or the matrix $\mathbf{G}$ is nonsingular, and

2. the magnitude of the control $\mathbf{u}(t)$ is constrained as

$$U^- \leq \mathbf{u}(t) \leq U^+ \quad \longrightarrow \quad |\mathbf{u}(t)| \leq \mathbf{U} \qquad (7.1.12)$$

or component wise

$$|u_j(t)| \leq U_j, \quad j = 1, 2, \cdots, r. \tag{7.1.13}$$

Here, U^+ and U^- are the upper and lower bounds of $\mathbf{U}$. But, the constraint relation (7.1.12) can also be written more conveniently (by absorbing the magnitude $\mathbf{U}$ into the matrix $\mathbf{B}$) as

$$-1 \leq \mathbf{u}(t) \leq +1 \quad \longrightarrow \quad |\mathbf{u}(t)| \leq 1 \tag{7.1.14}$$

or component wise,

$$|u_j(t)| \leq 1, \tag{7.1.15}$$

3. the initial state is $\mathbf{x}(t_0)$ and the final (target) state is $\mathbf{0}$.

The problem statement is: *Find the (optimal) control $\mathbf{u}^*(t)$ which satisfies the constraint (7.1.15) and drives the system (7.1.10) from the initial state $\mathbf{x}(t_0)$ to the origin $\mathbf{0}$ in* **minimum time**.

7.1.3 Solution of the TOC System

We develop the solution to this time-optimal control (TOC) system stated previously under the following steps. First let us list all the steps here and then discuss the same in detail.

- **Step 1:** *Performance Index*

- **Step 2:** *Hamiltonian*

- **Step 3:** *State and Costate Equations*

- **Step 4:** *Optimal Condition*

- **Step 5:** *Optimal Control*

- **Step 6:** *Types of Time-Optimal Controls*

- **Step 7:** *Bang-Bang Control Law*

- **Step 8:** *Conditions for Normal Time Optimal Control System*

- **Step 9:** *Uniqueness of Optimal Control*

- **Step 10:** *Number of Switchings*

- **Step 1:** *Performance Index:* For the minimum-time system formulation specified by (7.1.10) and by the control constraint relation (7.1.14), the performance index (PI) becomes

$$J(\mathbf{u}(t)) = \int_{t_0}^{t_f} V\left[\mathbf{x}(t), \mathbf{u}(t), t\right] dt = \int_{t_0}^{t_f} 1\, dt = t_f - t_0 \quad (7.1.16)$$

where, t_0 is fixed and t_f is *free*. If the final time t_f is *fixed*, trying to minimize a fixed quantity makes no sense.

- **Step 2:** *Hamiltonian:* We form the Hamiltonian $\mathcal{H}$ for the problem described by the system (7.1.10) and the PI (7.1.16) as

$$\mathcal{H}(\mathbf{x}(t), \boldsymbol{\lambda}(t), \mathbf{u}(t)) = 1 + \boldsymbol{\lambda}'(t)\left[\mathbf{A}\mathbf{x}(t) + \mathbf{B}\mathbf{u}(t)\right],$$
$$= 1 + \left[\mathbf{A}\mathbf{x}(t)\right]'\boldsymbol{\lambda}(t) + \mathbf{u}'(t)\mathbf{B}'\boldsymbol{\lambda}(t) \quad (7.1.17)$$

where, $\boldsymbol{\lambda}(t)$ is the costate variable.

- **Step 3:** *State and Costate Equations:* Let us assume the optimal values $\mathbf{u}^*(t)$, $\mathbf{x}^*(t)$, and $\boldsymbol{\lambda}^*(t)$. Then, the state $\mathbf{x}^*(t)$ and the costate $\boldsymbol{\lambda}^*(t)$ are given by

$$\dot{\mathbf{x}}^*(t) = +\left(\frac{\partial \mathcal{H}}{\partial \boldsymbol{\lambda}}\right)_* = \mathbf{A}\mathbf{x}^*(t) + \mathbf{B}\mathbf{u}^*(t), \quad (7.1.18)$$

$$\dot{\boldsymbol{\lambda}}^*(t) = -\left(\frac{\partial \mathcal{H}}{\partial \mathbf{x}}\right)_* = -\mathbf{A}'\boldsymbol{\lambda}^*(t) \quad (7.1.19)$$

with the boundary conditions

$$\mathbf{x}^*(t_0) = \mathbf{x}(t_0); \quad \mathbf{x}^*(t_f) = \mathbf{0} \quad (7.1.20)$$

where, we again note that t_f is free.

- **Step 4:** *Optimal Condition:* Now using Pontryagin Principle, we invoke the condition (7.1.6) for optimal control in terms of the Hamiltonian. Using (7.1.17) in (7.1.6), we have

$$1 + \left[\mathbf{A}\mathbf{x}^*(t)\right]'\boldsymbol{\lambda}^*(t) + \mathbf{u}^{*'}(t)\mathbf{B}'\boldsymbol{\lambda}^*(t)$$
$$\leq 1 + \left[\mathbf{A}\mathbf{x}^*(t)\right]'\boldsymbol{\lambda}^*(t) + \mathbf{u}'(t)\mathbf{B}'\boldsymbol{\lambda}^*(t) \quad (7.1.21)$$

which can be simplified to

$$\mathbf{u}^{*'}(t)\mathbf{B}'\boldsymbol{\lambda}^*(t) \leq \mathbf{u}'(t)\mathbf{B}'\boldsymbol{\lambda}^*(t),$$
$$\mathbf{u}^{*'}(t)\mathbf{q}^*(t) \leq \mathbf{u}'(t)\mathbf{q}^*(t),$$
$$= \min_{|\mathbf{u}(t)| \leq 1} \left\{\mathbf{u}'(t)\mathbf{q}^*(t)\right\} \quad (7.1.22)$$

where $\mathbf{q}^*(t) = \mathbf{B}'\boldsymbol{\lambda}^*(t)$, and $\mathbf{q}^*(t)$ is not to be confused as the vector version of the weighting matrix $\mathbf{Q}$ used in quadratic performance measures.

- **Step 5:** *Optimal Control*: We now derive the optimal sequence for $\mathbf{u}^*(t)$. From the optimal condition (7.1.21)

    1. if $\mathbf{q}^*(t)$ is *positive*, the optimal control $\mathbf{u}^*(t)$ must be the *smallest* admissible control value -1 so that

    $$\min_{|\mathbf{u}(t)|\leq 1} \left\{\mathbf{u}'(t)\mathbf{q}^*(t)\right\} = -\mathbf{q}^*(t) = -|\mathbf{q}^*(t)|, \quad (7.1.23)$$

    2. and on the other hand, if $\mathbf{q}^*(t)$ is *negative*, the optimal control $\mathbf{u}^*(t)$ must be the *largest* admissible value $+1$ so that

    $$\min_{|\mathbf{u}(t)|\leq 1} \left\{\mathbf{u}'(t)\mathbf{q}^*(t)\right\} = +\mathbf{q}^*(t) = -|\mathbf{q}^*(t)|. \quad (7.1.24)$$

In other words, the previous two relations can be written in a compact form (for either $\mathbf{q}^*(t)$ is positive or negative) as

$$\min_{|\mathbf{u}(t)|\leq 1} \left\{\mathbf{u}'(t)\mathbf{q}^*(t)\right\} = -|\mathbf{q}^*(t)|. \quad (7.1.25)$$

Also, the combination of (7.1.23) and (7.1.24) means that

$$\mathbf{u}^*(t) = \begin{cases} +1 & \text{if } \mathbf{q}^*(t) < 0, \\ -1 & \text{if } \mathbf{q}^*(t) > 0, \\ \text{indeterminate} & \text{if } \mathbf{q}^*(t) = 0. \end{cases} \quad (7.1.26)$$

Now, using the *signum* function (see Figure 7.1) defined between input f_i and output f_0, written as $f_o = sgn\{f_i\}$ as

$$f_0 = \begin{cases} +1 & \text{if } f_i > 0 \\ -1 & \text{if } f_i < 0 \\ \text{indeterminate} & \text{if } f_i = 0. \end{cases}$$

The engineering realization of the *signum* function is an *ideal relay*.

Then, we can write the control algorithm (7.1.26) in a compact form as

$$\boxed{\mathbf{u}^*(t) = -SGN\{\mathbf{q}^*(t)\}} \quad (7.1.27)$$

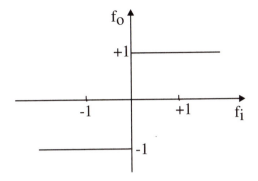

Figure 7.1   Signum Function

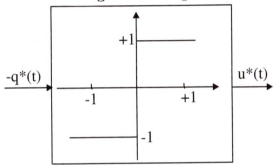

Figure 7.2   Time-Optimal Control

where the relation between the time-optimal control $\mathbf{u}^*(t)$ and the function $\mathbf{q}^*(t)$ is shown in Figure 7.2.

In terms of component wise,

$$
\begin{aligned}
u_j^*(t) &= -sgn\{q_j^*(t)\}\\
&= -sgn\{\mathbf{b}_j'\boldsymbol{\lambda}^*(t)\}
\end{aligned}
\tag{7.1.28}
$$

where, $\mathbf{b}_j, j = 1, 2, \ldots, r$ denote the column vectors of the input matrix $\mathbf{B}$. From the time-optimal control relation (7.1.27), note that the optimal control $\mathbf{u}^*(t)$ depends on the *costate* function $\boldsymbol{\lambda}^*(t)$.

- **Step 6:** *Types of Time-Optimal Controls*: We now have two types of time-optimal controls, depending upon the nature of the function $\mathbf{q}^*(t)$.

  1. *Normal Time-Optimal Control* (NTOC) System: Suppose that during the interval $[t_0, t_f^*]$, there exists a set of times

$t_1, t_2, \ldots, t_{\gamma j} \in [t_0, t_f], \quad \gamma = 1, 2, 3, \ldots, j = 1, 2, \ldots, r$ such that

$$q_j^*(t) = \mathbf{b}_j' \boldsymbol{\lambda}^*(t) = \begin{cases} 0, & \text{if and only if } t = t_{\gamma j} \\ \text{nonzero}, & \text{otherwise}, \end{cases}$$

$$(7.1.29)$$

then we have a *normal time-optimal control* (NTOC) system. The situation is depicted in Figure 7.3. Here, the

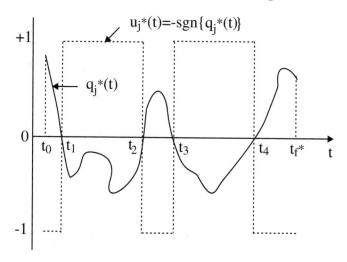

Figure 7.3   Normal Time-Optimal Control System

function $q_j^*(t)$ is zero only at four instants of time, and the time optimal control is *piecewise constant function* with simple *switchings* at t_1, t_2, t_3, and t_4. Thus, the optimal control $u_j^*(t)$ switches *four* times, or the number of switchings is four.

2. *Singular Time-Optimal Control (STOC) System:* Suppose that during the interval $[t_0, t_f^*]$, there is one (or more) subintervals $[T_1, T_2]$, such that

$$q_j^*(t) = 0 \quad \forall \, t \in [T_1, T_2] \qquad (7.1.30)$$

then, we have a *singular time-optimal control (STOC) system*, and the interval $[T_1, T_2]$ is called *singularity intervals*. The situation is shown in Figure 7.4. During this singularity intervals, the time-optimal control is not defined.

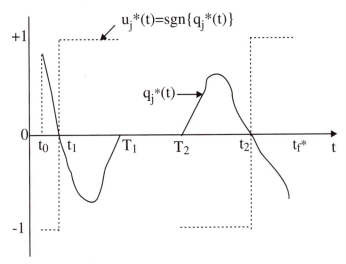

Figure 7.4 Singular Time-Optimal Control System

- **Step 7:** *Bang-Bang Control Law:* For a *normal* time-optimal system, the optimal control, given by (7.1.27)

$$\boxed{\mathbf{u}^*(t) = -SGN\{\mathbf{q}^*(t)\} = -SGN\{\mathbf{B}'\boldsymbol{\lambda}^*(t)\}} \quad (7.1.31)$$

for all $t \in [t_0, t_f^*]$, is a piecewise constant function of time (i.e., *bang-bang*).

- **Step 8:** *Conditions for NTOC System:* Here, we derive the conditions necessary for the system to be not *singular*, thereby obtaining the conditions for the system to be *normal*. First of all, the solution of the costate equation (7.1.19) is

$$\boldsymbol{\lambda}^*(t) = \epsilon^{-\mathbf{A}'t}\boldsymbol{\lambda}^*(0) \quad (7.1.32)$$

and assume that the costate initial condition $\boldsymbol{\lambda}^*(0)$ must be a nonzero vector. With this solution for $\boldsymbol{\lambda}^*(t)$, the control law (7.1.31) becomes

$$\mathbf{u}^*(t) = -SGN\{\mathbf{B}'\epsilon^{-\mathbf{A}'t}\boldsymbol{\lambda}^*(0)\} \quad (7.1.33)$$

or component wise,

$$u_j^*(t) = -sgn\{q_j^*(t)\},$$
$$= -sgn\left\{\mathbf{b}_j'\epsilon^{-\mathbf{A}'t}\boldsymbol{\lambda}^*(0)\right\}. \quad (7.1.34)$$

Let us suppose that there is an interval of time $[T_1, T_2]$ during which the function $\mathbf{q}^*(t)$ is zero. Then, it follows that during the time interval $[T_1, T_2]$ all the derivatives of $\mathbf{q}^*(t)$ must be zero. That is

$$q_j^*(t) = \mathbf{b}_j' \epsilon^{-\mathbf{A}'t} \boldsymbol{\lambda}^*(0) = 0$$
$$\dot{q}^*(t) = \mathbf{b}_j' \mathbf{A}' \epsilon^{-\mathbf{A}'t} \boldsymbol{\lambda}^*(0) = 0$$
$$\ddot{q}^*(t) = \mathbf{b}_j' \mathbf{A}'^2 \epsilon^{-\mathbf{A}'t} \boldsymbol{\lambda}^*(0) = 0$$
$$\cdots\cdots\cdots\cdots$$
$$q^{(n-1)*}(t) = \mathbf{b}_j' \mathbf{A}'^{(n-1)} \epsilon^{-\mathbf{A}'t} \boldsymbol{\lambda}^*(0) = 0 \qquad (7.1.35)$$

which in turn can be written in a compact form as

$$\mathbf{G}_j' \epsilon^{-\mathbf{A}'t} \boldsymbol{\lambda}^*(0) = 0 \qquad (7.1.36)$$

where,

$$\mathbf{G}_j = \left[\mathbf{b}_j \ \vdots \ \mathbf{A}\mathbf{b}_j \ \vdots \ \mathbf{A}^2\mathbf{b}_j \ \vdots \ \cdots \vdots \ \mathbf{A}^{n-1}\mathbf{b}_j \right]$$
$$= \left[\mathbf{B} \ \vdots \ \mathbf{A}\mathbf{B} \ \vdots \ \mathbf{A}^2\mathbf{B} \ \vdots \cdots \vdots \ \mathbf{A}^{n-1}\mathbf{B} \right]. \qquad (7.1.37)$$

In the condition (7.1.36), we know that $\epsilon^{-\mathbf{A}'t}$ is *nonsingular*, and $\boldsymbol{\lambda}^*(0) \neq 0$, and hence the matrix $\mathbf{G}_j$ must be *singular*. Hence, for the STOC system, $\mathbf{G}_j$ must be *singular*. Or for the NTOC system, $\mathbf{G}_j$ must be *nonsingular*. We know that the matrix $\mathbf{G}_j$ is *nonsingular* if and only if the original system (7.1.10) is completely *controllable*. This leads us to say that the time-optimal control system is *normal* if the matrix $\mathbf{G}_j$ is *nonsingular* or if the system is *completely controllable*. These results are stated as follows (the proofs are found in books such as [6]).

THEOREM 7.1

The necessary and sufficient conditions for the time-optimal control system to be normal is that the matrix $\mathbf{G}_j, j = 1, 2, ..., r$, is nonsingular or that the system is completely controllable.

THEOREM 7.2

The necessary and sufficient conditions for the time-optimal control system to be singular is that the matrix $\mathbf{G}_j, j = 1, 2, ..., r$, is singular or that the system is uncontrollable.

Thus, for a *singular* interval to exist, it is necessary that the system is *uncontrollable*, conversely, if the system is completely *controllable*, a *singular* interval cannot exist.

- **Step 9:** *Uniqueness of Optimal Control:* If the time-optimal system is *normal*, then the time-optimal control is *unique*.

- **Step 10:** *Number of Switchings*: The result is again stated in the form of a theorem.

THEOREM 7.3

If the original system (7.1.10) is normal, and if all the n eigenvalues of the system are real, then the optimal control $\mathbf{u}^*(t)$ can switch (from $+1$ to -1 or from -1 to $+1$) at most $(n-1)$ times.

7.1.4   Structure of Time-Optimal Control System

We examine two natural structures, i.e., open-loop and closed-loop structures for implementation of time-optimal control system.

1. *Open-Loop Structure*: We repeat here again the time-optimal control system and summarize the result. For the *normal time-optimal control system*, where the system is described by

$$\dot{\mathbf{x}}(t) = \mathbf{A}\mathbf{x}(t) + \mathbf{B}\mathbf{u}(t) \tag{7.1.38}$$

with the constraint on the control as

$$|u_j(t)| \leq 1, \quad j = 1, 2, \ldots, r. \tag{7.1.39}$$

the time-optimal control is to find the control which drives the system (7.1.38) from any initial condition $\mathbf{x}(0)$ to target condition $\mathbf{0}$ in *minimum time* under the constraint (7.1.39). From the

previous discussion, we know that the optimal control is given
by

$$u_j^*(t) = -sgn\{b_j', \boldsymbol{\lambda}^*(t)\} \tag{7.1.40}$$

where, the costate function $\boldsymbol{\lambda}^*(t)$ is

$$\boldsymbol{\lambda}^*(t) = \epsilon^{-\mathbf{A}'t}\boldsymbol{\lambda}^*(0). \tag{7.1.41}$$

Let us note that the initial condition $\boldsymbol{\lambda}^*(0)$ is not specified and
hence arbitrary, and hence we have to adopt an iterative proce-
dure. Thus, the steps involved in obtaining the optimal control
are given as follows.

(a) Assume a value for the initial condition $\boldsymbol{\lambda}^*(0)$.

(b) Using the initial value in (7.1.41), compute the costate $\boldsymbol{\lambda}^*(t)$.

(c) Using the costate $\boldsymbol{\lambda}^*(t)$, evaluate the control (7.1.40).

(d) Using the control $\mathbf{u}^*(t)$, solve the system relation (7.1.38).

(e) Monitor the solution $\mathbf{x}^*(t)$ and find if there is a time t_f
such that the system goes to zero, i.e., $\mathbf{x}(t_f) = 0$. Then
the corresponding control computed previously is the time-
optimal control. If not, then change the initial value of $\boldsymbol{\lambda}^*(0)$
and repeat the previous steps until $\mathbf{x}(t_f) = 0$.

A schematic diagram showing the open-loop, time-optimal control structure is shown in Figure 7.5. The relay shown in the

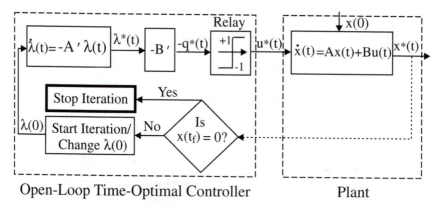

Open-Loop Time-Optimal Controller   Plant

Figure 7.5   Open-Loop Structure for Time-Optimal Control System

figure is an engineering realization of *signum function*. It gives
the required control sequences $+1$ or -1 depending on its input.
However, we note the following:

(a) The adjoint system (7.1.19) has *unstable* modes for a stable
system (7.1.10). This makes the already an iterative procedure much more tedious.

(b) We all know the obvious disadvantages of the open-loop implementation of a control system.

One should try for *closed-loop* implementation of the time-optimal
control system, which is discussed next.

2. *Closed-Loop Structure*: Intuitively, we can feel the relation between the control $\mathbf{u}^*(t)$ and the state $\mathbf{x}^*(t)$ recalling the results
of Chapters 3 and 4 where we used Riccati transformation $\boldsymbol{\lambda}^*(t) = \mathbf{P}(t)\mathbf{x}^*(t)$ to express the optimal control $\mathbf{u}^*(t)$, which was a function of the costate $\boldsymbol{\lambda}^*(t)$, as a function of the state $\mathbf{x}^*(t)$. Thus, we
assume that at any time there is a time-optimal control $\mathbf{u}^*(t)$ as a
function of the state $\mathbf{x}^*(t)$. That is, there is a switching function
$\mathbf{h}(\mathbf{x}^*(t))$ such that

$$\boxed{\mathbf{u}^*(t) = -SGN\{\mathbf{h}(\mathbf{x}^*(t))\}} \qquad (7.1.42)$$

where an analytical and/or computational algorithm

$$\mathbf{h}(\mathbf{x}^*(t)) = \mathbf{B}'\boldsymbol{\lambda}^*(\mathbf{x}^*(t)). \qquad (7.1.43)$$

needs to be developed as shown in the example to follow. Then,
the optimal control law (7.1.42) is implemented as shown in Figure 7.6. The relay implements the optimal control depending on
its input which in turn is decided by the feedback of the states.
The determination of the switching functions $\mathbf{h}[\mathbf{x}^*(t)]$ is the important aspect of the implementation of the control law. In the
next section, we demonstrate the way we try to obtain the closed-
loop structure for time-optimal control system of a second order
(double integral) system.

7.2   *TOC of a Double Integral System*

Here we examine the time-optimal control (TOC) of a classical double integral system. This simple example demonstrates some of the
important features of the TOC system [6].

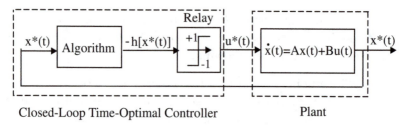

Closed-Loop Time-Optimal Controller   Plant

Figure 7.6   Closed-Loop Structure for Time-Optimal Control System

7.2.1   *Problem Formulation and Statement*

Consider a simple motion of an inertial load in a frictionless environment. The motion is described by

$$m\ddot{y}(t) = f(t) \tag{7.2.1}$$

where, m is the mass of a body (system or plant), $y(t), \dot{y}(t)$, and $\ddot{y}(t)$ are the position, velocity and acceleration, respectively, and $f(t)$ is the external force applied to the system. Defining a set of state variables as

$$x_1(t) = y(t); \quad x_2(t) = \dot{y}(t) \tag{7.2.2}$$

we have the double integral system described as

$$\dot{x}_1(t) = x_2(t)$$
$$\dot{x}_2(t) = u(t) \tag{7.2.3}$$

where, $u(t) = f(t)/m$. Let us assume that the control (input) $u(t)$ to the system is constrained as

$$|u(t)| \leq 1 \; \forall t \in [t_0, t_f]. \tag{7.2.4}$$

This constraint on the control is due to physical limitations such as current in a circuit or thrust of an engine.

Problem Statement: Given the double integral system (7.2.3) and the constraint on the control (7.2.4), find the admissible control that forces the system from any initial state $[x_1(0), x_2(0)]$ to the origin in *minimum time*.

Let us assume that we are dealing with *normal* system and no *singular* controls are allowed. Now, we attempt to solve the system following the procedure described in the previous section.

7.2.2   *Problem Solution*

Our problem solution consists of the list of the following steps with the details following.

- **Step 1:** *Performance Index*

- **Step 2:** *Hamiltonian*

- **Step 3:** *Minimization of Hamiltonian*

- **Step 4:** *Costate Solutions*

- **Step 5:** *Time-Optimal Control Sequences*

- **Step 6:** *State Trajectories*

- **Step 7:** *Switch Curve*

- **Step 8:** *Phase Plane Regions*

- **Step 9:** *Control Law*

- **Step 10:** *Minimum Time*

- **Step 1:** *Performance Index*: For minimum-time system, the performance index (7.1.16) is easily seen to be

$$J = \int_{t_0}^{t_f} 1 \, dt = t_f - t_0 \qquad (7.2.5)$$

where, t_0 is fixed and t_f is *free*.

- **Step 2:** *Hamiltonian*: From the system (7.2.3) and the PI (7.2.5), form the Hamiltonian (7.1.17) as

$$\mathcal{H}(\mathbf{x}(t), \boldsymbol{\lambda}(t), u(t)) = 1 + \lambda_1(t)x_2(t) + \lambda_2(t)u(t). \qquad (7.2.6)$$

- **Step 3:** *Minimization of Hamiltonian*: According to the Pontryagin Principle, we need to minimize the Hamiltonian as

$$\mathcal{H}(\mathbf{x}^*(t), \boldsymbol{\lambda}^*(t), u^*(t)) \leq \mathcal{H}(\mathbf{x}^*(t), \boldsymbol{\lambda}^*(t), u(t),)$$
$$= \min_{|u| \leq 1} \mathcal{H}(\mathbf{x}^*(t), \boldsymbol{\lambda}^*(t), u(t)). \qquad (7.2.7)$$

Using the Hamiltonian (7.2.6) in the condition (7.2.7), we have

$$1 + \lambda_1^*(t)x_2^*(t) + \lambda_2^*(t)u^*(t)$$
$$\leq 1 + \lambda_1^*(t)x_2^*(t) + \lambda_2^*(t)u(t) \qquad (7.2.8)$$

which leads to

$$\lambda_2^*(t)u^*(t) \leq \lambda_2^*(t)u(t). \qquad (7.2.9)$$

Using the result of the previous section, we have the optimal control (7.1.27) given in terms of the *signum* function as

$$u^*(t) = -sgn\{\lambda_2^*(t)\}. \qquad (7.2.10)$$

Now to know the nature of the optimal control, we need to solve for the costate function $\lambda_2^*(t)$.

- **Step 4:** *Costate Solutions*: The costate equations (7.1.19) along with the Hamiltonian (7.2.6) are

$$\dot{\lambda}_1^*(t) = -\frac{\partial \mathcal{H}}{\partial x_1^*} = 0,$$

$$\dot{\lambda}_2^*(t) = -\frac{\partial \mathcal{H}}{\partial x_2^*} = -\lambda_1^*(t). \qquad (7.2.11)$$

Solving the previous equations, we get the costates as

$$\lambda_1^*(t) = \lambda_1^*(0),$$
$$\lambda_2^*(t) = \lambda_2^*(0) - \lambda_1(0)t. \qquad (7.2.12)$$

- **Step 5:** *Time-Optimal Control Sequences*: From the solutions of the costates (7.2.12), we see that $\lambda_2^*(t)$ is a straight line, and that there are four possible (assuming initial conditions $\lambda_1(0)$ and $\lambda_2(0)$ to be nonzero) solutions as shown in Figure 7.7. Also shown are the four possible *optimal* control sequences

$$\{+1\}, \quad \{-1\}, \quad \{+1, -1\}, \quad \{-1, +1\} \qquad (7.2.13)$$

that satisfy the optimal control relation (7.2.10). Let us reiterate that the admissible *optimal* control sequences are the ones given by (7.2.13). That is, a control sequence like $\{+1, -1, +1\}$ is not an optimal control sequence. Also, the control sequence

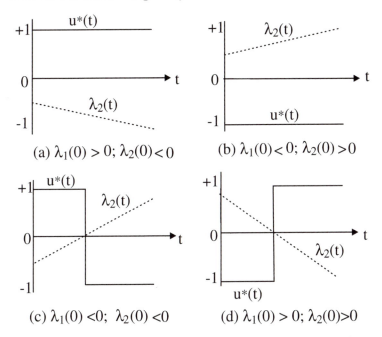

(a) $\lambda_1(0) > 0$; $\lambda_2(0) < 0$

(b) $\lambda_1(0) < 0$; $\lambda_2(0) > 0$

(c) $\lambda_1(0) < 0$; $\lambda_2(0) < 0$

(d) $\lambda_1(0) > 0$; $\lambda_2(0) > 0$

Figure 7.7   Possible Costates and the Corresponding Controls

$\{+1, -1, +1\}$ requires two switchings which is in violation of the earlier result (Theorem 7.3) that a second (nth) order system will have *at most* 1 $(n-1)$ switchings. From Figure 7.7, we see that the time-optimal control for the second order (double integral) system is a piecewise constant and can switch at most *once*. In order to arrive at *closed-loop* realization of the optimal control, we need to find the phase $(x_1(t), x_2(t))$ plane (state) trajectories.

- **Step 6:** *State Trajectories*: Solving the state equations (7.2.3), we have

$$x_1^*(t) = x_1^*(0) + x_2^*(0)t + \frac{1}{2}Ut^2,$$
$$x_2^*(t) = x_2^*(0) + Ut, \qquad (7.2.14)$$

where, $U = u^*(t) = \pm 1$. For phase plane plots, we need to eliminate t from solutions (7.2.14) for the states. Thus, (for simplicity, we omit * since we are now dealing with all optimal functions only and write $x_1(0) = x_{10}, x_2(0) = x_{20}$)

$$t = (x_2(t) - x_{20})/U,$$
$$x_1(t) = x_{10} - \frac{1}{2}Ux_{20}^2 + \frac{1}{2}Ux_2^2(t), \qquad (7.2.15)$$

where, we used $U = \pm 1 = 1/U$. If

$$u = U = +1, \quad \begin{cases} t & = x_2(t) - x_{20} \\ x_1(t) = x_{10} - \frac{1}{2}x_{20}^2 + \frac{1}{2}x_2^2(t) = C_1 + \frac{1}{2}x_2^2(t) \end{cases}$$

$$(7.2.16)$$

and if

$$u = U = -1, \quad \begin{cases} t & = x_{20} - x_2(t), \\ x_1(t) = x_{10} + \frac{1}{2}x_{20}^2 - \frac{1}{2}x_2^2(t) = C_2 - \frac{1}{2}x_2^2(t) \end{cases}$$

$$(7.2.17)$$

where, $C_1 = x_{10} - \frac{1}{2}x_{20}^2$ and $C_2 = x_{10} + \frac{1}{2}x_{20}^2$ are constants. Now, we can easily see that the relations (7.2.16) and (7.2.17) represent a family of *parabolas* in (x_1, x_2) plane (or phase plane) as shown in Figure 7.8. The arrow indicates the direction of motion for increasing (positive) time. Our aim is to drive the system from

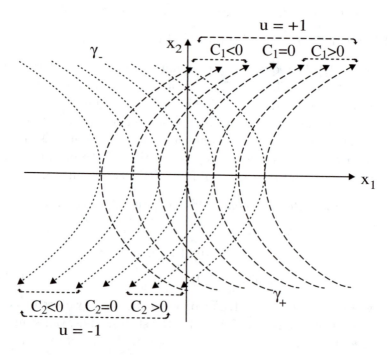

Figure 7.8 Phase Plane Trajectories for $u = +1$ (dashed lines) and $u = -1$ (dotted lines)

any initial state $(x_1(0), x_2(0))$ to origin $(0,0)$ in minimum time.

Then, from (7.2.15), we find that at $t = t_f$

$$x_1(t = t_f) = 0; \qquad x_2(t = t_f) = 0. \qquad (7.2.18)$$

With this, (7.2.15) becomes

$$0 = x_{10} - \frac{1}{2}Ux_{20}^2 + 0 \longrightarrow x_{10} = \frac{1}{2}Ux_{20}^2. \qquad (7.2.19)$$

Now, rewriting this for any initial state $x_1 = x_{10}, x_2 = x_{20}$, we have

$$x_1 = \frac{1}{2}Ux_2^2. \qquad (7.2.20)$$

Note that x_1 and x_2 in (7.2.19) are any initial states and not to be confused with $x_1(t)$ and $x_2(t)$ in (7.2.15) which are the states at any time t.

Now we can restate our problem as to find the time-optimal control sequence to drive the system from any initial state (x_1, x_2) to the origin $(0, 0)$ in minimum time.

- **Step 7:** *Switch Curve:* From Figure 7.8, we see that there are two curves labeled γ_+ and γ_- which transfer any initial state (x_1, x_2) to the origin $(0, 0)$.

  1. *The γ_+ curve is the locus of all (initial) points (x_1, x_2) which can be transferred to the final point $(0, 0)$ by the control $u = +1$. That is*

  $$\gamma_+ = \left\{ (x_1, x_2) : \quad x_1 = \frac{1}{2}x_2^2, \quad x_2 \le 0 \right\}. \qquad (7.2.21)$$

  2. *The γ_- curve is the locus of all (initial) points (x_1, x_2) which can be transferred to the final point $(0, 0)$ by the control $u = -1$. That is*

  $$\gamma_- = \left\{ (x_1, x_2) : \quad x_1 = -\frac{1}{2}x_2^2, \quad x_2 \ge 0 \right\}. \qquad (7.2.22)$$

  3. *The complete switch curve, i.e., the γ curve, is defined as the union (either or) of the partial switch curves γ_+ and γ_-. That is*

  $$\gamma = \left\{ (x_1, x_2) : \quad x_1 = -\frac{1}{2}x_2 |x_2| \right\},$$

  $$= \gamma_+ \cup \gamma_- \qquad (7.2.23)$$

  where, $\cup$ means the union operation.

The switch curve γ is shown in Figure 7.9

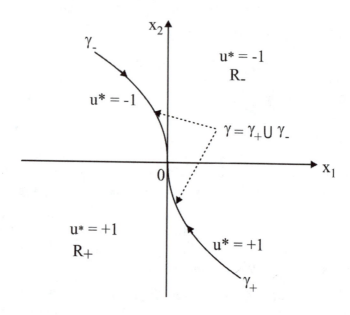

Figure 7.9   Switch Curve for Double Integral Time-Optimal Control System

- **Step 8:** *Phase Plane Regions:* Let us now define the regions in which we need to apply the control $u = +1$ or $u = -1$.

  1. *Let R_+ be the region of the points such that*

$$R_+ = \left\{ (x_1, x_2): \quad x_1 < -\frac{1}{2} x_2 \, |x_2| \right\}. \qquad (7.2.24)$$

  That is R_+ consists of the region of the points to the left of the switch curve γ.

  2. *Let R_- be the region of the points such that*

$$R_- = \left\{ (x_1, x_2): \quad x_1 > -\frac{1}{2} x_2 \, |x_2| \right\}. \qquad (7.2.25)$$

  That is R_- consists of the region of the points to the right of the switch curve γ.

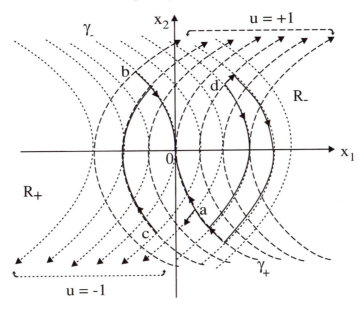

Figure 7.10   Various Trajectories Generated by Four Possible
Control Sequences

Figure 7.10 shows four possible control sequences (7.2.13) which
drive the system from any initial condition to the origin.

1. If the system is initially anywhere (say a) on the γ_+ curve,
   the optimal control is $u = +1$ to drive the system to origin
   in minimum time t_f.

2. If the system is at rest anywhere (say b) on the γ_- curve,
   the optimal control is $u = -1$ to drive the system to origin
   in minimum time t_f.

3. If the system is initially anywhere (say c) in the R_+ region,
   the optimal control sequence is $u = \{+1, -1\}$ to drive the
   system to origin in minimum time t_f.

4. If the system is initially anywhere (say d) in the R_- region,
   the optimal control sequence is $u = \{-1, +1\}$ to drive the
   system to origin in minimum time t_f.

   If we start at d and use the control $u = +1$ and use the
   optimal control sequence $u = \{-1, +1\}$, we certainly drive
   the system to origin but

(a) we then have a control sequence $\{+1, -1, +1\}$ which is not a member of the optimal control sequence (7.2.13), and

(b) the time t_f taken for the system using the control sequence $\{+1, -1, +1\}$ is higher than the corresponding time t_f taken for the system with control sequence $\{-1, +1\}$.

- **Step 9:** *Control Law:* We now reintroduce * to indicate the optimal values. The time-optimal control u^* as a function of the state $[x_1, x_2]$ is given by

$$u^* = u^*(x_1, x_2) = +1 \quad \text{for all} \quad (x_1, x_2) \in \gamma_+ \cup R_+$$
$$u^* = u^*(x_1, x_2) = -1 \quad \text{for all} \quad (x_1, x_2) \in \gamma_- \cup R_-. \quad (7.2.26)$$

Alternatively, if we define $z = x_1 + \frac{1}{2}x_2|x_2|$, then if

$$z > 0, \ u^* = -1, \quad \text{and}$$
$$z < 0, \ u^* = +1. \quad (7.2.27)$$

- **Step 10:** *Minimum Time:* We can easily calculate the time taken for the system starting at any position in state space and ending at the origin. We use the set of equations (7.2.15) for each portion of the trajectory. It can be shown that the minimum time t_f^* for the system starting from (x_1, x_2) and arriving at $(0, 0)$ is given by [6]

$$t_f^* = \begin{cases} x_2 + \sqrt{4x_1 + 2x_2^2} & \text{if} \quad (x_1, x_2) \in R_- \quad \text{or} \quad x_1 > -\frac{1}{2}x_2|x_2| \\ -x_2 + \sqrt{-4x_1 + 2x_2^2} & \text{if} \quad (x_1, x_2) \in R_+ \quad \text{or} \quad x_1 < -\frac{1}{2}x_2|x_2| \\ |x_2| & \text{if} \quad (x_1, x_2) \in \gamma \quad \text{or} \quad x_1 = -\frac{1}{2}x_2|x_2| \end{cases}$$
$$(7.2.28)$$

7.2.3 Engineering Implementation of Control Law

Figure 7.11 shows the implementation of the optimal control law (7.2.26).

1. If the system is initially at $(x_1, x_2) \in R_-$, then $x_1 > -\frac{1}{2}x_2|x_2|$, which means $z > 0$ and hence the output of the relay is $u^* = -1$.

2. On the other hand, if the system is initially at $(x_1, x_2) \in R_+$, then $x_1 < -\frac{1}{2}x_2|x_2|$, which means $z < 0$ and hence the output of the relay is $u^* = +1$.

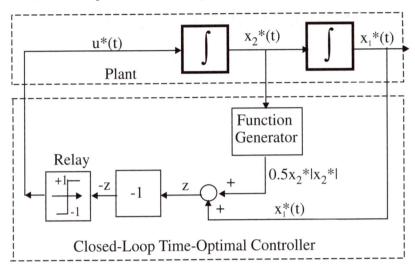

Figure 7.11   Closed-Loop Implementation of Time-Optimal Control Law

Let us note that the closed-loop (feedback) optimal controller is *non-linear* (control u^* is a nonlinear function of x_1^* and x_2^*) although the system is *linear*. On the other hand, we found in Chapters 3 and 4 for *unconstrained* control, the optimal control u^* is a *linear* function of the state x^*.

7.2.4   *SIMULINK© Implementation of Control Law*

The SIMULINK© implementation of time-optimal control law is very easy and convenient. The controller is easily obtained by using *abs* and *signum* function blocks as shown in Figure 7.12. Using different initial conditions, one can get the phase-plane (x_1 and x_2 plane) trajectories belonging to γ_+, γ_-, R_+ and R_- shown in Figures 7.13, 7.14, 7.15, and 7.16, respectively.

7.3   *Fuel-Optimal Control Systems*

Fuel-optimal control systems arise often in aerospace systems where the vehicles are controlled by thrusts and torques. These inputs like thrusts are due to the burning of fuel or expulsion of mass. Hence, the natural question is weather we can control the vehicle to minimize the fuel consumption. Another source of fuel-optimal control systems is

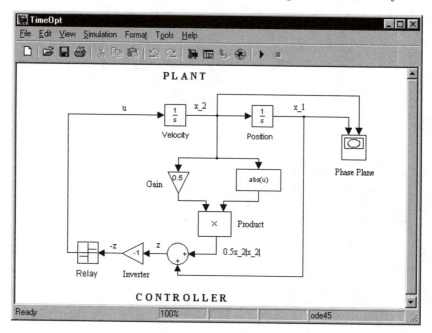

Figure 7.12 SIMULINK© Implementation of Time-Optimal Control Law

nuclear reactor control systems where fuel remains within the system and not expelled out of the system like in aerospace systems.

An interesting historical account is found in [59] regarding fuel-optimal control as applicable to the terminal phase of the lunar landing problem [100] of Apollo 11 mission, in which astronauts Neil Armstrong and Edwin Aldrin soft-landed the Lunar Excursion Module (LEM) "Eagle" on the lunar surface on July 20, 1969, while astronaut Michael Collins was in the orbit with Apollo Command Module "Columbia".

7.3.1 Fuel-Optimal Control of a Double Integral System

In this section, we formulate the fuel-optimal control system and obtain a solution to the system.

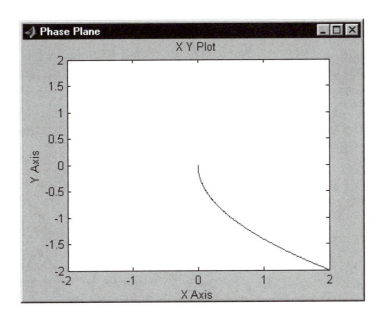

Figure 7.13 Phase-Plane Trajectory for γ_+: Initial State (2,-2) and Final State (0,0)

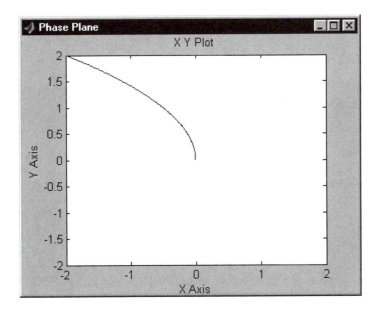

Figure 7.14 Phase-Plane Trajectory for γ_-: Initial State (-2,2) and Final State (0,0)

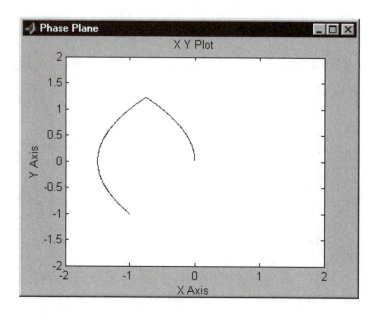

Figure 7.15　Phase-Plane Trajectory for R_+: Initial State (-1,-1) and Final State (0,0)

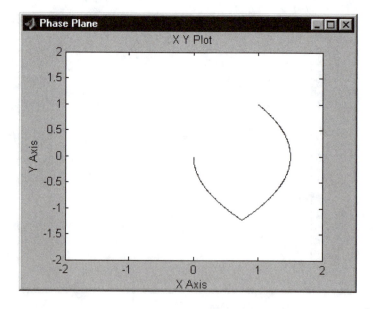

Figure 7.16　Phase-Plane Trajectory for R_-: Initial State (1,1) and Final State (0,0)

7.3.2   Problem Formulation and Statement

Consider a body with a unit mass undergoing translational motion

$$\dot{x}_1(t) = x_2(t)$$
$$\dot{x}_2(t) = u(t), \qquad |u(t)| \le 1 \qquad\qquad (7.3.1)$$

where, $x_1(t)$ is the position, $x_2(t)$ is the velocity, and $u(t)$ is the thrust force. Let us assume that the thrust (i.e., the control) is proportional to $\phi(t)$, the rate of fuel consumption. Then, the total fuel consumed becomes

$$J = \int_{t_0}^{t_f} \phi(t)dt. \qquad\qquad (7.3.2)$$

Let us further assume that

1. the mass of fuel consumed is small compared with the total mass of the body,

2. the rate of fuel, $\phi(t)$ is proportional to the *magnitude* of the thrust, $u(t)$, and

3. the final time t_f is *free* or *fixed*.

Then from (7.3.2), the performance index can be formulated as

$$J(u) = \int_{t_0}^{t_f} |u(t)|\, dt. \qquad\qquad (7.3.3)$$

The fuel-optimal control problem may be stated as follows: *Find the control u(t) which forces the system (7.3.1) from any initial state $(x_1(0),$ $x_2(0) = x_{10}, x_{20})$ to the origin in a certain unspecified final time t_f while minimizing the fuel consumption (7.3.3).*

Note that in case the final time t_f is *fixed* then that final time t_f must be greater than the minimum time t_f^* required to drive the system from (x_{10}, x_{20}) to the origin.

7.3.3   Problem Solution

The solution to the fuel-optimal system is provided first under the following list of steps and then explained in detail.

- **Step 1:** *Hamiltonian*

- **Step 2:** *Optimal Condition*

- **Step 3:** *Optimal Control*

- **Step 4:** *Costate Solutions*

- **Step 5:** *State Trajectories*

- **Step 6:** *Minimum Fuel*

- **Step 7:** *Switching Sequences*

- **Step 8:** *Control Law*

- **Step 1:** *Hamiltonian:* Let us formulate the Hamiltonian as

$$\mathcal{H}(\mathbf{x}(t), \boldsymbol{\lambda}(t), u(t)) = |u(t)| + \lambda_1(t)x_2(t) + \lambda_2(t)u(t). \quad (7.3.4)$$

- **Step 2:** *Optimal Condition:* According to the Minimum Principle, the optimal condition is

$$\mathcal{H}(\mathbf{x}^*(t), \boldsymbol{\lambda}^*(t), u^*(t)) \le \mathcal{H}(\mathbf{x}^*(t), \boldsymbol{\lambda}^*(t), u(t)),$$
$$= \min_{|u(t)| \le 1} \{\mathcal{H}(\mathbf{x}^*(t), \boldsymbol{\lambda}^*(t), u(t))\}. \quad (7.3.5)$$

Using (7.3.4) in (7.3.5), we have

$$|u^*(t)| + \lambda_1^*(t)x_2^*(t) + \lambda_2^*(t)u^*(t)$$
$$\le |u(t)| + \lambda_1^*(t)x_2^*(t) + \lambda_2^*(t)u(t), \quad (7.3.6)$$

which reduces to

$$|u^*(t)| + u^*(t)\lambda_2^*(t) \le |u(t)| + u(t)\lambda_2^*(t). \quad (7.3.7)$$

- **Step 3:** *Optimal Control:* Let us note at this point that

$$\min_{|u(t)| \le 1} \{|u(t)| + u(t)\lambda_2^*(t)\} = |u^*(t)| + u^*(t)\lambda_2^*(t) \quad (7.3.8)$$

and

$$|u(t)| = \begin{cases} +u(t) & \text{if} \quad u(t) \ge 0, \\ -u(t) & \text{if} \quad u(t) \le 0. \end{cases} \quad (7.3.9)$$

Hence, we have

$$\min_{|u(t)| \leq 1} \{|u(t)| + u(t)\lambda_2^*(t)\}$$

$$= \begin{cases} \min_{|u(t)| \leq 1} \{[+1 + \lambda_2^*(t)]\, u(t)\} \text{ if } u(t) \geq 0 \\ \min_{|u(t)| \leq 1} \{[-1 + \lambda_2^*(t)]\, u(t)\} \text{ if } u(t) \leq 0. \end{cases} \qquad (7.3.10)$$

Let us now explore all the possible values of $\lambda_2^*(t)$ and the corresponding optimal values of $u^*(t)$. Thus, we have the following table.

Possible values of $\lambda_2^*(t)$	Resulting values of $u^*(t)$	Minimum value of $\{|u(t)| + u(t)\lambda_2^*(t)\}$
$\lambda_2^*(t) > +1$	$u^*(t) = -1$	$1 - \lambda_2^*(t)$
$\lambda_2^*(t) < -1$	$u^*(t) = +1$	$1 + \lambda_2^*(t)$
$\lambda_2^*(t) = +1$	$-1 \leq u^*(t) \leq 0$	0
$\lambda_2^*(t) = -1$	$0 \leq u^*(t) \leq +1$	0
$-1 < \lambda_2^*(t) < 1$	$u^*(t) = 0$	0
Possible values of $\lambda_2^*(t)$	Resulting values of $u^*(t)$	Maximum value of $\{|u(t)| + u(t)\lambda_2^*(t)\}$
$\lambda_2^*(t) = 0$	$u^*(t) = +1 \text{ or } -1$	$+1$
$\lambda_2^*(t) > 0$	$u^*(t) = +1$	$1 + \lambda_2^*(t)$
$\lambda_2^*(t) < 0$	$u^*(t) = -1$	$1 - \lambda_2^*(t)$

These relations are also exhibited in Figure 7.17

The previous tabular relations are also written as

$$u^*(t) = \begin{cases} 0 \text{ if } -1 < \lambda_2^*(t) < +1 \\ +1 \text{ if } \qquad \lambda_2^*(t) < -1 \\ -1 \text{ if } \qquad \lambda_2^*(t) > +1 \\ \end{cases} \qquad (7.3.11)$$
$$0 \leq u^*(t) \leq +1 \qquad \text{if} \qquad \lambda_2^*(t) = -1$$
$$-1 \leq u^*(t) \leq 0 \qquad \text{if} \qquad \lambda_2^*(t) = +1.$$

The previous relation is further rewritten as

$$u^*(t) = \begin{cases} 0 & \text{if } |\lambda_2^*(t)| < 1 \\ -sgn\,\{\lambda_2^*(t)\} & \text{if } |\lambda_2^*(t)| > 1 \\ \text{undetermined} & \text{if } |\lambda_2^*(t)| = 1 \end{cases} \qquad (7.3.12)$$

where, sgn is already defined in the previous section on time-optimal control systems. In order to write the relation (7.3.12) in

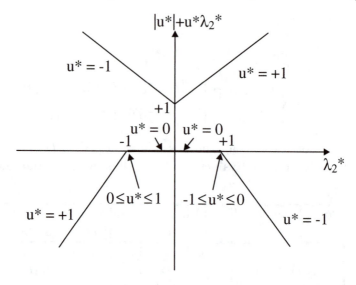

Figure 7.17   Relations Between $\lambda_2^*(t)$ and $|u^*(t)| + u^*(t)\lambda_2^*(t)$

a more compact form, let us define a *dead-zone function* between input function f_i and output function f_o, denoted by $dez\{\ \}$, as $f_o = dez\{f_i\}$ means that

$$
f_o = \begin{cases} 0 & \text{if } |f_i| < 1 \\ sgn\{f_i\} & \text{if } |f_i| > 1 \end{cases}
$$
$$
\begin{array}{ll} 0 \le f_o \le 1 & \text{if } f_i = +1 \\ -1 \le f_o \le 0 & \text{if } f_i = -1. \end{array}
$$
(7.3.13)

The *dead-zone function* is illustrated in Figure 7.18.

Using the definition of the *dez* function (7.3.13), we write the control strategy (7.3.12) as

$$
\boxed{u^*(t) = -dez\{\lambda_2^*(t)\}.}
$$
(7.3.14)

Using the previous definition of *dead-zone* function (7.3.13), the optimal control (7.3.14) is illustrated by Figure 7.19.

- **Step 4:** *Costate Solutions:* Using the Hamiltonian (7.3.4), the

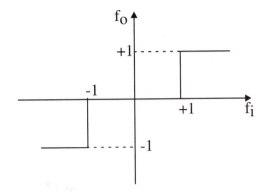

Figure 7.18   Dead-Zone Function

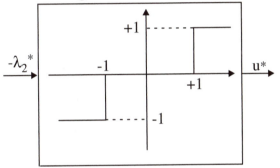

Figure 7.19   Fuel-Optimal Control

costates are described by

$$\dot{\lambda}_1^*(t) = -\frac{\partial \mathcal{H}}{\partial x_1^*} = 0,$$

$$\dot{\lambda}_2^*(t) = -\frac{\partial \mathcal{H}}{\partial x_2^*} = -\lambda_1^*(t), \qquad (7.3.15)$$

the solutions of which become

$$\lambda_1^*(t) = \lambda_1(0); \quad \lambda_2^*(t) = -\lambda_1(0)t + \lambda_2(0). \qquad (7.3.16)$$

From Figure 7.19), depending upon the values of $\lambda_1(0) \neq 0$ and $\lambda_2(0)$, there are 9 *admissible* fuel-optimal control sequences:

$$\{0\}, \ \{+1\}, \ \{-1\}, \ \{-1,0\}, \ \{0,+1\}, \ \{+1,0\}, \ \{0,-1\}$$
$$\{-1,0,+1\}, \ \{+1,0,-1\}. \qquad (7.3.17)$$

- **Step 5:** *State Trajectories:* The solutions of the state equations
  (7.3.1), already obtained in (7.2.15) under time-optimal control
  system, are (omitting * for simplicity)

$$x_1(t) = x_{10} - \frac{1}{2}Ux_{20}^2 + \frac{1}{2}Ux_2^2(t),$$
$$t = [x_2(t) - x_{20}]/U \tag{7.3.18}$$

for the control sequence $u(t) = U = \pm 1$. The switching curve
is the same as shown in Figure 7.9 (for time-optimal control
of a double integral system) which is repeated here in Figure 7.20.
For the control sequence $u(t) = U = 0$, we have from (7.3.1)

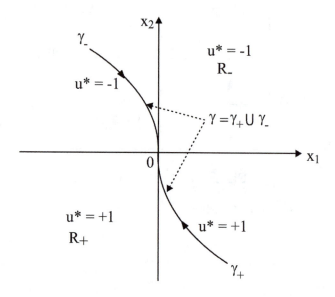

Figure 7.20   Switching Curve for a Double Integral Fuel-Optimal
Control System

$$x_1(t) = x_{10} + x_{20}t,$$
$$x_2(t) = x_{20},$$
$$t = (x_1(t) - x_{10})/x_{20}. \tag{7.3.19}$$

These trajectories for $u(t) = 0$ are shown in Figure 7.21. Here,
we cannot drive the system from any initial state to the origin by

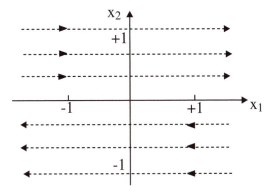

Figure 7.21 Phase-Plane Trajectories for $u(t) = 0$

means of the zero control. For example, if the system is on the x_1 axis at $(1,0)$, it continues to stay there for ever. Or if the system is at $(0,1)$ or $(0,-1)$, it travels along the trajectory with constant x_{20} towards the right or left.

- **Step 6:** *Minimum Fuel:* If there is a control $u(t)$ which drives the system from any initial condition (x_{10}, x_{20}) to the origin $(0,0)$, then the minimum fuel satisfies the relation

$$J^*(x_{10}, x_{20}) = J^* \geq |x_{20}| \qquad (7.3.20)$$

and hence

$$\boxed{J^* = |x_{20}|.} \qquad (7.3.21)$$

Proof: Solving the state equation (7.3.1), we have

$$x_2(t) = x_{20} + \int_0^t u(\tau)d\tau. \qquad (7.3.22)$$

Since we must reach the origin at t_f, it follows from the previous equation

$$x_2(t_f) = 0 = x_{20} + \int_0^{t_f} u(t)dt \qquad (7.3.23)$$

which yields

$$x_{20} = -\int_0^{t_f} u(t)dt. \qquad (7.3.24)$$

Or using the well-known inequality,

$$|x_{20}| = \left| \int_0^{t_f} u(t)dt \right| \leq \int_0^{t_f} |u(t)|\, dt = J. \qquad (7.3.25)$$

Hence, $|x_{20}| = J^*$. Note, if the initial state is $(x_{10}, 0)$, the fuel consumed $J = 0$, which implies that $u(t) = 0$ for all $t \in [0, t_f]$. In other words, a minimum-fuel solution does not exist for the initial state $(x_{10}, 0)$.

- **Step 7:** *Switching Sequences:* Now let us define the various regions in state space. (See Figure 7.22.)

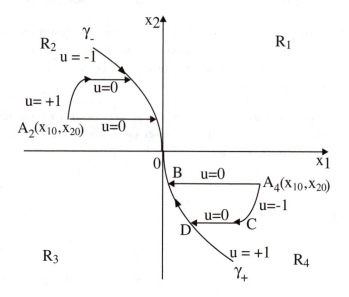

Figure 7.22   Fuel-Optimal Control Sequences

1. The R_1 (R_3) is the region to the right (left) of γ curve and for the positive (negative) values of x_2.

2. The R_2 (R_4) is the region to the left (right) of γ curve and for the positive (negative) values of x_2.

Now, depending upon the initial position of the system, we have a particular optimal control sequence (see Figure 7.22).

1. γ_+ *and* γ_- *Curves:* If the initial condition $(x_{10}, x_{20}) \in \gamma_+ (\gamma_-)$, then the control $u(t) = +1 (u(t) = -1)$.

2. R_2 *and* R_4 *Regions:* If the initial condition is $(x_{10}, x_{20}) \in R_4$, then the control sequence $\{0, +1\}$ forces the system to $(0,0)$, through A_4 to B, and then to 0, and hence is fuel-optimal. Although the control sequence $\{-1, 0, +1\}$ also drives the system to origin through $A_4 C D 0$, it is not *optimal*. Similarly, in the region R_2, the optimal control sequence is $\{0, -1\}$.

3. R_1 *and* R_3 *Regions:* Let us position the system at $A_1 (x_{10}, x_{20})$ in the region R_1, as shown in Figure 7.23. As seen, staying

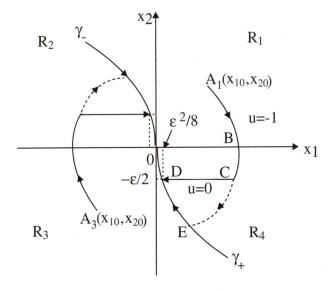

Figure 7.23 ϵ-Fuel-Optimal Control

in region R_1 there is no way one can drive the system at A_1 to origin, as the control sequence $u^*(t) = 0$ drives the system towards right (or away from the origin).

Thus, there is no *fuel-optimal* solution for the system for region R_1. However, given any $\epsilon > 0$, there is a control sequence $\{-1, 0, +1\}$, which forces the system to origin. Then, the fuel consumed is

$$J_\epsilon = |x_{20}| + \left|\frac{-\epsilon}{2}\right| + \left|\frac{\epsilon}{2}\right| = |x_{20}| + \epsilon = J^* + \epsilon \geq J^*.$$

(7.3.26)

We call such a control ϵ *fuel-optimal*. Similarly, for the region R_3, we have the ϵ fuel-optimal control sequence given as $\{+1, 0, -1\}$. Note that the control sequence $\{-1, +1\}$ through $A_1 BCEO$ is not an allowable optimal control sequence (7.3.17) and also consumes more fuel than the ϵ-fuel optimal through $A_1 BCDO$. Also, we like to *make ϵ as small as possible* and apply the control $\{0\}$ as soon as the trajectory enters the region R_4.

- **Step 8:** *Control Law:* The fuel-optimal control law for driving the system from any initial state (x_1, x_2) to the origin, can be stated as follows:

$$u^*(t) = \begin{cases} +1 & \forall \ (x_1, x_2) \in \gamma_+ \\ -1 & \forall \ (x_1, x_2) \in \gamma_- \\ 0 & \forall \ (x_1, x_2) \in R_2 \cup R_4. \end{cases}$$

(7.3.27)

If $(x_1, x_2) \in R_1 \cup R_3$, there is no fuel-optimal control. However, there is $\epsilon-$ fuel-optimal control as described above.

7.4   Minimum-Fuel System: LTI System

7.4.1   Problem Statement

Let us consider a linear, time-invariant system

$$\dot{\mathbf{x}}(t) = \mathbf{A}\mathbf{x}(t) + \mathbf{B}\mathbf{u}(t)$$

(7.4.1)

where, $\mathbf{x}(t)$ and $\mathbf{u}(t)$ are $n-$ and $r-$ dimensional state and control vectors, respectively. Let us assume that the control $\mathbf{u}(t)$ is constrained as

$$-1 \leq \mathbf{u}(t) \leq +1 \quad \text{or} \quad |\mathbf{u}(t)| \leq 1$$

(7.4.2)

or component wise,

$$|u_j(t)| \leq 1 \quad j = 1, 2, \ldots, r.$$

(7.4.3)

Our problem is to find the optimal control $\mathbf{u}^*(t)$ which transfers the system (7.4.1) from any initial condition $\mathbf{x}(0)$ to a given final state (usually the origin) and minimizes the performance measure

$$J(\mathbf{u}) = \int_0^{t_f} \sum_{j=1}^r |u_j(t)| dt. \qquad (7.4.4)$$

7.4.2 *Problem Solution*

We present the solution to this fuel-optimal system under the following steps. First let us list the steps.

- **Step 1:** *Hamiltonian*

- **Step 2:** *Optimal Condition*

- **Step 3:** *Costate Functions*

- **Step 4:** *Normal Fuel-Optimal Control System*

- **Step 5:** *Bang-off-Bang Control Law*

- **Step 6:** *Implementation*

- **Step 1:** *Hamiltonian:* Let us formulate the Hamiltonian for the system (7.4.1) and the performance measure (7.4.4) as

$$\mathcal{H}(\mathbf{x}(t), \mathbf{u}(t), \boldsymbol{\lambda}(t)) = \sum_{j=1}^r |u_j(t)| + \boldsymbol{\lambda}'(t)\mathbf{A}\mathbf{x}(t) + \boldsymbol{\lambda}'(t)\mathbf{B}\mathbf{u}(t).$$

$$(7.4.5)$$

- **Step 2:** *Optimal Condition:* According to the Pontryagin Principle, the optimal condition is given by

$$\mathcal{H}(\mathbf{x}^*(t), \boldsymbol{\lambda}^*(t), \mathbf{u}^*(t)) \le \mathcal{H}(\mathbf{x}^*(t), \boldsymbol{\lambda}^*(t), \mathbf{u}(t)),$$
$$= \min_{|\mathbf{u}(t)| \le 1} \{\mathcal{H}(\mathbf{x}^*(t), \boldsymbol{\lambda}^*(t), \mathbf{u}(t))\}. \quad (7.4.6)$$

Using (7.4.5) in (7.4.6), we have

$$\sum_{j=1}^r |u_j^*(t)| + \boldsymbol{\lambda}^{*\prime}(t)\mathbf{A}\mathbf{x}^*(t) + \boldsymbol{\lambda}^{*\prime}(t)\mathbf{B}\mathbf{u}^*(t)$$

$$\le \sum_{j=1}^r |u_j(t)| + \boldsymbol{\lambda}^{*\prime}(t)\mathbf{A}\mathbf{x}^*(t) + \boldsymbol{\lambda}^{*\prime}(t)\mathbf{B}\mathbf{u}(t) \qquad (7.4.7)$$

which in turn yields

$$\sum_{j=1}^{r} |u_j^*(t)| + \boldsymbol{\lambda}^{*\prime}(t)\mathbf{B}\mathbf{u}^*(t) \le \sum_{j=1}^{r} |u_j(t)| + \boldsymbol{\lambda}^{*\prime}(t)\mathbf{B}\mathbf{u}(t)$$

or transposing

$$\sum_{j=1}^{r} |u_j^*(t)| + \mathbf{u}^{*\prime}(t)\mathbf{B}'\boldsymbol{\lambda}^*(t) \le \sum_{j=1}^{r} |u_j(t)| + \mathbf{u}'(t)\mathbf{B}'\boldsymbol{\lambda}^*(t). \quad (7.4.8)$$

Considering the various possibilities as before for the double integral system, we have

$$\mathbf{q}^*(t) = \mathbf{B}'\boldsymbol{\lambda}^*(t). \quad (7.4.9)$$

Using the earlier relations (7.3.11) and (7.3.12) for the dead-zone function, we can write the condition (7.4.6) as

$$\boxed{\mathbf{u}^*(t) = -DEZ\{\mathbf{q}^*(t)\} = -DEZ\{\mathbf{B}'\boldsymbol{\lambda}^*(t)\}} \quad (7.4.10)$$

or component wise,

$$u_j^*(t) = -dez\{q_j^*(t)\} = -dez\{\mathbf{b}_j'\boldsymbol{\lambda}^*(t)\} \quad (7.4.11)$$

where, $j = 1, 2, \ldots, r$. The optimal control (7.4.10) in terms of the dead-zone (dez) function is shown in Figure 7.24.

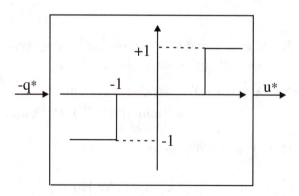

Figure 7.24   Optimal Control as Dead-Zone Function

- **Step 3:** *Costate Functions:* The costate functions $\boldsymbol{\lambda}^*(t)$ are given in terms of the Hamiltonian as

$$\dot{\boldsymbol{\lambda}}^*(t) = -\frac{\partial \mathcal{H}}{\partial \mathbf{x}} = -\mathbf{A}'\boldsymbol{\lambda}^*(t) \tag{7.4.12}$$

the solution of which is

$$\boldsymbol{\lambda}^*(t) = \epsilon^{-\mathbf{A}'t}\boldsymbol{\lambda}(0). \tag{7.4.13}$$

Depending upon the nature of the function $\mathbf{q}^*(t)$, we can classify it as *normal fuel-optimal control* (NFOC) system, if $|\mathbf{q}^*(t)| = 1$ only at switch times as shown in Figure 7.25 or *singular fuel-optimal control* (SFOC) system, if $|\mathbf{q}^*(t)| = 1$ as shown in Figure 7.26, for some $t \in [T_1, T_2]$.

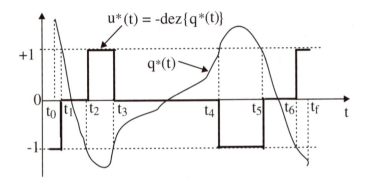

Figure 7.25   Normal Fuel-Optimal Control System

- **Step 4:** *Normal Fuel-Optimal Control System:* We first derive the *necessary* conditions for the fuel-optimal system to be *singular* and then translate these into *sufficient* conditions for the system to be *normal*, that is, the negation of the conditions for singular is taken as that for normal.

For the fuel-optimal system to be singular, it is necessary that in the system interval $[0, t_f]$, there is at least one subinterval $[T_1, T_2]$ for which

$$\mathbf{q}^*(t) = 1 \ \forall \ t \in [T_1, T_2]. \tag{7.4.14}$$

Using (7.4.9), the previous condition becomes

$$|\mathbf{q}^*(t)| = |\mathbf{B}'\boldsymbol{\lambda}^*(t)| = 1. \tag{7.4.15}$$

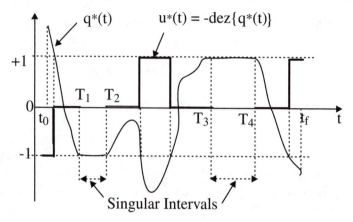

Figure 7.26   Singular Fuel-Optimal Control System

This means that the function $\mathbf{q}^*(t)$ is constant and hence all its time derivatives must vanish. By repeated differentiation of (7.4.15) and using (7.4.12), we have

$$(\mathbf{Ab}_j)'\boldsymbol{\lambda}^*(t) = 0,$$
$$(\mathbf{A}^2\mathbf{b}_j)'\boldsymbol{\lambda}^*(t) = 0,$$
$$\dotsb\dotsb\dotsb$$
$$(\mathbf{A}^{n-1}\mathbf{b}_j)'\boldsymbol{\lambda}^*(t) = 0,$$
$$(\mathbf{A}^n\mathbf{b}_j)'\boldsymbol{\lambda}^*(t) = 0, \qquad\qquad (7.4.16)$$

for all $t \in [T_1, T_2]$, where $j = 1, 2, \ldots, r$. We can rewrite the previous set of equations as

$$[\mathbf{AG}_j]'\boldsymbol{\lambda}^*(t) = 0 \qquad\qquad (7.4.17)$$

where,

$$\mathbf{G}_j = [\mathbf{b}_j,\ \mathbf{Ab}_j, \cdots, \mathbf{A}^{n-1}\mathbf{b}_j]. \qquad\qquad (7.4.18)$$

The condition (7.4.17) can further be rewritten as

$$\mathbf{G}_j'\mathbf{A}'\boldsymbol{\lambda}^*(t) = 0 \ \forall\ t \in [T_1, T_2]. \qquad\qquad (7.4.19)$$

But the condition (7.4.15) implies that $\boldsymbol{\lambda}^*(t) \neq 0$. Then, for (7.4.19) to hold, it is necessary that the matrix $\mathbf{G}_j'\mathbf{A}'$ must be *singular*. This means that

$$\det\{\mathbf{G}_j'\mathbf{A}'\} = \det \mathbf{A} \det \mathbf{G_j} = 0. \qquad\qquad (7.4.20)$$

Thus, the sufficient condition for the system to be *normal* is that

$$\det\{\mathbf{G}'_j\mathbf{A}'\} \neq 0 \quad \forall \quad j = 1, 2, \cdots, r. \qquad (7.4.21)$$

Thus, if the system (7.4.1) is *normal* (that is also *controllable*), and if the matrix $\mathbf{A}$ is *nonsingular*, then the fuel-optimal system is *normal*.

- **Step 5:** *Bang-off-Bang Control Law:* If the linear, time-invariant system (7.4.1) is *normal* and $\mathbf{x}^*(t)$ and $\boldsymbol{\lambda}^*(t)$ are the state and costate trajectories, then the optimal control law $\mathbf{u}^*(t)$ given by (7.4.10) is repeated here as

$$\mathbf{u}^*(t) = -DEZ\left\{\mathbf{B}'\boldsymbol{\lambda}^*(t)\right\} \qquad (7.4.22)$$

  for all $t \in [t_0, t_f]$. In other words, if the fuel-optimal system is *normal*, the components of the fuel-optimal control are piecewise constant functions of time. The fuel-optimal control can switch between $+1, 0$ and -1 and hence is called the *bang-off-bang* control (or principle).

- **Step 6:** *Implementation:* As before in time-optimal control system, the fuel-optimal control law can be implemented either in *open-loop* configuration as shown in Figure 7.27. Here, an iterative procedure is to be used to finally drive the state to origin. On the other hand, we can realize *closed-loop* configuration as

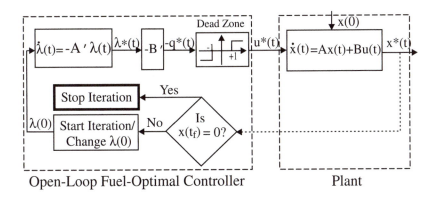

Figure 7.27 Open-Loop Implementation of Fuel-Optimal Control System

shown in Figure 7.28, where the current initial state is used to realize the fuel-optimal control law (7.4.22).

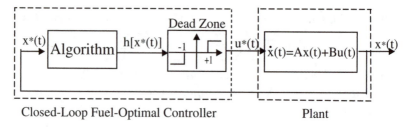

Closed-Loop Fuel-Optimal Controller   Plant

Figure 7.28   Closed-Loop Implementation of Fuel-Optimal Control System

7.4.3   *SIMULINK© Implementation of Control Law*

The SIMULINK© implementation of fuel-optimal control law for the double integral system described in the previous section is very convenient. The controller is obtained by using *abs, signum*, and *dead-zone* function blocks as shown in Figure 7.29. Further note that since the relay with dead-zone function block required for fuel-optimal control, as shown in Figure 7.19, is not readily available in SIMULINK© library, the function block is realized by combining *dead-zone* and *sign3* function blocks [6]. Using different initial conditions one can get the phase-plane (x_1 and x_2 plane) trajectories belonging to γ_+, γ_-, R_1, R_3, R_2 and R_4 shown in Figures 7.30, 7.31, 7.32, 7.33, 7.34, and 7.35, respectively. In particular note the trajectories belonging to R_1 and R_3 regions showing ϵ-fuel-optimal condition.

7.5   *Energy-Optimal Control Systems*

In minimum-energy (energy-optimal) systems with *constraints*, we often formulate the performance measure as the energy of an electrical (or mechanical) system. For example, if $u(t)$ is the voltage input to a field circuit in a typical constant armature-current, field controlled positional control system, with negligible field inductance and a unit field resistance, the total energy to the field circuit is (power is $u^2(t)/R_f$, where, $R_f = 1$ is the field resistance)

$$J = \int_{t_0}^{t_f} u^2(t)dt \qquad (7.5.1)$$

and the field voltage $u(t)$ is constrained by $|u(t)| \leq 110$. This section is based on [6, 89].

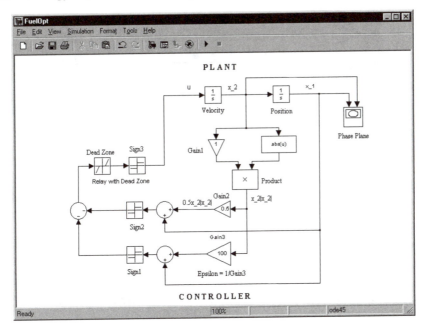

Figure 7.29   SIMULINK© Implementation of Fuel-Optimal
Control Law

7.5.1   *Problem Formulation and Statement*

Let us now formulate the energy-optimal control (EOC) system with
magnitude-constrained control. Consider a linear, time-varying, fully
controllable system

$$\dot{\mathbf{x}}(t) = \mathbf{A}(t)\mathbf{x}(t) + \mathbf{B}(t)\mathbf{u}(t) \tag{7.5.2}$$

where, $\mathbf{x}(t)$ and $\mathbf{u}(t)$ are $n-$ and $r-$dimensional state and control vectors, respectively, and the energy cost functional

$$J = \frac{1}{2}\int_{t_0}^{t_f} \mathbf{u}'(t)\mathbf{R}(t)\mathbf{u}(t)dt. \tag{7.5.3}$$

Let us assume that the control $\mathbf{u}(t)$ is *constrained* as

$$-1 \le \mathbf{u}(t) \le +1 \quad \text{or} \quad |\mathbf{u}(t)| \le 1 \tag{7.5.4}$$

or component wise,

$$|u_j(t)| \le 1 \quad j = 1, 2, \ldots, r. \tag{7.5.5}$$

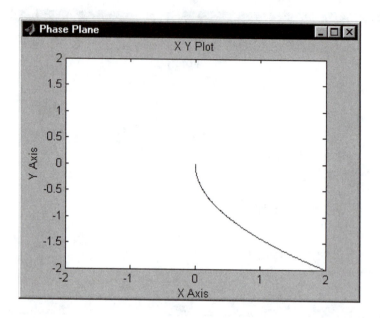

Figure 7.30   Phase-Plane Trajectory for γ_+: Initial State (2,-2) and Final State (0,0)

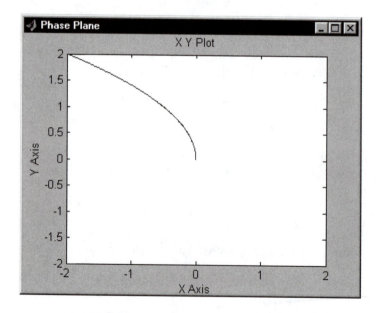

Figure 7.31   Phase-Plane Trajectory for γ_-: Initial State (-2,2) and Final State (0,0)

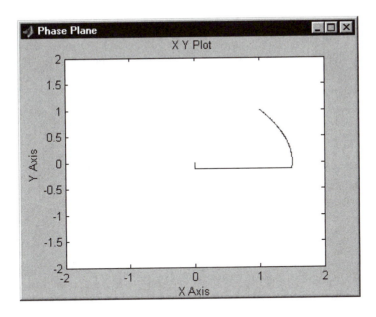

Figure 7.32   Phase-Plane Trajectory for R_1: Initial State (1,1) and
Final State (0,0)

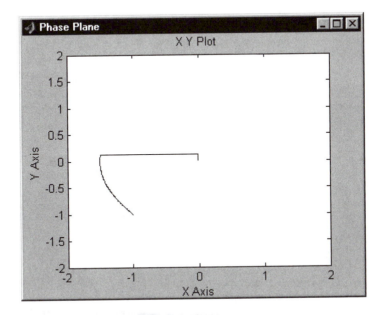

Figure 7.33   Phase-Plane Trajectory for R_3: Initial State (-1,-1)
and Final State (0,0)

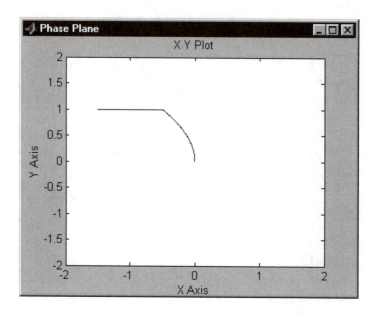

Figure 7.34   Phase-Plane Trajectory for R_2: Initial State (-1.5,1) and Final State (0,0)

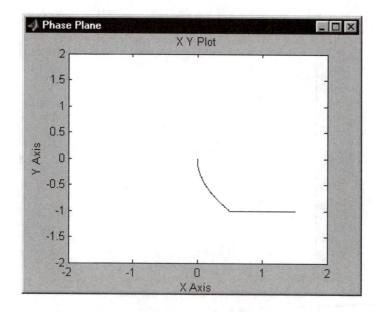

Figure 7.35   Phase-Plane Trajectory for R_4: Initial State (1.5,-1) and Final State (0,0)

Problem Statement

The energy-optimal control system is to transfer the system (7.5.2) from any initial state $\mathbf{x}(t = t_0) = \mathbf{x}(t_0) \neq 0$ to the origin in time t_f and at the same time minimize the energy cost functional (7.5.3) with the constraint relation (7.5.4).

7.5.2   Problem Solution

We present the solution to this energy-optimal system under the following steps. But first let us list the various steps involved.

- **Step 1:** *Hamiltonian*

- **Step 2:** *State and Costate Equations*

- **Step 3:** *Optimal Condition*

- **Step 4:** *Optimal Control*

- **Step 5:** *Implementation*

- **Step 1:** *Hamiltonian:* Let us formulate the Hamiltonian for the system (7.5.2) and the PI (7.5.3) as

$$\mathcal{H}(\mathbf{x}(t), \mathbf{u}(t), \boldsymbol{\lambda}(t)) = \frac{1}{2}\mathbf{u}'(t)\mathbf{R}(t)\mathbf{u}(t) + \boldsymbol{\lambda}'(t)\mathbf{A}\mathbf{x}(t) + \boldsymbol{\lambda}'(t)\mathbf{B}\mathbf{u}(t)$$

$$(7.5.6)$$

  where, $\boldsymbol{\lambda}(t)$ is the costate variable.

- **Step 2:** *State and Costate Equations:* Let us assume optimal values $\mathbf{u}^*(t)$, $\mathbf{x}^*(t)$, and $\boldsymbol{\lambda}^*(t)$. Then, the state $\mathbf{x}^*(t)$ and the costate $\boldsymbol{\lambda}^*(t)$ optimal values are given in terms of the Hamiltonian as

$$\dot{\mathbf{x}}^*(t) = + \left(\frac{\partial \mathcal{H}}{\partial \boldsymbol{\lambda}}\right)_* = \mathbf{A}(t)\mathbf{x}^*(t) + \mathbf{B}(t)\mathbf{u}^*(t)$$

$$\dot{\boldsymbol{\lambda}}^*(t) = - \left(\frac{\partial \mathcal{H}}{\partial \mathbf{x}}\right)_* = -\mathbf{A}'(t)\boldsymbol{\lambda}^*(t) \qquad (7.5.7)$$

  with the boundary conditions

$$\mathbf{x}(t_0) = \mathbf{x}(t_0); \quad \mathbf{x}(t_f) = 0 \qquad (7.5.8)$$

  where, we again note that t_f is either fixed or free.

- **Step 3:** *Optimal Condition:* Now using Pontryagin Principle, we invoke the condition for optimal control in terms of the Hamiltonian, that is,

$$\mathcal{H}(\mathbf{x}^*(t), \boldsymbol{\lambda}^*(t), \mathbf{u}^*(t)) \le \mathcal{H}(\mathbf{x}^*(t), \boldsymbol{\lambda}^*(t), \mathbf{u}(t))$$
$$= \min_{|\mathbf{u}(t)| \le 1} \mathcal{H}(\mathbf{x}^*(t), \boldsymbol{\lambda}^*(t), \mathbf{u}(t)). \quad (7.5.9)$$

Using (7.5.6) in (7.5.9), we have

$$\frac{1}{2}\mathbf{u}^{*\prime}(t)\mathbf{R}(t)\mathbf{u}^*(t) + \boldsymbol{\lambda}^{*\prime}(t)\mathbf{A}(t)\mathbf{x}^*(t) + \boldsymbol{\lambda}^{*\prime}(t)\mathbf{B}(t)\mathbf{u}^*(t)$$
$$\le \frac{1}{2}\mathbf{u}^{\prime}(t)\mathbf{R}(t)\mathbf{u}(t) + \boldsymbol{\lambda}^{*\prime}(t)\mathbf{A}(t)\mathbf{x}^*(t) + \boldsymbol{\lambda}^{*\prime}(t)\mathbf{B}(t)\mathbf{u}(t) \quad (7.5.10)$$

which becomes

$$\frac{1}{2}\mathbf{u}^{*\prime}(t)\mathbf{R}(t)\mathbf{u}^*(t) + \boldsymbol{\lambda}^{*\prime}(t)\mathbf{B}(t)\mathbf{u}^*(t)$$
$$\le \frac{1}{2}\mathbf{u}^{\prime}(t)\mathbf{R}(t)\mathbf{u}(t) + \boldsymbol{\lambda}^{*\prime}(t)\mathbf{B}(t)\mathbf{u}(t)$$
$$= \min_{|\mathbf{u}^{\prime}(t)| \le 1} \left\{ \frac{1}{2}\mathbf{u}^{\prime}(t)\mathbf{R}(t)\mathbf{u}(t) + \boldsymbol{\lambda}^{*\prime}(t)\mathbf{B}(t)\mathbf{u}(t) \right\}. \quad (7.5.11)$$

- **Step 4:** *Optimal Control:* Let us denote

$$\mathbf{q}^*(t) = \mathbf{R}^{-1}(t)\mathbf{B}^{\prime}(t)\boldsymbol{\lambda}^*(t) \quad (7.5.12)$$

and write

$$\boldsymbol{\lambda}^{*\prime}(t)\mathbf{B}(t)\mathbf{u}^*(t) = \mathbf{u}^{*\prime}(t)\mathbf{B}^{\prime}(t)\boldsymbol{\lambda}^*(t) = \mathbf{u}^{*\prime}(t)\mathbf{R}(t)\mathbf{q}^*(t). \quad (7.5.13)$$

Using (7.5.12) and (7.5.13) in (7.5.11), we get

$$\frac{1}{2}\mathbf{u}^{*\prime}(t)\mathbf{R}(t)\mathbf{u}^*(t) + \mathbf{u}^{*\prime}(t)\mathbf{R}(t)\mathbf{q}^*(t)$$
$$\le \frac{1}{2}\mathbf{u}^{\prime}(t)\mathbf{R}(t)\mathbf{u}(t) + \mathbf{u}^{\prime}(t)\mathbf{R}(t)\mathbf{q}^*(t). \quad (7.5.14)$$

Now, adding

$$\frac{1}{2}\mathbf{q}^{*\prime}(t)\mathbf{R}(t)\mathbf{q}^*(t) = \frac{1}{2}\boldsymbol{\lambda}^{*\prime}(t)\mathbf{B}(t)\mathbf{R}^{-1}(t)\mathbf{B}^{\prime}(t)\boldsymbol{\lambda}^*(t) \quad (7.5.15)$$

to both sides of (7.5.14), we get

$$[\mathbf{u}^*(t) + \mathbf{q}^*(t)]'\mathbf{R}(t)[\mathbf{u}^*(t) + \mathbf{q}^*(t)]$$
$$\leq [\mathbf{u}(t) + \mathbf{q}^*(t)]'\,\mathbf{R}(t)\,[\mathbf{u}(t) + \mathbf{q}^*(t)]. \qquad (7.5.16)$$

That is

$$\mathbf{w}^{*\prime}(t)\mathbf{R}(t)\mathbf{w}^*(t) \leq \mathbf{w}'(t)\mathbf{R}(t)\mathbf{w}(t)$$
$$= \min_{|\mathbf{u}(t)|\leq 1} \{\mathbf{w}'(t)\mathbf{R}(t)\mathbf{w}(t)\}, \qquad (7.5.17)$$

where,

$$\mathbf{w}(t) = \mathbf{u}(t) + \mathbf{q}^*(t) = \mathbf{u}(t) + \mathbf{R}^{-1}(t)\mathbf{B}'(t)\boldsymbol{\lambda}^*(t)$$
$$\mathbf{w}^*(t) = \mathbf{u}^*(t) + \mathbf{q}^*(t) = \mathbf{u}^*(t) + \mathbf{R}^{-1}(t)\mathbf{B}'(t)\boldsymbol{\lambda}^*(t). \qquad (7.5.18)$$

The relation (7.5.17) implies that $\mathbf{w}'(t)$ attains its minimum value at $\mathbf{w}^*(t)$.

Now we know that

1. if $\mathbf{R}(t)$ is *positive definite* for all t,
   its eigenvalues $d_1(t), d_2(t), \ldots, d_r(t)$ are *positive*,

2. if $\mathbf{D}(t)$ is the diagonal matrix of
   the eigenvalues $d_1(t), d_2(t), \ldots, d_r(t)$ of $\mathbf{R}(t)$, then

3. there is an *orthogonal* matrix $\mathbf{M}$ such that

$$\mathbf{M}'\mathbf{M} = \mathbf{I} \rightarrow \mathbf{M}' = \mathbf{M}^{-1} \qquad (7.5.19)$$

and

$$\mathbf{D} = \mathbf{M}'\mathbf{R}\mathbf{M} \rightarrow \mathbf{M}\mathbf{D}\mathbf{M}' = \mathbf{R}. \qquad (7.5.20)$$

Now, using (7.5.20) along with (7.5.17), we have

$$\mathbf{w}'(t)\mathbf{R}\mathbf{w}(t) = \mathbf{w}'(t)\mathbf{M}\mathbf{D}\mathbf{M}'\mathbf{w}(t)$$
$$= \mathbf{v}'(t)\mathbf{D}\mathbf{v}(t) = \sum_{j=1}^{r} d_j(t)v_j^2(t) \qquad (7.5.21)$$

where, $\mathbf{v}(t) = \mathbf{M}'\mathbf{w}(t)$ and note that $d_j > 0$. Since both $\mathbf{M}'$ and $\mathbf{M}$ are orthogonal, we know that

$$\mathbf{v}'(t)\mathbf{v}(t) = \mathbf{w}'(t)\mathbf{M}\mathbf{M}'\mathbf{w}(t) = \mathbf{w}'(t)\mathbf{w}(t) \qquad (7.5.22)$$

where, we used $\mathbf{M}'\mathbf{M} = \mathbf{I}$. We can equivalently write (7.5.22) component wise as

$$\sum_{j=1}^{r} v_j^2(t) = \sum_{j=1}^{r} w_j^2(t). \tag{7.5.23}$$

Now, (7.5.17) implies that (using (7.5.21))

$$\min_{|\mathbf{u}(t)|\leq 1} \left\{ \mathbf{w}'(t)\mathbf{R}(t)\mathbf{w}(t) \right\} = \min_{|\mathbf{u}(t)|\leq 1} \left\{ \sum_{j=1}^{r} d_j(t)v_j^2(t) \right\}$$

$$= \sum_{j=1}^{r} \min_{v_j(t)} \left\{ v_j^2(t) \right\}. \tag{7.5.24}$$

This implies that if $\mathbf{w}^*(t)$ minimizes $\mathbf{w}'(t)\mathbf{R}(t)\mathbf{w}(t)$, then the components $v_j(t)$ also minimize $\mathbf{v}'(t)\mathbf{v}(t)$. This fact is also evident from (7.5.22). In other words, we have established that

$$\text{if} \quad \mathbf{w}^{*\prime}(t)\mathbf{R}(t)\mathbf{w}^*(t) \leq \mathbf{w}'(t)\mathbf{R}(t)\mathbf{w}(t)$$
$$\text{then} \quad \mathbf{w}^{*\prime}(t)\mathbf{w}^*(t) \leq \mathbf{w}'(t)\mathbf{w}(t) \tag{7.5.25}$$

and the converse is also true. Or the effect of $\mathbf{R}(t)$ is nullified in the minimization process. Thus,

$$\min_{|\mathbf{u}(t)|\leq 1} \left\{ \mathbf{w}'(t)\mathbf{R}(t)\mathbf{w}(t) \right\} = \min_{|\mathbf{u}(t)|\leq 1} \left\{ \mathbf{w}'(t)\mathbf{w}(t) \right\},$$

$$= \sum_{j=1}^{r} \min_{\mathbf{w}(t)} \left\{ w_j^2(t) \right\},$$

$$= \sum_{j=1}^{r} \min_{|\mathbf{u}(t)|\leq 1} \left\{ [u_j(t) + q_j^*(t)]^2 \right\}. \tag{7.5.26}$$

A careful examination of (7.5.26) reveals that to minimize the *positive* quantity $[u_j(t) + q_j^*(t)]^2$, we must select

$$u^*(t) = \begin{cases} -q_j^*(t) & \text{if } |q_j^*(t)| \leq 1, \\ +1 & \text{if } q_j^*(t) < -1, \\ -1 & \text{if } q_j^*(t) > +1. \end{cases} \tag{7.5.27}$$

First, let us define $sat\{\ \}$ as the *saturation* function between the input f_i and the output f_o (see Figure 7.36) as $f_o = sat\{f_i\}$ means that

$$f_o = \begin{cases} f_i & \text{if } |f_i| \le 1 \\ sgn\{f_i\} & \text{if } |f_i| > 1. \end{cases} \tag{7.5.28}$$

The *sgn* function is already defined in Section 7.1.1. Then the

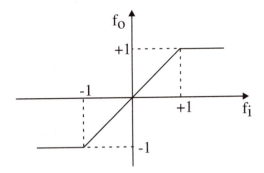

Figure 7.36   Saturation Function

relation (7.5.27) can be conveniently written as

$$u^*(t) = \begin{cases} -q_j^*(t) & \text{if } |q_j^*(t)| \le 1 \\ -sgn\{q_j^*(t)\} & \text{if } |q_j^*(t)| > 1, \end{cases} \tag{7.5.29}$$

or more compactly component-wise as

$$u_j^*(t) = -sat\{q_j^*(t)\}, \tag{7.5.30}$$

or in vector form as

$$\boxed{\mathbf{u}^*(t) = -SAT\{\mathbf{q}^*(t)\} = -SAT\{\mathbf{R}^{-1}(t)\mathbf{B}'(t)\boldsymbol{\lambda}^*(t)\}} \tag{7.5.31}$$

shown in Figure 7.37.

The following notes are in order.

1. The constrained minimum-energy control law (7.5.31) is valid only if $\mathbf{R}(t)$ is *positive definite*.

2. The *energy-optimal* control law (7.5.31), described by *saturation (SAT)* function, which is different from the *signum (SGN)* function for *time-optimal* control and *dead-zone (DEZ)*

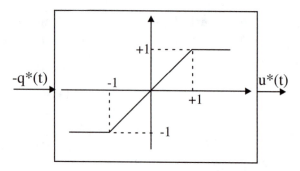

Figure 7.37 Energy-Optimal Control

function for *fuel-optimal* control functions, is a *well-defined* (determinate) function. Hence, the minimum-energy system has no option to be *singular*.

3. In view of the above, it also follows that the optimal control $\mathbf{u}^*(t)$ is a *continuous* function of time which again is different from the *piece-wise constant* functions of time for time-optimal and fuel-optimal control systems discussed earlier in this chapter.

4. If the minimum-energy system described by the system (7.5.2) and the PI (7.5.3) has *no constraint* (7.5.4) on the control, then by the results of Chapter 3, we obtain the optimal control $\mathbf{u}^*(t)$ by using the Hamiltonian (7.5.6) and the condition

$$\frac{\partial \mathcal{H}}{\partial \mathbf{u}} = 0 \longrightarrow \mathbf{R}(t)\mathbf{u}_n^*(t) + \mathbf{B}'(t)\boldsymbol{\lambda}^*(t) = 0 \longrightarrow$$
$$\mathbf{u}_n^*(t) = -\mathbf{R}^{-1}(t)\mathbf{B}'(t)\boldsymbol{\lambda}^*(t) = -\mathbf{q}^*(t), \qquad (7.5.32)$$

where, $\mathbf{u}_n^*(t)$ refers to *unconstraint* control. Comparing the relation (7.5.32) with (7.5.29), we see that

$$\mathbf{u}_n^*(t) = -\mathbf{q}^*(t) = \mathbf{u}^*(t) \text{ if } |\mathbf{q}^*(t)| \le 1 \qquad (7.5.33)$$

where $\mathbf{u}^*(t)$ refers to *constrained* control. Thus, if $\mathbf{q}^*(t) \le 1$, the *constrained* optimal control $\mathbf{u}^*(t)$ and the *unconstrained* optimal control $\mathbf{u}^*(t)$ are the *same*.

5. For the *constrained* energy-optimal control system, using optimal control (7.5.31), the state and costate system (7.5.7)

becomes

$$\dot{\mathbf{x}}^*(t) = \mathbf{A}\mathbf{x}^*(t) - \mathbf{B}SAT\left\{\mathbf{R}^{-1}\mathbf{B}'\boldsymbol{\lambda}^*(t)\right\}$$

$$\dot{\boldsymbol{\lambda}}^*(t) = -\mathbf{A}'\boldsymbol{\lambda}^*(t). \tag{7.5.34}$$

We notice that this is a set of $2n$ *nonlinear* differential equations and can only be solved by using numerical simulations.

- **Step 5:** *Implementation:* The implementation of the energy-optimal control law (7.5.31) can be performed in *open-loop* or *closed-loop* configuration. In the open-loop case (Figure 7.38), it becomes iterative to try different values of initial conditions for $\boldsymbol{\lambda}(0)$ to satisfy the final condition of driving the state to origin. On the other hand, the closed-loop case shown in Figure 7.39 becomes more attractive.

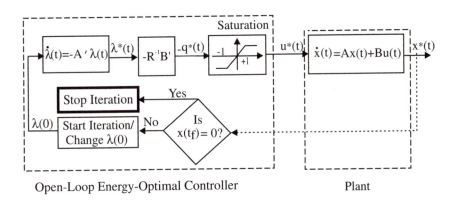

Open-Loop Energy-Optimal Controller   Plant

Figure 7.38   Open-Loop Implementation of Energy-Optimal
Control System

A more general *constrained* minimum-energy control system is where the performance measure (7.5.3) contains additional weighting terms $\mathbf{x}'(t)\mathbf{Q}(t)\mathbf{x}(t)$ and $2\mathbf{x}(t)\mathbf{S}(t)\mathbf{u}(t)$ [6].

Example 7.1

Consider a simple scalar system

$$\dot{x}(t) = ax(t) + u(t), \quad a < 0 \tag{7.5.35}$$

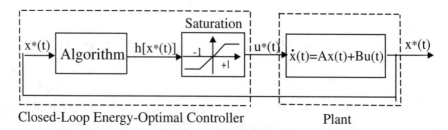

Closed-Loop Energy-Optimal Controller   Plant

Figure 7.39   Closed-Loop Implementation of Energy-Optimal
Control System

to be transferred from an arbitrary initial state $x(t = 0) = x_0$ to
the origin that minimize the performance index

$$J = \int_0^{t_f} u^2(t)dt \qquad (7.5.36)$$

where, the final time t_f is free and the control $u(t)$ is *constrained*
as

$$|u(t)| \leq 1. \qquad (7.5.37)$$

Discuss the resulting optimal control system.

Solution: Comparing the system (7.5.35) and the performance
measure (7.5.36) with the general formulations of the corresponding system (7.5.2) and the performance index (7.5.3), we easily
see that $\mathbf{A}(t) = a, \mathbf{B}(t) = b = 1, \mathbf{R}(t) = r = 2$. Then using the
step-by-step procedure in the last section, we get the following.

- **Step 1:** Form the Hamiltonian (7.5.6) as

$$\mathcal{H}(x(t), \lambda(t), u(t)) = \frac{1}{2}x2u^2(t) + \lambda(t)ax(t) + \lambda(t)u(t).$$

$$(7.5.38)$$

- **Step 2:** The state and costate relations (7.5.7) are

$$\dot{x}^*(t) = + \left(\frac{\partial \mathcal{H}}{\partial \lambda}\right)_* = ax^*(t) + u^*(t)$$

$$\dot{\lambda}^*(t) = - \left(\frac{\partial \mathcal{H}}{\partial x}\right)_* = -a\lambda^*(t). \qquad (7.5.39)$$

The solution of the costate function $\lambda^*(t)$ is easily seen to be

$$\lambda^*(t) = \lambda(0)\epsilon^{-at}. \qquad (7.5.40)$$

- **Step 3:** The optimal control (7.5.30) becomes

$$u^*(t) = -sat\{q_i^*(t)\}$$
$$= -sat\{r^{-1}b\lambda^*(t)\}$$
$$= -sat\{0.5\lambda^*(t)\}. \tag{7.5.41}$$

In other words,

$$u^*(t) = \begin{cases} +1.0 & \text{if } 0.5\lambda^*(t) \le -1 \text{ or } \lambda^*(t) \le -2, \\ -1.0 & \text{if } 0.5\lambda^*(t) \ge +1 \text{ or } \lambda^*(t) \ge +2, \\ -0.5\lambda^*(t) & \text{if } |0.5\lambda^*(t)| \le 1 \text{ or } |\lambda^*(t)| \le +2. \end{cases}$$
$$\tag{7.5.42}$$

The previous relationship between the optimal control $u^*(t)$ and the optimal costate $\lambda^*(t)$ is shown in Figure 7.40.

We note from (7.5.42) that the condition $u^*(t) = -\frac{1}{2}\lambda^*(t)$ is also obtained from the results of *unconstrained* control using the Hamiltonian (7.5.38) and the condition

$$\frac{\partial H}{\partial u} = 0 \longrightarrow 2u^*(t) + \lambda^*(t) = 0 \longrightarrow u^*(t) = -\frac{1}{2}\lambda^*(t). \tag{7.5.43}$$

Consider the costate function $\lambda^*(t)$ in (7.5.40). The condition $\lambda^*(0) = 0$ is not admissible because then according to (7.5.41), $u^*(t) = 0$ for $t \in [0, t_f]$, and the state $x^*(t) = x(0)\epsilon^{at}$ in (7.5.39) will never reach the origin in time t_f for an arbitrarily given initial state $x(0)$.

Then, the costate $\lambda^*(t) = \lambda(0)\epsilon^{-at}$ has four possible solutions depending upon the initial values $(0 < \lambda(0) < 2, \lambda(0) > 2, -2 < \lambda(0) < 0, \lambda(0) < -2)$ as shown in Figure 7.41.

1. $0 < \lambda(0) < 2$: For this case, Figure 7.41, curve (a),

$$u^*(t) = \left\{-\frac{1}{2}\lambda^*(t)\right\} \text{ or } \left\{-\frac{1}{2}\lambda^*(t), -1\right\} \tag{7.5.44}$$

   depending upon whether the system reaches the origin before or after time t_a, the function $\lambda^*(t)$ reaches the value of $+2$.

2. $\lambda(0) > 2$: In this case, Figure 7.41, curve (b), since $\lambda^*(t) > +2$, the optimal control $u^*(t) = \{-1\}$.

3. $-2 < \lambda(0) < 0$: Depending on whether the state reaches the origin before or after time t_c, the function $\lambda^*(t)$ reaches the value -2, the optimal control is (Figure 7.41, curve (c))

$$u^*(t) = \left\{-\frac{1}{2}\lambda^*(t)\right\} \text{ or } \left\{-\frac{1}{2}\lambda^*(t), +1\right\}. \tag{7.5.45}$$

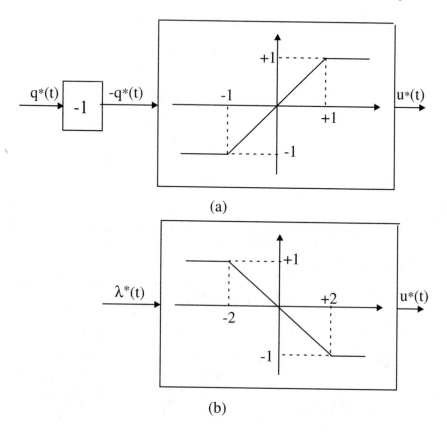

Figure 7.40   Relation between Optimal Control $u^*(t)$ vs (a) $q^*(t)$
and (b) $0.5\lambda^*(t)$

    4. $\lambda(0) < -2$: Here, Figure 7.39, curve (d), since $\lambda^*(t) < -2$,
the optimal control $u^*(t) = \{+1\}$.

The previous discussion refers to the *open-loop implementation* in
the sense that depending upon the values of the costate variable
$\lambda^*(t)$. However, in this scalar case, it may be possible to obtain
closed-loop implementation.

- **Step 4:** *Closed-Loop Implementation:* In this scalar case, it may
  be easy to get a closed-loop optimal control. First, let us note
  that if the final time t_f is *free* and the Hamiltonian (7.5.38) does
  not contain time t explicitly, then we know that

$$H(x^*(t), \lambda^*(t), u^*(t)) = 0 \;\; \forall \;\; t \in [0, t_f] \tag{7.5.46}$$

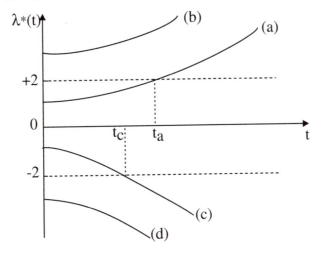

Figure 7.41   Possible Solutions of Optimal Costate $\lambda^*(t)$

which means that

$$u^{*^2}(t) + \lambda^*(t)[ax^*(t) + u^*(t)] = 0. \qquad (7.5.47)$$

Solving for the optimal state

$$x^*(t) = \frac{u^*(t)}{-a}\left[\frac{u^*(t)}{\lambda^*(t)} + 1\right]. \qquad (7.5.48)$$

Let us now discuss two situations.

1. *Saturated Region:* (i) At time $t = t_a$ (Figure 7.41(a)), $\lambda^*(t_a) = 2$, $u^*(t_a) = -1$, then the optimal state (7.5.48) becomes

$$x^*(t_a) = \frac{1}{2a}, \quad \text{and since} \quad a < 0, \quad x^*(t_a) < 0. \;(7.5.49)$$

Next, for time $t \in [t_a, t_f]$, $u^*(t) = -1$ and $\lambda^*(t) > 2$ and the relation (7.5.48) reveals that $x^*(t) < x^*(t_a)$. Combining this with (7.5.49), we have

$$x^*(t) < x^*(t_a) < 0 \qquad (7.5.50)$$

(ii) At time $t = t_c$, (Figure 7.41(c)), $\lambda^*(t_c) = -2$, $u^*(t_c) = +1$, then the optimal state (7.5.48) becomes

$$x^*(t_c) = -\frac{1}{2a}, \quad \text{and since} \quad a < 0, \quad x^*(t_c) > 0. \;(7.5.51)$$

Next, for time $t \in [t_c, t_f]$, $u^*(t) = +1$ and $\lambda^*(t) < -2$ and the relation (7.5.48) reveals that $x^*(t) > x^*(t_c)$. Combining this with (7.5.51), we have

$$x^*(t) > x^*(t_c) > 0. \qquad (7.5.52)$$

2. *Unsaturated Region:* During the unsaturated region,

$$|\lambda^*(t)| \le 1 \quad \text{and} \quad u^*(t) = -\frac{1}{2}\lambda^*(t) \qquad (7.5.53)$$

and using this, the Hamiltonian condition (7.5.47) becomes

$$u^{*2}(t) + \lambda^*(t)[ax^*(t) + u^*(t)] = 0$$

$$\frac{1}{4}\lambda^{*2}(t) + a\lambda^*(t)x^*(t) - \frac{1}{2}\lambda^{*(2)}(t) = 0 \longrightarrow$$

$$\lambda^*(t)\left[\frac{1}{4}\lambda^*(t) - ax^*(t)\right] = 0 \quad (7.5.54)$$

solution of which becomes

$$\lambda^*(t) = 0 \quad \text{or} \quad \lambda^*(t) = 4ax^*(t). \qquad (7.5.55)$$

Here, $\lambda^*(t) = 0$ is not admissible because then the optimal control (7.5.44) becomes zero. For $\lambda^*(t) = 4ax^*(t)$, the optimal control (7.5.44) becomes

$$u^*(t) = -2ax^*(t), \quad a < 0. \qquad (7.5.56)$$

The previous relation also means that

$$\begin{aligned}
&\text{If } x^*(t) > 0, \quad \text{then } u^*(t) = +1 \\
&\text{If } x^*(t) < 0, \quad \text{then } u^*(t) = -1 \\
&\text{If } x^*(t) = 0, \quad \text{then } u^*(t) = 0.
\end{aligned} \qquad (7.5.57)$$

Control Law: Combining the previous relations for unsaturated region and for the saturated region, we finally get the control law for the entire region as

$$u^*(t) = \begin{cases}
-1, & \text{if } x^*(t) < +\frac{1}{2a} < 0, \\
+1, & \text{if } x^*(t) > -\frac{1}{2a} > 0, \\
-2ax^*(t), & \text{if } x^*(t) > -\frac{1}{2a} > 0, \\
-2ax^*(t), & \text{if } x^*(t) < +\frac{1}{2a} < 0, \\
0, & \text{if } x^*(t) = 0
\end{cases} \qquad (7.5.58)$$

and the implementation of the energy-optimal control law is shown in Figure 7.42.

Further, for a combination of time-optimal and fuel-optimal control systems and other related problems with control constraints, see excellent texts [6, 116].

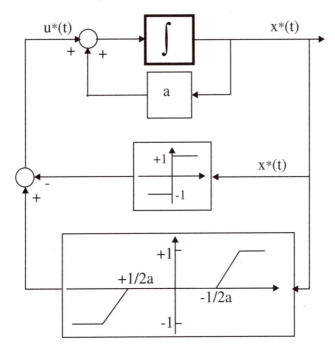

Figure 7.42   Implementation of Energy-Optimal Control Law

7.6   *Optimal Control Systems with State Constraints*

In the previous sections, we discussed the optimal control systems with *control constraints*. In this section, we address the optimal control systems with *state constraints* [79, 120].

Optimal control systems with state constraints (Constrained Optimal Control) has been of great interest to engineers. Some examples of state-constrained problems are the solution of the minimum time-to-climb problem for an aircraft that is required to start within a specified flight envelope, the determination of the best control policy for an industrial mechanical robot subject to path constraints, and the speed of an electric motor which cannot exceed a certain value without damaging some of the mechanical components such as bearings and shaft. There have been several methods proposed to handle state variable inequality constraints. In general, there are three methods for handling these systems [49]:

1. slack variables,

2. penalty functions, and

3. interior-point constraints.

Let us first consider the penalty function method.

7.6.1 *Penalty Function Method*

Let us consider the system as

$$\dot{\mathbf{x}}(t) = \mathbf{f}(\mathbf{x}(t), \mathbf{u}(t), t) \tag{7.6.1}$$

and the performance index as

$$J = \int_{t_0}^{t_f} V(\mathbf{x}(t), \mathbf{u}(t), t) \tag{7.6.2}$$

where, $\mathbf{x}(t)$ and $\mathbf{u}(t)$ are n and r dimensional state and control vectors, respectively. Let the inequality constraints on the states be expressed as

$$\mathbf{g}(\mathbf{x}(t), t) \geq \mathbf{0} \tag{7.6.3}$$

or

$$
\begin{aligned}
g_1(x_1(t), x_2(t), \ldots, x_n(t), t) &\geq 0 \\
g_2(x_1(t), x_2(t), \ldots, x_n(t), t) &\geq 0 \\
\cdots\cdots\cdots\cdots\cdots\cdots \\
g_p(x_1(t), x_2(t), \ldots, x_n(t), t) &\geq 0
\end{aligned}
\tag{7.6.4}
$$

where, $\mathbf{g}$ is a $p \leq n$ vector function of the states and assumed to have continuous first and second partial derivatives with respect to state $\mathbf{x}(t)$. There are several methods of solving this system where the *inequality* constraints (7.6.3) are converted to *equality* constraints. One such methodology is described below. Let us define a new variable $x_{n+1}(t)$ by

$$
\begin{aligned}
\dot{x}_{n+1}(t) &\cong f_{n+1}(\mathbf{x}(t), t), \\
&= [g_1(\mathbf{x}(t), t]^2 H(g_1) + [g_2(\mathbf{x}(t), t)]^2 H(g_2) + \cdots \\
&\quad + [g_p(\mathbf{x}(t), t)] H(g_p),
\end{aligned}
\tag{7.6.5}
$$

where, $H(g_i)$ is a unit Heaviside step function defined by

$$H(g_i) = \begin{cases} 0, & \text{if } g_i(\mathbf{x}(t), t) \geq 0, \\ 1, & \text{if } g_i(\mathbf{x}(t), t) < 0, \end{cases} \tag{7.6.6}$$

for $i = 1, 2, \ldots, p$. The relations (7.6.6) and (7.6.5) mean that $\dot{x}_{n+1}(t) = 0$ for all t when the constraint relation (7.6.3) is satisfied and $\dot{x}_{n+1}(t) \geq 0$ for all t due to the square terms in (7.6.5). Further, let us require that the new variable $x_{n+1}(t)$ has the boundary conditions

$$x_{n+1}(t_0) = 0, \quad \text{and} \quad x_{n+1}(t_f) = 0 \tag{7.6.7}$$

such that

$$
\begin{aligned}
x_{n+1}(t) &= \int_{t_0}^{t} \dot{x}_{n+1}(t) dt \\
&= \int_{t_0}^{t} \Big\{ [g_1(\mathbf{x}(t), t]^2 \, H(g_1) + [g_2(\mathbf{x}(t), t)]^2 \, H(g_2) + \cdots \\
&\quad + [g_p(\mathbf{x}(t), t)]^2 \, H(g_p) \Big\} dt.
\end{aligned}
\tag{7.6.8}
$$

Now we use the Hamiltonian approach to minimize the PI (7.6.2) subject to the system equation (7.6.1) and the state inequality constraint (7.6.3). Let us define the Hamiltonian as

$$
\begin{aligned}
\mathcal{H}(\mathbf{x}(t), &\mathbf{u}(t), \boldsymbol{\lambda}(t), \lambda_{n+1}(t), t) \\
&= V(\mathbf{x}(t), \mathbf{u}(t), t) + \boldsymbol{\lambda}'(t) \mathbf{f}(\mathbf{x}(t), \mathbf{u}(t), t) \\
&\quad + \lambda_{n+1}(t) \Big\{ [g_1(\mathbf{x}(t), t]^2 \, H(g_1) + [g_2(\mathbf{x}(t), t)]^2 \, H(g_2) + \cdots \\
&\quad + [g_p(\mathbf{x}(t), t)]^2 \, H(g_m) \Big\}, \\
&= V(\mathbf{x}(t), \mathbf{u}(t), t) + \boldsymbol{\lambda}'(t) \mathbf{f}(\mathbf{x}(t), \mathbf{u}(t), t) \\
&\quad + \lambda_{n+1}(t) f_{n+1}(\mathbf{x}(t), t).
\end{aligned}
\tag{7.6.9}
$$

Thus, the previous Hamiltonian is formed with $n+1$ costates and $n+1$ states. Note that the Hamiltonian (7.6.9) does not explicitly contain the new state variable $x_{n+1}(t)$. Now, we apply the necessary optimality conditions for the state as

$$\dot{\mathbf{x}}^*(t) = \frac{\partial \mathcal{H}}{\partial \boldsymbol{\lambda}} = \mathbf{f}(\mathbf{x}^*(t), \mathbf{u}^*(t), t),$$

$$\dot{x}_{n+1}^*(t) = \frac{\partial \mathcal{H}}{\partial \lambda_{n+1}} = f_{n+1}(\mathbf{x}^*(t), t), \tag{7.6.10}$$

for the costate as

$$\dot{\boldsymbol{\lambda}}^*(t) = -\frac{\partial \mathcal{H}}{\partial \mathbf{x}},$$

$$\dot{\lambda}_{n+1}^*(t) = -\frac{\partial \mathcal{H}}{\partial x_{n+1}} = 0 \qquad (7.6.11)$$

and for the control as

$$\boxed{\mathcal{H}(\mathbf{x}^*(t), \mathbf{u}(t), \boldsymbol{\lambda}^*(t), \lambda_{n+1}^*(t), t) \le \mathcal{H}(\mathbf{x}^*(t), \mathbf{u}(t), \boldsymbol{\lambda}^*(t), \lambda_{n+1}^*(t), t).}$$

$$(7.6.12)$$

or

$$\boxed{\min_{|\mathbf{u}(t)| \le \mathbf{U}} \left\{ \mathcal{H}(\mathbf{x}^*(t), \mathbf{u}(t), \boldsymbol{\lambda}^*(t), \lambda_{n+1}^*(t), t) \right\} = \mathcal{H}(\mathbf{x}^*(t), \mathbf{u}(t), \boldsymbol{\lambda}^*(t), \lambda_{n+1}^*(t), t).}$$

$$(7.6.13)$$

Note that in the above, $\dot{\lambda}_{n+1}^*(t) = 0$ because the Hamiltonian (7.6.9) does not contain $x_{n+1}(t)$ explicitly (see Table 7.1).

Let us now illustrate the previous method by an example.

Example 7.2

Consider a second order system

$$\dot{x}_1(t) = x_2(t)$$
$$\dot{x}_2(t) = u(t), \qquad (7.6.14)$$

and the performance index

$$J(u) = \frac{1}{2} \int_{t_0}^{t_f} \left[x_1^2(t) + u^2(t) \right] dt, \qquad (7.6.15)$$

where, time t_f is free and the final state $\mathbf{x}(t_f)$ is free. The control $u(t)$ is constrained as

$$-1 \le u(t) \le +1 \quad \text{or} \quad |u(t)| \le +1 \text{ for } t \in [t_0, t_f], \qquad (7.6.16)$$

and the state $x_2(t)$ is constrained as

$$-3 \le x_2(t) \le +3 \quad \text{or} \quad |x_2(t)| \le +3 \text{ for } t \in [t_0, t_f]. \qquad (7.6.17)$$

Find the optimal control.

Table 7.1   Procedure Summary of Optimal Control Systems with State Constraints

A. Statement of the Problem
Given the system as $\dot{\mathbf{x}}(t) = \mathbf{f}(\mathbf{x}(t), \mathbf{u}(t), t)$, the performance index as $J = S(\mathbf{x}(t_f), t_f) + \int_{t_0}^{t_f} V(\mathbf{x}(t), \mathbf{u}(t), t) dt$, the state constraints as $\mathbf{g}(\mathbf{x}(t), t) \geq 0$, and the boundary conditions as $\mathbf{x}(t_0) = \mathbf{x}_0$ and t_f and $\mathbf{x}(t_f) = \mathbf{x}_f$ are free, find the optimal control.

B. Solution of the Problem	
Step 1	Form the Pontryagin $\mathcal{H}$ function $\mathcal{H}(\mathbf{x}(t), \mathbf{u}(t), \boldsymbol{\lambda}(t), \lambda_{n+1}(t), t) = V(\mathbf{x}(t), \mathbf{u}(t), t)$ $+ \boldsymbol{\lambda}'(t) \mathbf{f}(\mathbf{x}(t), \mathbf{u}(t), t) + \lambda_{n+1}(t) f_{n+1}(\mathbf{x}(t), t)$.
Step 2	Solve the set of $2n + 2$ differential equations $\dot{\mathbf{x}}^*(t) = \left(\dfrac{\partial \mathcal{H}}{\partial \boldsymbol{\lambda}}\right)_*, \quad \dot{x}_{n+1}^*(t) = \left(\dfrac{\partial \mathcal{H}}{\partial \lambda_{n+1}}\right)_*,$ and $\dot{\boldsymbol{\lambda}}^*(t) = -\left(\dfrac{\partial \mathcal{H}}{\partial \mathbf{x}}\right)_*, \quad \dot{\lambda}_{n+1}^*(t) = -\left(\dfrac{\partial \mathcal{H}}{\partial x_{n+1}}\right)_* = 0$ with boundary conditions $\mathbf{x}_0, x_{n+1}(t_0) = 0, x_{n+1}(t_f) = 0$ and $\left[\mathcal{H} + \dfrac{\partial S}{\partial t}\right]_{*t_f} \delta t_f + \left[\dfrac{\partial S}{\partial \mathbf{x}} - \boldsymbol{\lambda}\right]'_{*t_f} \delta \mathbf{x}_f = 0$.
Step 3	Minimize $\mathcal{H}$ w.r.t. $\mathbf{u}(t) (\leq \mathbf{U})$ $\mathcal{H}(\mathbf{x}^*(t), \mathbf{u}^*(t), \boldsymbol{\lambda}^*(t), \lambda_{n+1}^*(t), t)$ $\qquad \leq \mathcal{H}(\mathbf{x}^*(t), \mathbf{u}(t), \boldsymbol{\lambda}^*(t), \lambda_{n+1}^*(t), t)$.

Solution: To express the state constraint (7.6.17) as state inequality constraint (7.6.3), let us first note

$$[x_2(t) + 3] \geq 0, \quad \text{and} \tag{7.6.18}$$
$$[3 - x_2(t)] \geq 0 \tag{7.6.19}$$

and then

$$g_1(\mathbf{x}(t)) = [x_2(t) + 3] \geq 0,$$
$$g_2(\mathbf{x}(t)) = [3 - x_2(t)] \geq 0. \tag{7.6.20}$$

- **Step 1:** First formulate the Hamiltonian as

$$\mathcal{H}(\mathbf{x}(t), u(t), \boldsymbol{\lambda}(t), \lambda_3(t))$$
$$= \frac{1}{2}x_1^2(t) + \frac{1}{2}u^2(t) + \lambda_1(t)x_2(t) - \lambda_2(t)u(t)$$
$$+ \lambda_3(t)\left\{[x_2(t) + 3]^2 H(x_2(t) + 3)\right.$$
$$\left. + [3 - x_2(t)]^2 H(3 - x_2(t))\right\}. \tag{7.6.21}$$

- **Step 2:** The *necessary* condition for the state (7.6.10) becomes

$$\dot{x}_1^*(t) = x_2^*(t),$$
$$\dot{x}_2^*(t) = u^*(t),$$
$$\dot{x}_3^*(t) = [x_2(t) + 3]^2 H(x_2(t) + 3) + [3 - x_2(t)]^2 H(3 - x_2(t)), \tag{7.6.22}$$

and for the costate (7.6.11)

$$\dot{\lambda}_1^*(t) = -\frac{\partial \mathcal{H}}{\partial x_1} = -x_1^*(t),$$

$$\dot{\lambda}_2^*(t) = -\frac{\partial \mathcal{H}}{\partial x_2} = -\lambda_1^*(t) - 2\lambda_3^*(t)[x_2^*(t) + 3]H(x_2^*(t) + 3)$$
$$+ 2\lambda_3^*(t)[3 - x_2^*(t)]H(3 - x_2^*(t))$$

$$\dot{\lambda}_3^*(t) = -\frac{\partial \mathcal{H}}{\partial x_3} = 0 \longrightarrow \lambda_3^*(t) = \text{constant}. \tag{7.6.23}$$

- **Step 3:** Minimize $\mathcal{H}$ w.r.t. the control (7.6.13)

$$\mathcal{H}(\mathbf{x}^*(t), u^*(t), \boldsymbol{\lambda}^*(t), \lambda_3^*(t)) \leq \mathcal{H}(\mathbf{x}^*(t), u(t), \boldsymbol{\lambda}^*(t), \lambda_3^*(t)). \tag{7.6.24}$$

Using (7.6.21) in the condition (7.6.24) and taking out the terms not containing the control $u(t)$ explicitly, we get

$$\frac{1}{2}u^{*2}(t) + \lambda_2^*(t)u^*(t) \leq \frac{1}{2}u^2(t) + \lambda_2^*(t)u(t)$$

$$= \min_{|u| \leq 1}\left\{\frac{1}{2}u^2(t) + \lambda_2^*(t)u(t)\right\}. \tag{7.6.25}$$

By simple calculus, we see that the expression $\frac{1}{2}u^2(t) + \lambda_2^*(t)u(t)$ will attain the *optimum* value for

$$u^*(t) = -\lambda_2^*(t) \qquad (7.6.26)$$

when the control $u^*(t)$ is *unconstrained*. This can also be seen alternatively by using the relation

$$\frac{\partial \mathcal{H}}{\partial u} = 0 \longrightarrow u^*(t) + \lambda_2^*(t) = 0 \longrightarrow u^*(t) = -\lambda_2^*(t). \qquad (7.6.27)$$

But, for the present *constrained* control situation (7.6.14), we see from (7.6.25) or (7.6.26) that

$$u^*(t) = \begin{cases} -1, & \text{if} \quad \lambda_2^*(t) > +1 \\ +1, & \text{if} \quad \lambda_2^*(t) < -1. \end{cases} \qquad (7.6.28)$$

Combining the *unsaturated* or *unconstrained* control (7.6.26) with the *saturated* or *constrained* control (7.6.28), we have

$$u^*(t) = \begin{cases} +1, & \text{if} \quad \lambda_2^*(t) < -1 \\ -1, & \text{if} \quad \lambda_2^*(t) > +1 \\ -\lambda_2^*(t), & \text{if} \quad -1 \le \lambda_2^*(t) \le +1. \end{cases} \qquad (7.6.29)$$

Using the definition of *saturation* function (7.5.28), the previous optimal control strategy can be written as

$$u_j^*(t) = -sat\left\{\lambda_2^*(t)\right\}. \qquad (7.6.30)$$

The situation is shown in Figure 7.43. Thus, one has to solve for the costate function $\lambda_2^*(t)$ completely to find the optimal control $u^*(t)$ from (7.6.29) to get *open-loop optimal control implementation*.

Note: In obtaining the optimal control strategy *in general*, one cannot obtain the *unconstrained* or *unsaturated* control first and then just extend the same for *constrained* or *saturated* region. Instead, one has to really use the Hamiltonian relation (7.6.13) to obtain the optimal control. Although, in this chapter, we considered control constraints and state constraints separately, we can combine both of them and have a situation with constraints as

$$\mathbf{g}(\mathbf{x}(t), \mathbf{u}(t), t) \le 0. \qquad (7.6.31)$$

For further details, see the recent book [61] and the survey article [62].

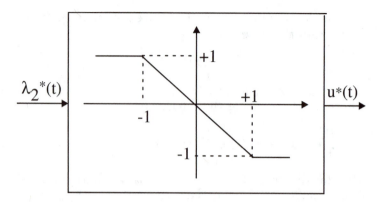

Figure 7.43   Relation between Optimal Control $u^*(t)$ and Optimal
Costate $\lambda_2^*(t)$

7.6.2   Slack Variable Method

The slack variable approach [68, 137], often known as Valentine's method,
transforms the given *inequality* state (path) constraint into an *equality*
state (path) constraint by introducing a slack variable. For the sake of
completeness, let us restate the state constraint problem.

Consider the optimal control system

$$\dot{\mathbf{x}}(t) = \mathbf{f}(\mathbf{x}(t), \mathbf{u}(t), t), \quad \mathbf{x}(t = t_o) = \mathbf{x}_o \tag{7.6.32}$$

which minimizes the performance index

$$\mathbf{J} = F(\mathbf{x}(t_f), t_f) + \int_{t_o}^{t_f} V(\mathbf{x}(t), \mathbf{u}(t), t) dt \tag{7.6.33}$$

subject to the state-variable inequality constraint

$$S(\mathbf{x}(t), t) \leq 0. \tag{7.6.34}$$

Here, $\mathbf{x}(t)$ is an n-dimensional order state vector, $\mathbf{u}(t)$ is an r-dimensional
control vector and the constraint S is of pth order in the sense that the
pth derivative of S contains the control $\mathbf{u}(t)$ explicitly.

The state-constrained, optimal control problem is solved by con-
verting the given *inequality* constrained problem into an *equality* constrained one by introducing a "slack variable," as [68, 137]

$$S(\mathbf{x}(t), t) + \frac{1}{2}\alpha^2(t) = 0. \tag{7.6.35}$$

Differentiating (7.6.35) p times with respect to time t, we obtain

$$S_1(\mathbf{x}(t), t) + \alpha \alpha_1 = 0$$
$$S_2(\mathbf{x}(t), t) + \alpha_1^2 + \alpha \alpha_2 = 0$$
$$\dots$$

$$S_p(\mathbf{x}(t), \underline{u}(t), t) + \{\text{terms involving } \alpha(t), \alpha_1)(t), , ..., \alpha_p(t)\} = 0$$

$$(7.6.36)$$

where, the subscripts on S and α denote the time derivatives, that is,

$$S_1 = \frac{dS}{dt} = \left(\frac{\partial S}{\partial \mathbf{x}}\right)\left(\frac{d\mathbf{x}}{dt}\right) + \frac{\partial S}{\partial t} \text{ and } \alpha_1 = \frac{d\alpha}{dt}. \qquad (7.6.37)$$

Since the control $\mathbf{u}(t)$ is explicitly present in the pth derivative equation, we can solve for the control to obtain

$$\mathbf{u}(t) = g(\mathbf{x}(t), \alpha(t), \alpha_1)(t), , ..., \alpha_p(t), t). \qquad (7.6.38)$$

Substituting the control (7.6.38) in the plant (7.6.32) and treating the various $\alpha, ..., \alpha_{p-1}$ as additional state variables, the new unconstrained control becomes α_p. Thus,

$$\dot{\mathbf{x}}(t) = f(\mathbf{x}(t), g(\mathbf{x}(t), \alpha(t), \alpha)1(t), , ..., \alpha_p(t), t), \quad x(t = t_0) = \mathbf{x}_0$$
$$\dot{\alpha} = \alpha_1, \quad \alpha(t = t_0) = \alpha(t_0)$$
$$\dot{\alpha}_1 = \alpha_2, \quad \alpha_1(t = t_0) = \alpha_1(t_0)$$
$$\dot{\alpha}_{p-1} = \alpha_p, \quad \alpha_{p-1}(t = t_0) = \alpha_{p-1}(t_0). \qquad (7.6.39)$$

The new cost functional is then given by

$$J = F(\mathbf{x}(t_f), t_f) + \int_{t_o}^{t_f} V(\mathbf{x}(t), g(\mathbf{x}(t), \alpha(t), \alpha_)(t), , ..., \alpha_p(t), t), t)dt$$

$$(7.6.40)$$

The new initial conditions $\alpha(t_0), ..., \alpha_{p-1}(t_0)$ are required to satisfy (7.6.35) and (7.6.36), so after some algebraic manipulations, we get

$$\alpha(t_0) = \pm\sqrt{-2S(\mathbf{x}(t_0), t_0)}$$
$$\alpha_1(t_0) = -S_1(\mathbf{x}(t_0), t_0)/\alpha(t_0)$$
$$\alpha_2(t_0) = -[S_2(\mathbf{x}(t_0), t_0) + \alpha_1^2(t_0)]/\alpha(t_0)$$
$$\dots$$

$$(7.6.41)$$

With this choice of boundary conditions, the original relations w.r.t. the constraints (7.6.35) and (7.6.36) are satisfied for all t for any control function $\alpha_p(\cdot)$. In other words, any function $\alpha_p(\cdot)$ will produce an admissible trajectory. Thus, the original *constrained* problem is transformed into an *unconstrained* problem.

Now we apply the Pontryagin Principle to this unconstrained problem. [68, 27, 40, 62]. In general terms, we define a new $n + p$th state vector

$$\mathcal{Z}(t) = [\mathbf{x}(t), \alpha(t), \cdots \alpha_{p-1}]' \tag{7.6.42}$$

then, the new plant (7.6.39) becomes

$$\dot{\mathcal{Z}} = \mathcal{F}(\mathcal{Z}(t), \alpha_p(t), t) \tag{7.6.43}$$

where the $(n+p)$-dimensional vector function $\mathcal{F}$ represents the right-hand side of (7.6.39). Next, we define the Hamiltonian as

$$\mathcal{H} = V + \lambda \mathcal{F} \tag{7.6.44}$$

where λ is an $n + p$-dimensional Lagrange multiplier. Then, for the state

$$\dot{\mathcal{Z}}(t) = \mathcal{H}_\lambda, \quad \mathcal{Z}(t_0), \tag{7.6.45}$$

for the costate

$$-\dot{\lambda}(t) = \mathcal{H}_{\mathcal{Z}}, \quad \lambda(t_f) = [F_x, 0, \cdots 0]', \tag{7.6.46}$$

and for the control

$$\frac{\partial \mathcal{H}}{\partial \alpha_p} = \mathcal{H}_{\alpha_p} = 0, \tag{7.6.47}$$

where the subscripts in $\mathcal{H}_\lambda$, $\mathcal{H}_{\mathcal{Z}}$, and F_x denote the partial derivative with respect to the subscripted variable.

The previous set of equations for the states (7.6.45) and costates (7.6.46) and their initial and final conditions constitute a two point boundary value problem (TPBVP). Such problems can be solved, depending on the difficulty, with a closed solution, or highly nonlinear problems must be solved with specialized software [68, 129, 130, 63].

7.7 Problems

1. Make reasonable assumptions wherever necessary.

2. Use MATLAB© wherever possible to solve the problems and plot all the optimal controls and states for all problems. Provide the relevant MATLAB© m files.

Problem 7.1 Derive the expressions for minimum time given by (7.2.28) for a double integral system.

Problem 7.2 A second order system, described by

$$\dot{x}_1(t) = x_2(t)$$
$$\dot{x}_2(t) = -2x_2(t) + x_1(t) + u(t)$$

where, the initial and final states are specified, is to minimize the performance index

$$J = \frac{1}{2} \int_0^1 \left[2x_1^2(t) + x_2^2(t) + u^2(t) \right] dt.$$

Find the optimal control $u^*(t)$ for
(a) $u(t)$ unconstrained, and
(b) $u(t)$ constrained as $|u(t)| \leq 1$.

Problem 7.3 Find the optimal control law for transferring the second order linear system

$$\dot{x}_1(t) = x_2(t)$$
$$\dot{x}_2(t) = u(t)$$

where, (a) the control $u(t)$ is unconstrained and (b) the control $|u(t)| \leq 1$, from any arbitrary initial state to the final state $[2, 2]$ in minimum time.

Problem 7.4 For the second order, linear system

$$\dot{x}_1(t) = -x_1(t) - u(t)$$
$$\dot{x}_2(t) = -3x_2(t) - 2u(t)$$

to be transferred from any arbitrary initial state to origin in minimum time, find the optimal control law if the control $u(t)$ is (a) unconstrained and (b) constrained as $|u(t)| \leq 1$.

Problem 7.5 Given a second order linear system

$$\dot{x}_1(t) = -x_1(t) - u(t)$$
$$\dot{x}_2(t) = -3x_2(t) - 2u(t), \quad |u(t)| \leq 1$$

find the expression for minimum time to transfer the above system from any initial state to the origin.

Problem 7.6 For a first order system

$$\dot{x}(t) = u(t), \quad |u(t)| \leq 1$$

find the optimal control law to minimize the performance index

$$J = \int_0^{t_f} |u(t)| dt \quad t_f \text{ is free}$$

so that the system is driven from $x(0) = x_0$ to origin.

Problem 7.7 Formulate and solve Problem 7.4 as fuel-optimal control problem.

Problem 7.8 A second order system

$$\dot{x}_1(t) = x_2(t)$$
$$\dot{x}_2(t) = -ax_2(t) + u(t), \quad a > 0$$

with control constraint as $|u(t)| \leq 1$, discuss the optimal control strategy to transfer the system to origin and at the same time minimize the performance index

$$J = \int_0^{t_f} [\beta + |u(t)|] dt$$

where, the final time t_f is free and $\beta > 0$.

Problems

Problem 7.9 For a double integral plant

$$\dot{x}_1(t) = x_2(t)$$
$$\dot{x}_2(t) = u(t)$$

with control constraint $|u(t)| \le 1$, find the optimal control which transfers the plant from initial condition $x_1(0) = 1, x_2(0) = 1$ to the final condition $x_1(t_f) = x_2(t_f) = 0$ in such a way so as to minimize the performance measure

$$J = \int_0^4 u^2(t)dt$$

and calculate the minimum value J^*.

Problem 7.10 For a second-order system

$$\dot{x}_1(t) = x_2(t)$$
$$\dot{x}_2(t) = -2x_2(t) + 3x_2(t) + 5u(t)$$

with control constraint $|u(t)| \le 1$, find the optimal control which transfers the plant from initial condition $x_1(0) = 1, x_2(0) = 1$ to the final condition $x_1(t_f) = x_2(t_f) = 0$ in such a way so as to minimize the performance measure

$$J = \int_0^5 u^2(t)dt$$

and calculate the minimum value J^*.

Problem 7.11 The double integral plant

$$\dot{x}_1(t) = x_2(t)$$
$$\dot{x}_2(t) = u(t)$$

is to be transferred from any state to the origin in minimum time with the state and control constraints as $|u(t)| \le 1$ and $|x_2(t)| \le 2$. Determine the optimal control law.

Problem 7.12 For the liquid-level control system described in Problem 1.2, formulate the time-optimal control problem and find the optimal control law.

Problem 7.13 For the D.C. motor speed control system described in Problem 1.1, formulate the minimum-energy problem and find the optimal control law if the control input is constrained as $u(t) \leq 1$.

Problem 7.14 For the mechanical control system described in Problem 1.4, formulate the minimum-energy problem and find the optimal control.

Problem 7.15 For the automobile suspension system described in Problem 1.5, formulate the minimum-energy problem and find the optimal control.

Problem 7.16 For the chemical control system described in Problem 1.6, formulate the minimum-energy problem and find the optimal control.

@@@@@@@@@@@@@

Appendix A

Vectors and Matrices

The main purpose of this appendix is to provide a brief summary of the results on matrices, vectors and matrix algebra to serve as a review of these topics rather than any in depth treatment of the topics. For more details on this subject, the reader is referred to [54, 10, 13].

A.1 Vectors

Vector

A *vector* $\mathbf{x}$, generally considered as a *column vector*, is an arrangement of n elements in a column as

$$\mathbf{x} = \begin{bmatrix} x_1 \\ x_2 \\ \cdot \\ \cdot \\ \cdot \\ x_n \end{bmatrix}, \tag{A.1.1}$$

. The number n is also referred to as *order, size* or *dimensions* of the vector. We can also write the vector $\mathbf{x}$ as

$$\mathbf{x} = [x_1 \, x_2 \, ... \, x_n]' \tag{A.1.2}$$

where, $'$ denotes the transpose as defined below.

Transpose of a Vector

The *transpose* of a vector $\mathbf{x}$ is the interchange of the column vector into a row vector. Thus

$$\mathbf{x}' = \begin{bmatrix} x_1 & x_2 & \cdots & x_n \end{bmatrix}. \tag{A.1.3}$$

Norm of a Vector

The norm of a vector $\mathbf{x}$, written as $||\mathbf{x}||$, is a measure of the size or length of the vector. Further,

1. $||\mathbf{x}|| > 0$ for all $\mathbf{x}$ and $||\mathbf{x}|| = 0$ only if $\mathbf{x} = 0$.

2. $||\alpha\mathbf{x}|| = a||\mathbf{x}||$ for any scalar α and for all $\mathbf{x}$.

3. $||\mathbf{x} + \mathbf{y}|| \leq ||\mathbf{x}|| + ||\mathbf{y}||$ for all $\mathbf{x}$ and $\mathbf{y}$, called the Schwartz inequality.

The norm is calculated by either of the following ways

1.

$$||\mathbf{x}||^2 = <\mathbf{x}, \mathbf{x}> = \mathbf{x}'\mathbf{x}, \quad \text{or}$$

$$||\mathbf{x}|| = \left[\sum_{i=1}^{n} x_i^2 \right]^{1/2}, \quad \text{called the Euclidean norm} \tag{A.1.4}$$

2.

$$||\mathbf{x}|| = \max|x_i| \tag{A.1.5}$$

3.

$$||\mathbf{x}|| = \sum_{i=1}^{n} |x_i| \tag{A.1.6}$$

Multiplication of Vectors

The multiplication of two vectors is done by transposing one of the vectors and then multiplying this vector with the other vector. Thus,

$$\mathbf{x}'\mathbf{y} = <\mathbf{x}, \mathbf{y}> = \sum_{i}^{n} x_i y_i = x_1 y_1 + x_2 y_2 + \cdots + x_n y_n. \tag{A.1.7}$$

This product $< \mathbf{x}, \mathbf{y} >$ which is a scalar, is often called the *inner product* of these two vectors.

On the other hand, the *outer product* $\mathbf{x} >< \mathbf{y}$ of two vectors is defined as

$$
\mathbf{x} >< \mathbf{y} = \mathbf{x}\mathbf{y}' = \begin{bmatrix} x_1 y_1 & x_1 y_2 & \cdots & x_1 y_n \\ x_2 y_1 & x_2 y_2 & \cdots & x_2 y_n \\ \cdots & \cdots & \cdots & \cdots \\ x_n y_1 & x_n y_2 & \cdots & x_n y_n \end{bmatrix},
\tag{A.1.8}
$$

which is a *matrix* defined next.

A.2   Matrices

Matrix

An nxm *matrix* $\mathbf{A}$ is an arrangement of nm elements a_{ij} $(i = 1, 2, ..., n;$ $j = 1, 2, ..., m)$ into n rows and m columns as

$$
\mathbf{A} = \begin{bmatrix} a_{11} & a_{12} & \cdots & a_{1m} \\ a_{21} & a_{22} & \cdots & a_{2m} \\ \cdots & \cdots & \cdots & \cdots \\ a_{n1} & a_{n2} & \cdots & a_{nm} \end{bmatrix}.
\tag{A.2.1}
$$

The nxn of the matrix $\mathbf{A}$ is also referred to as *order*, *size* or *dimension* of the matrix.

Square Matrix

If the number of rows and columns is the same, that is, if $m = n$ in the matrix $\mathbf{A}$ of (A.2.1), then it is called a *square* matrix.

Unity Matrix

A *unity* matrix $\mathbf{I}$ is defined as the matrix having values 1 for all diagonal elements and having the rest of the elements as zero as

$$
\mathbf{I} = \begin{bmatrix} 1 & 0 & 0 & \cdots & 0 \\ 0 & 1 & 0 & \cdots & 0 \\ \cdots & \cdots & \cdots & \cdots & \cdots \\ 0 & 0 & 0 & \cdots & 1 \end{bmatrix}.
\tag{A.2.2}
$$

Addition/Subtraction of Matrices

The addition (or subtraction) of two matrices $\mathbf{A}$ and $\mathbf{B}$ is simply the addition of the corresponding elements in a particular row and column, and hence, obviously these two matrices should be of the same size or order. Thus we get a new matrix $\mathbf{C}$ as

$$\mathbf{C} = \mathbf{A} + \mathbf{B} \tag{A.2.3}$$

where, $c_{ij} = a_{ij} + b_{ij}$. The addition of two matrices is *commutative* as

$$\mathbf{A} + \mathbf{B} = \mathbf{B} + \mathbf{A}. \tag{A.2.4}$$

Multiplication of a Matrix by a Scalar

The scalar multiplication of two matrices of the same order, and the addition or subtraction is easily seen to be

$$\mathbf{C} = \alpha_1 \mathbf{A} + \alpha_2 \mathbf{B} \tag{A.2.5}$$

where, α_1 and α_2 are scalars, and

$$c_{ij} + \alpha_1 a_{ij} + \alpha_2 b_{ij}. \tag{A.2.6}$$

Multiplication of Matrices

The product of nxp matrix $\mathbf{A}$ and pxm matrix $\mathbf{B}$ is defined as

$$\mathbf{C} = \mathbf{AB}, \quad \text{where}$$
$$c_{ij} = \sum_{k=1}^{p} a_{ik} b_{kj}. \tag{A.2.7}$$

Note that the element c_{ij} is formed by summing the multiplication of elements in the i *row* of matrix $\mathbf{A}$ and with the elements in the j *column* of the matrix $\mathbf{B}$. Obviously, the columns of $\mathbf{A}$ should be the same as the rows of $\mathbf{B}$ so that the resultant matrix $\mathbf{C}$ has the same rows of $\mathbf{A}$ and the same columns of $\mathbf{B}$.

The product of two or more matrices is defined as

$$\mathbf{D} = \mathbf{ABC} = (\mathbf{AB})\,\mathbf{C} = \mathbf{A}\,(\mathbf{BC}). \tag{A.2.8}$$

However, note that in general

$$\mathbf{AB} \neq \mathbf{BA}. \tag{A.2.9}$$

Thus, the multiplication process is *associative*, but not generally *commutative*. However, with *unity* matrix $\mathbf{I}$, we have

$$\mathbf{AI} = \mathbf{A} = \mathbf{IA}. \tag{A.2.10}$$

Transpose of a Matrix

A *transpose* of a matrix $\mathbf{A}$, denoted as $\mathbf{A}'$, is obtained by interchanging the rows and columns. Thus, the transpose $\mathbf{B}'$ of the matrix $\mathbf{A}$ is written as

$$\mathbf{B} = \mathbf{A}' \quad \text{such that} \quad b_{ij} = a_{ji}. \tag{A.2.11}$$

Also, it can be easily seen that the transpose of the *sum* of two matrices is

$$(\mathbf{A} + \mathbf{B})' = \mathbf{A}' + \mathbf{B}'. \tag{A.2.12}$$

The transpose of the *product* of two or more matrices is defined as

$$(\mathbf{AB})' = \mathbf{B}'\mathbf{A}'$$
$$(\mathbf{ABC})' = \mathbf{C}'\mathbf{B}'\mathbf{A}'. \tag{A.2.13}$$

Symmetric Matrix

A *symmetric* matrix is one whose row elements are the same as the corresponding column elements. Thus, $a_{ij} = a_{ji}$. In other words, if $\mathbf{A} = \mathbf{A}'$, then the matrix $\mathbf{A}$ is symmetric.

Norm of a Matrix

For matrices, the various norms are defined as

1. $||\mathbf{Ax}|| \leq ||\mathbf{A}||.||\mathbf{x}||,$

2. $||\mathbf{A} + \mathbf{B}|| \leq ||\mathbf{A}|| + ||\mathbf{B}||,$ called the Schwartz inequality,

3. $||\mathbf{AB}|| \leq ||\mathbf{A}||.||\mathbf{B}||$

where . denotes multiplication.

Determinant

The *determinant* $|\mathbf{A}|$ of an *nxn* matrix $\mathbf{A}$ is evaluated in many ways. One of the ways for a 3x3 matrix $\mathbf{A}$ is as follows.

$$\mathbf{A} = \begin{bmatrix} a_{11} & a_{12} & a_{13} \\ a_{21} & a_{22} & a_{23} \\ a_{31} & a_{32} & a_{33} \end{bmatrix} ;$$

$$|\mathbf{A}| = a_{11}(-1)^{1+1}\begin{vmatrix} a_{22} & a_{23} \\ a_{32} & a_{33} \end{vmatrix} + a_{12}(-1)^{1+2}\begin{vmatrix} a_{21} & a_{23} \\ a_{31} & a_{33} \end{vmatrix}$$

$$+ a_{13}(-1)^{1+3}\begin{vmatrix} a_{21} & a_{22} \\ a_{31} & a_{32} \end{vmatrix}. \tag{A.2.14}$$

Note that in (A.2.14), the sub-determinant associated with a_{11} is formed by deleting the row and column containing a_{11}. Thus, the 3x3 determinant $|\mathbf{A}|$ is expressed in terms of the 2x2 sub-determinants. Once again, this 2x2 sub-determinant can be written, for example, as

$$\begin{vmatrix} a_{22} & a_{23} \\ a_{32} & a_{33} \end{vmatrix} = a_{22}a_{33} - a_{23}a_{32}. \tag{A.2.15}$$

Some useful results on determinants are

$$|\mathbf{A}| = |\mathbf{A}'|$$
$$|\mathbf{AB}| = |\mathbf{A}|.|\mathbf{B}|$$
$$|\mathbf{I} + \mathbf{AB}| = |\mathbf{I} + \mathbf{BA}|. \tag{A.2.16}$$

Cofactor of a Matrix

A *cofactor* of an element in the ith row and jth column of a matrix $\mathbf{A}$ is $(-1)^{i+j}$ times the determinant of the matrix formed by deleting the ith row and jth column. Thus, the determinant given by (A.2.14) can be written in terms of cofactors as

$$|\mathbf{A}| = a_{11}[\text{cofactor of } a_{11}] + a_{12}[\text{cofactor of } a_{12}]$$
$$+ + a_{13}[\text{cofactor of } a_{13}]$$

$$\tag{A.2.17}$$

Adjoint of a Matrix

The *adjoint* of a matrix $\mathbf{A}$, denoted as adj$\mathbf{A}$, is obtained by replacing each element by its cofactor and transposing.

Singular Matrix

A matrix $\mathbf{A}$ is called *singular* if its determinant is zero, that is, if $|A| = 0$. $\mathbf{A}$ is said to be *nonsingular* if $|\mathbf{A}| \neq 0$.

Rank of a Matrix

The *rank* or *full rank* of a matrix $\mathbf{A}$ of order $n \times n$ is defined as

1. the number of linearly independent columns or rows of $\mathbf{A}$, or

2. the greatest order of nonzero determinant of submatrices of $\mathbf{A}$.

If the rank of $\mathbf{A}$ is n, it means that the matrix $\mathbf{A}$ is nonsingular.

Inverse of a Matrix

If we have a relation

$$\mathbf{PA} = \mathbf{I}, \qquad \text{where } \mathbf{I} \text{ is an identity matrix,} \qquad (A.2.18)$$

then $\mathbf{P}$ is called the *inverse* of the matrix $\mathbf{A}$ denoted as $\mathbf{A}^{-1}$. The inverse of a matrix can be calculated in several ways. Thus,

$$\mathbf{A}^{-1} = \frac{\text{adj}\mathbf{A}}{|\mathbf{A}|}. \qquad (A.2.19)$$

It can be easily seen that

$$\left(\mathbf{A}^{-1}\right)' = (\mathbf{A}')^{-1}$$
$$(\mathbf{AB})^{-1} = \mathbf{B}^{-1}\mathbf{A}^{-1}. \qquad (A.2.20)$$

Further, the inverse of sum of matrices is given as

$$[\mathbf{A} + \mathbf{BCD}]^{-1} = \mathbf{A}^{-1} - \mathbf{A}^{-1}\mathbf{B}\left[\mathbf{DA}^{-1}\mathbf{B} + \mathbf{C}^{-1}\right]^{-1}\mathbf{DA}^{-1} \quad (A.2.21)$$

where, $\mathbf{A}$ and $\mathbf{C}$ are nonsingular matrices, the matrix $[\mathbf{A} + \mathbf{BCD}]$ can be formed and is nonsingular and the matrix $[\mathbf{DA}^{-1}\mathbf{B} + \mathbf{C}^{-1}]$ is nonsingular. As a special case

$$\left[\mathbf{I} - \mathbf{F}\left[s\mathbf{I} - A\right]^{-1}\mathbf{B}\right]^{-1} = \mathbf{I} + \mathbf{F}\left[s\mathbf{I} - \mathbf{A} - \mathbf{BF}\right]^{-1}\mathbf{B}. \quad (A.2.22)$$

If a matrix $\mathbf{A}$ consists of submatrices as

$$\mathbf{A} = \begin{bmatrix} \mathbf{A}_{11} & \mathbf{A}_{12} \\ \mathbf{A}_{21} & \mathbf{A}_{22} \end{bmatrix} \qquad (A.2.23)$$

then

$$|\mathbf{A}| = |\mathbf{A}_{11}|.|\mathbf{A}_{22} - \mathbf{A}_{21}\mathbf{A}_{11}^{-1}\mathbf{A}_{12}|$$

$$= |\mathbf{A}_{22}|.|\mathbf{A}_{11} - \mathbf{A}_{12}\mathbf{A}_{22}^{-1}\mathbf{A}_{21}|$$

$$\mathbf{A}^{-1} = \begin{bmatrix} \mathbf{E}_1^{-1} & -\mathbf{E}_1^{-1}\mathbf{A}_{12}\mathbf{A}_{22}^{-1} \\ -\mathbf{A}_{22}^{-1}\mathbf{A}_{21}\mathbf{E}_1^{-1} & \mathbf{E}_2^{-1} \end{bmatrix} \qquad (A.2.24)$$

where, the inverses of $\mathbf{A}_{11}$ and $\mathbf{A}_{22}$ exist and

$$\mathbf{E}_1 = \left[\mathbf{A}_{11} - \mathbf{A}_{12}\mathbf{A}_{22}^{-1}\mathbf{A}_{21}\right], \qquad \mathbf{E}_2 = \left[\mathbf{A}_{22} - \mathbf{A}_{21}\mathbf{A}_{11}^{-1}\mathbf{A}_{12}\right] \quad (A.2.25)$$

Powers of a Matrix

The *m power* of a square matrix $\mathbf{A}$ denoted as $\mathbf{A}^m$, is defined as

$$\mathbf{A}^m = \mathbf{A}\mathbf{A}\cdots\mathbf{A} \qquad \text{upto } m \text{ terms.} \qquad (A.2.26)$$

Exponential of a Matrix

The *exponential* of a square matrix $\mathbf{A}$ can be expressed as

$$exp(\mathbf{A}) = e^{\mathbf{A}} = \mathbf{I} + \mathbf{A} + \frac{1}{2!}\mathbf{A}^2 + \frac{1}{3!}\mathbf{A}^3 + \cdots \qquad (A.2.27)$$

Differentiation and Integration

Differentiation of a Scalar w.r.t. a Vector

If a scalar f is a function of a (column) vector $\mathbf{x}$, then the derivative of f w.r.t. the $\mathbf{x}$ becomes

$$\frac{df}{d\mathbf{x}} = \nabla_{\mathbf{x}}f = \begin{bmatrix} \frac{\partial f}{\partial x_1} \\ \frac{\partial f}{\partial x_2} \\ \cdots \\ \frac{\partial f}{\partial x_n} \end{bmatrix}. \qquad (A.2.28)$$

This is also called a *gradient* of the function f w.r.t. the vector $\mathbf{x}$.

The second derivative (also called *Hessian*) of f w.r.t. the vector $\mathbf{x}$ is

$$f_{\mathbf{xx}} \triangleq \frac{\partial^2 f}{\partial \mathbf{x}^2} = \left[\frac{\partial^2 f}{\partial \mathbf{x}\partial \mathbf{x}}\right]. \qquad (A.2.29)$$

Differentiation of a Vector w.r.t. a Scalar

If a vector $\mathbf{x}$ of dimension n is a function of scalar t, then the derivative of $\mathbf{x}$ w.r.t. t is

$$\frac{d\mathbf{x}}{dt} = \begin{bmatrix} \frac{dx_1}{dt} \\ \frac{dx_2}{dt} \\ \cdots \\ \frac{dx_n}{dt} \end{bmatrix}. \tag{A.2.30}$$

Differentiation of a Vector w.r.t. a Vector

The derivative of an mth order vector function $\mathbf{f}$ w.r.t. an nth vector $\mathbf{x}$ is written as

$$\frac{d\mathbf{f}'}{d\mathbf{x}} = \mathbf{G} = \frac{\partial \mathbf{f}'}{\partial \mathbf{x}} = \begin{bmatrix} \frac{\partial f_1}{\partial x_1} & \frac{\partial f_2}{\partial x_1} & \cdots & \frac{\partial f_m}{\partial x_1} \\ \frac{\partial f_1}{\partial x_2} & \frac{\partial f_2}{\partial x_2} & \cdots & \frac{\partial f_m}{\partial x_2} \\ \cdots & \cdots & \cdots & \cdots \\ \frac{\partial f_1}{\partial x_n} & \frac{\partial f_2}{\partial x_n} & \cdots & \frac{\partial f_m}{\partial x_n} \end{bmatrix} \tag{A.2.31}$$

where, $\mathbf{G}$ is a matrix of order $n \times m$. Note that

$$\mathbf{G}' = \left[\frac{\partial \mathbf{f}'}{\partial \mathbf{x}} \right]' \qquad \text{which is written as} \quad \frac{\partial \mathbf{f}}{\partial \mathbf{x}}. \tag{A.2.32}$$

This is also called Jacobian matrix denoted as

$$\mathbf{J_x}(\mathbf{f}(\mathbf{x})) = \frac{d\mathbf{f}}{d\mathbf{x}} = \frac{\partial \mathbf{f}}{\partial \mathbf{x}} = \left[\frac{\partial f_i}{\partial x_j} \right]$$

$$= \begin{bmatrix} \frac{\partial f_1}{\partial x_1} & \frac{\partial f_1}{\partial x_2} & \cdots & \frac{\partial f_1}{\partial x_n} \\ \frac{\partial f_2}{\partial x_1} & \frac{\partial f_2}{\partial x_2} & \cdots & \frac{\partial f_2}{\partial x_2} \\ \cdots & \cdots & \cdots & \cdots \\ \frac{\partial f_m}{\partial x_1} & \frac{\partial f_n}{\partial x_2} & \cdots & \frac{\partial f_m}{\partial x_n} \end{bmatrix}. \tag{A.2.33}$$

Thus, the total differential of $\mathbf{f}$ is

$$d\mathbf{f} = \frac{\partial \mathbf{f}}{\partial \mathbf{x}} d\mathbf{x}. \tag{A.2.34}$$

Differentiation of a Scalar w.r.t. Several Vectors

For a scalar f as a function of two vectors $\mathbf{x}$ and $\mathbf{y}$, we have

$$f = f(\mathbf{x}, \mathbf{y})$$

$$df = \left[\frac{\partial f}{\partial \mathbf{x}} \right]' d\mathbf{x} + \left[\frac{\partial f}{\partial \mathbf{y}} \right]' d\mathbf{y} \tag{A.2.35}$$

where, df is the total differential. For a scalar function

$$f = f(\mathbf{x}(t), \mathbf{y}(t), t), \quad \mathbf{y}(t) = \mathbf{y}(\mathbf{x}(t), t)$$

$$\frac{df}{d\mathbf{x}} = \left[\frac{\partial \mathbf{y}'}{\partial \mathbf{x}}\right] \frac{\partial f}{\partial \mathbf{y}} + \frac{\partial f}{\partial \mathbf{x}}$$

$$\frac{df}{dt} = \left[\left\{\frac{\partial \mathbf{y}'}{\partial \mathbf{x}}\right\} \frac{\partial f}{\partial \mathbf{y}} + \frac{\partial f}{\partial \mathbf{x}}\right]' \frac{d\mathbf{x}}{dt} + \left[\frac{\partial f}{\partial \mathbf{y}}\right]' \frac{\partial \mathbf{y}}{\partial t} + \frac{\partial f}{\partial t} \quad \text{(A.2.36)}$$

Differentiation of a Vector w.r.t. Several Vectors

Similarly for a vector function $\mathbf{f}$, we have

$$\mathbf{f} = \mathbf{f}(\mathbf{y}, \mathbf{x}, t), \quad \mathbf{y} = \mathbf{y}(\mathbf{x}, t), \quad \mathbf{x} = \mathbf{x}(t)$$

$$\frac{d\mathbf{f}}{d\mathbf{x}} = \left[\frac{\partial \mathbf{f}'}{\partial \mathbf{y}}\right]' \left[\frac{\partial \mathbf{y}'}{\partial \mathbf{x}}\right]' + \frac{\partial \mathbf{f}}{\partial \mathbf{x}}$$

$$= \left[\frac{\partial \mathbf{f}}{\partial \mathbf{y}}\right] \left[\frac{\partial \mathbf{y}}{\partial \mathbf{x}}\right] + \frac{\partial \mathbf{f}}{\partial \mathbf{x}}$$

$$\frac{d\mathbf{f}}{dt} = \left[\frac{\partial \mathbf{f}'}{\partial \mathbf{y}}\right]' \left[\left\{\frac{\partial \mathbf{y}'}{\partial \mathbf{x}}\right\}' \frac{d\mathbf{x}}{dt} + \frac{\partial \mathbf{y}}{\partial t}\right] + \left[\frac{\partial \mathbf{f}'}{\partial \mathbf{x}}\right]' \frac{d\mathbf{x}}{dt} + \frac{\partial \mathbf{f}}{\partial t}$$

$$= \left[\frac{\partial \mathbf{f}}{\partial \mathbf{y}}\right] \left[\left\{\frac{\partial \mathbf{y}}{\partial \mathbf{x}}\right\} \frac{d\mathbf{x}}{dt} + \frac{\partial \mathbf{y}}{\partial t}\right] + \left[\frac{\partial \mathbf{f}}{\partial \mathbf{x}}\right] \frac{d\mathbf{x}}{dt} + \frac{\partial \mathbf{f}}{\partial t}. \quad \text{(A.2.37)}$$

Differentiation of a Matrix w.r.t. a Scalar

If each element of a matrix is a function of a scalar variable t, then the matrix $\mathbf{A}(t)$ is said to be a function of t. Then the derivative of the matrix $\mathbf{A}(t)$ is defined as

$$\frac{d\mathbf{A}(t)}{dt} = \begin{bmatrix} \frac{da_{11}}{dt} & \frac{da_{12}}{dt} & \cdots & \frac{da_{1m}}{dt} \\ \frac{da_{21}}{dt} & \frac{da_{22}}{dt} & \cdots & \frac{da_{2m}}{dt} \\ \cdots & \cdots & \cdots & \cdots \\ \frac{da_{n1}}{dt} & \frac{da_{n2}}{dt} & \cdots & \frac{da_{nm}}{dt} \end{bmatrix}. \quad \text{(A.2.38)}$$

It follows (from chain rule) that

$$\frac{d}{dt}[\mathbf{A}(t)\mathbf{B}(t)] = \frac{d\mathbf{A}(t)}{dt}\mathbf{B}(t) + \mathbf{A}(t)\frac{d\mathbf{B}(t)}{dt}. \quad \text{(A.2.39)}$$

It is obvious that

$$\frac{d}{dt}\left[e^{\mathbf{A}t}\right] = \mathbf{A}e^{\mathbf{A}t} = e^{\mathbf{A}t}\mathbf{A}$$

$$\frac{d}{dt}\left[\mathbf{A}^{-1}(t)\right] = -\mathbf{A}^{-1}(t)\frac{d\mathbf{A}(t)}{dt}\mathbf{A}^{-1}(t). \quad \text{(A.2.40)}$$

Differentiation of a Scalar w.r.t. a Matrix

Suppose a scalar f is a function of a matrix $\mathbf{A}$, then

$$\frac{df}{d\mathbf{A}} = \begin{bmatrix} \frac{df}{da_{11}} & \frac{df}{da_{12}} & \cdots & \frac{df}{da_{1m}} \\ \frac{df}{da_{21}} & \frac{df}{da_{22}} & \cdots & \frac{df}{da_{2m}} \\ \cdots & \cdots & \cdots & \cdots \\ \frac{df}{da_{n1}} & \frac{df}{da_{n2}} & \cdots & \frac{df}{da_{nm}} \end{bmatrix} \tag{A.2.41}$$

The *integration* process for matrices and vectors is similarly defined for all the previous cases. For example,

$$\int \mathbf{A} dt = \left[\int a_{ij} dt \right]. \tag{A.2.42}$$

Taylor Series Expansion

It is well known that the *Taylor series expansion* of a function f w.r.t. $\mathbf{x}$ about $\mathbf{x}_0$ is

$$f(\mathbf{x}) = f(\mathbf{x}_0) + \left[\frac{\partial f}{\partial \mathbf{x}} \right]' \bigg|_{\mathbf{x}_0} (\mathbf{x} - \mathbf{x}_0) + \frac{1}{2!} (\mathbf{x} - \mathbf{x}_0)' \frac{\partial^2 f}{\partial \mathbf{x}^2} \bigg|_{\mathbf{x}_0} (\mathbf{x} - \mathbf{x}_0)$$

$$+ \mathcal{O}(3) \tag{A.2.43}$$

where, $\mathcal{O}(3)$ indicates terms of order 3 and higher.

Trace of a Matrix

For a square matrix $\mathbf{A}$ of n dimension, the *trace* of $\mathbf{A}$ is defined as

$$\text{tr}\,[\mathbf{A}] = \sum_{i=1}^{n} a_{ii}. \tag{A.2.44}$$

Thus, the trace is the sum of the diagonal elements of a matrix. Also,

$$\text{tr}\,[\mathbf{A} + \mathbf{B}] = \text{tr}\,[\mathbf{A}] + \text{tr}\,[\mathbf{B}]$$
$$\text{tr}\,[\mathbf{A}\mathbf{B}] = \text{tr}\,[\mathbf{A}'\mathbf{B}'] = \text{tr}\,[\mathbf{B}'\mathbf{A}'] = \text{tr}\,[\mathbf{B}\mathbf{A}]. \tag{A.2.45}$$

Eigenvalues and Eigenvectors of a Square Matrix

For a square matrix $\mathbf{A}$ of order n, the roots (or zeros) of the *characteristic polynomial* equation in λ

$$|\lambda \mathbf{I} - \mathbf{A}| = 0 \tag{A.2.46}$$

are called *eigenvalues* of the matrix $\mathbf{A}$. If there is a nonzero vector $\mathbf{x}$ satisfying the equation

$$\mathbf{A}\mathbf{x} = \lambda_i \mathbf{x} \longrightarrow (\mathbf{A} - \lambda_i \mathbf{I})\mathbf{x} = 0 \qquad (\text{A.2.47})$$

for a particular eigenvalue λ_i, then the vector $\mathbf{x}$ is called the *eigenvector* of the matrix $\mathbf{A}$ corresponding to the particular eigenvalue λ_i. Also, note that the trace of a matrix is related as

$$\text{tr}\,[\mathbf{A}] = \sum_{i=1}^{n} \lambda_i. \qquad (\text{A.2.48})$$

Singular Values

Let $\mathbf{A}$ be an nxm matrix, then the *singular values* σ of the matrix $\mathbf{A}$ are defined as the square root values of the eigenvalues (λ) of the matrix $\mathbf{A}'\mathbf{A}$, that is

$$\sigma = \sqrt{\lambda(\mathbf{A}'\mathbf{A})}. \qquad (\text{A.2.49})$$

The singular values are usually arranged in the descending order of the magnitude.

A.3   Quadratic Forms and Definiteness

Quadratic Forms

Consider the inner product of a real symmetric matrix $\mathbf{P}$ and a vector $\mathbf{x}$ or the norm of vector $\mathbf{x}$ w.r.t. the real symmetric matrix $\mathbf{P}$ as

$$< \mathbf{x}, \mathbf{P}\mathbf{x} > = \mathbf{x}'\mathbf{P}\mathbf{x} = ||\mathbf{x}||_{\mathbf{P}}$$

$$= \begin{bmatrix} x_1 & x_2 & \cdots & x_n \end{bmatrix} \begin{bmatrix} p_{11} & p_{12} & \cdots & p_{1n} \\ p_{12} & p_{22} & \cdots & p_{2n} \\ \cdots & \cdots & \cdots & \cdots \\ p_{1n} & p_{2n} & \cdots & p_{nn} \end{bmatrix} \begin{bmatrix} x_1 \\ x_2 \\ \cdots \\ x_n \end{bmatrix}$$

$$= \sum_{i,j=1}^{n} p_{ij} x_i x_j. \qquad (\text{A.3.1})$$

The scalar quantity $\mathbf{x}'\mathbf{P}\mathbf{x}$ is called a *quadratic form* since it contains quadratic terms such as $x_1^2 p_{11}, x_1 x_2 p_{12}, \ldots$.

Definiteness

Let $\mathbf{P}$ be a real and symmetric matrix and $\mathbf{x}$ be a nonzero real vector, then

1. $\mathbf{P}$ is *positive definite* if the scalar quantity $\mathbf{x}'\mathbf{Px} > 0$ or is *positive*.

2. $\mathbf{P}$ is *positive semidefinite* if the scalar quantity $\mathbf{x}'\mathbf{Px} \geq 0$ or is *nonnegative*.

3. $\mathbf{P}$ is *negative definite* if the scalar quantity $\mathbf{x}'\mathbf{Px} < 0$ or is *negative*.

4. $\mathbf{P}$ is *negative semidefinite* if the scalar quantity $\mathbf{x}'\mathbf{Px} \leq 0$ or *nonpositive*.

A test for real symmetric matrix $\mathbf{P}$ to be positive definite is that all its *principal or leading minors* must be positive, that is,

$$p_{11} > 0, \quad \begin{vmatrix} p_{11} & p_{12} \\ p_{12} & p_{22} \end{vmatrix} > 0, \quad \begin{vmatrix} p_{11} & p_{12} & p_{13} \\ p_{12} & p_{22} & p_{23} \\ p_{13} & p_{23} & p_{33} \end{vmatrix} > 0 \qquad (A.3.2)$$

for a 3x3 matrix $\mathbf{P}$. The $>$ sign is changed accordingly for positive semidefinite ($\geq$), negative definite ($<$), and negative semidefinite ($\leq$ 0) cases. Another simple test for definiteness is using eigenvalues (all eigenvalues positive for positive definiteness, etc.).

Also, note that

$$[\mathbf{x}'\mathbf{Px}]' = \mathbf{x}'\mathbf{P}'\mathbf{x} = \mathbf{x}'\mathbf{Px}$$
$$\mathbf{P} = \sqrt{\mathbf{P}}\sqrt{\mathbf{P}'} = \sqrt{\mathbf{P}'}\sqrt{\mathbf{P}}. \qquad (A.3.3)$$

Derivative of Quadratic Forms

Some useful results in obtaining the derivatives of quadratic forms and related expressions are given below.

$$\frac{\partial}{\partial \mathbf{x}} (\mathbf{A}\mathbf{x}) = \mathbf{A}$$

$$\frac{\partial}{\partial \mathbf{y}} (\mathbf{x}'\mathbf{y}) = \frac{\partial}{\partial \mathbf{y}} (\mathbf{y}'\mathbf{x}) = \mathbf{x}$$

$$\frac{\partial}{\partial \mathbf{y}} (\mathbf{x}'\mathbf{A}\mathbf{y}) = \frac{\partial}{\partial \mathbf{y}} (\mathbf{y}'\mathbf{A}'\mathbf{x}) = \mathbf{A}'\mathbf{x}$$

$$\frac{\partial}{\partial \mathbf{x}} (\mathbf{x}'\mathbf{A}\mathbf{x}) = \mathbf{A}\mathbf{x} + \mathbf{A}'\mathbf{x}$$

$$\frac{\partial^2}{\partial \mathbf{x}^2} (\mathbf{x}'\mathbf{A}\mathbf{x}) = \mathbf{A} + \mathbf{A}'. \tag{A.3.4}$$

If there is a symmetric matrix $\mathbf{P}$, then

$$\frac{\partial}{\partial \mathbf{x}} (\mathbf{x}'\mathbf{P}\mathbf{x}) = 2\mathbf{P}\mathbf{x}$$

$$\frac{\partial^2}{\partial \mathbf{x}^2} (\mathbf{x}'\mathbf{P}\mathbf{x}) = 2\mathbf{P}. \tag{A.3.5}$$

Appendix B

State Space Analysis

The main purpose of this appendix is to provide a brief summary of the results on state space analysis to serve as a review of these topics rather than any in depth treatment of the topics. For more details on this subject, the reader is referred to [69, 147, 4, 41, 11, 35].

B.1 State Space Form for Continuous-Time Systems

A linear time-invariant (LTI), continuous-time, dynamical system is described by

$$\dot{\mathbf{x}}(t) = \mathbf{A}\mathbf{x}(t) + \mathbf{B}\mathbf{u}(t), \quad \text{state equation}$$
$$\mathbf{y}(t) = \mathbf{C}\mathbf{x}(t) + \mathbf{D}\mathbf{u}(t), \quad \text{output equation} \qquad \text{(B.1.1)}$$

with initial conditions $\mathbf{x}(t = t_0) = \mathbf{x}(t_0)$. Here, $\mathbf{x}(t)$ is an n-dimensional *state* vector, $\mathbf{u}(t)$ is an r dimensional *control* vector, and $\mathbf{y}(t)$ is a p dimensional *output* vector and the various matrices $\mathbf{A}, \mathbf{B}, ...,$ are of appropriate dimensionality. The Laplace transform (in terms of the Laplace variable s) of the preceding set of equations (B.1.1) is

$$s\mathbf{X}(s) - \mathbf{x}(t_0) = \mathbf{A}\mathbf{X}(s) + \mathbf{B}\mathbf{U}(s)$$
$$\mathbf{Y}(s) = \mathbf{C}\mathbf{X}(s) + \mathbf{D}\mathbf{U}(s) \qquad \text{(B.1.2)}$$

which becomes

$$\mathbf{X}(s) = [s\mathbf{I} - \mathbf{A}]^{-1}[\mathbf{x}(t_0) + \mathbf{B}\mathbf{U}(s)]$$
$$\mathbf{Y}(s) = \mathbf{C}[s\mathbf{I} - \mathbf{A}]^{-1}[\mathbf{x}(t_0) + \mathbf{B}\mathbf{U}(s)] + \mathbf{D}\mathbf{U}(s) \qquad \text{(B.1.3)}$$

where, $\mathbf{X}(s)$ =Laplace transform of $\mathbf{x}(t)$, etc. In terms of the transfer function $\mathbf{G}(s)$ with zero initial conditions $\mathbf{x}(t_0) = 0$, we have

$$\mathbf{G}(s) = \frac{\mathbf{Y}(s)}{\mathbf{U}(s)} = \mathbf{C}\left[s\mathbf{I} - \mathbf{A}\right]^{-1}\mathbf{B} + \mathbf{D}. \tag{B.1.4}$$

A linear time-varying (LTV), continuous-time, dynamical system is described by

$$\dot{\mathbf{x}}(t) = \mathbf{A}(t)\mathbf{x}(t) + \mathbf{B}(t)\mathbf{u}(t), \quad \text{state equation}$$
$$\mathbf{y}(t) = \mathbf{C}(t)\mathbf{x}(t) + \mathbf{D}(t)\mathbf{u}(t), \quad \text{output equation} \tag{B.1.5}$$

with initial conditions $\mathbf{x}(t = t_0) = \mathbf{x}(t_0)$. The solution of the continuous-time LTI system (B.1.1) is given by

$$\mathbf{x}(t) = \mathbf{\Phi}(t, t_0)\mathbf{x}(t_0) + \int_{t_0}^{t} \mathbf{\Phi}(t, \tau)\mathbf{B}\mathbf{u}(\tau)d\tau$$

$$\mathbf{y}(t) = \mathbf{C}\mathbf{\Phi}(t, t_0)\mathbf{x}(t_0) + \mathbf{C}\int_{t_0}^{t} \mathbf{\Phi}(t, \tau)\mathbf{B}\mathbf{u}(\tau)d\tau + \mathbf{D}\mathbf{u}(t) \tag{B.1.6}$$

where, $\mathbf{\Phi}(t, t_0)$, called the *state transition matrix* of the system (B.1.1), is given by

$$\mathbf{\Phi}(t, t_0) = e^{\mathbf{A}(t - t_0)} \tag{B.1.7}$$

having the properties

$$\mathbf{\Phi}(t_0, t_0) = \mathbf{I}, \quad \mathbf{\Phi}(t_2, t_1)\mathbf{\Phi}(t_1, t_0) = \mathbf{\Phi}(t_2, t_0). \tag{B.1.8}$$

Similarly, the solution of the continuous-time LTV system (B.1.5) is given by

$$\mathbf{x}(t) = \mathbf{\Phi}(t, t_0)\mathbf{x}(t_0) + \int_{t_0}^{t} \mathbf{\Phi}(t, \tau)\mathbf{B}(\tau)\mathbf{u}(\tau)d\tau$$

$$\mathbf{y}(t) = \mathbf{C}(t)\mathbf{\Phi}(t, t_0)\mathbf{x}(t_0) + \mathbf{C}(t)\int_{t_0}^{t} \mathbf{\Phi}(t, \tau)\mathbf{B}(\tau)\mathbf{u}(\tau)d\tau + \mathbf{D}(t)\mathbf{u}(t)$$

$$\tag{B.1.9}$$

where, $\mathbf{\Phi}(t, t_0)$, still called the *state transition matrix* of the system (B.1.5), cannot be easily computed analytically, but does satisfy the properties (B.1.8). However, in terms of a fundamental matrix $\mathbf{X}(t)$ satisfying

$$\dot{\mathbf{X}}(t) = \mathbf{A}(t)\mathbf{X}(t) \tag{B.1.10}$$

it can be written as [35]

$$\mathbf{\Phi}(t, t_0) = \mathbf{X}(t)\mathbf{X}^{-1}(t_0). \tag{B.1.11}$$

B.2   Linear Matrix Equations

A set of linear simultaneous equations for an unknown matrix $\mathbf{P}$ in terms of known matrices $\mathbf{A}$ and $\mathbf{Q}$, is written as

$$\mathbf{PA} + \mathbf{A'P} + \mathbf{Q} = 0. \tag{B.2.1}$$

In particular, if $\mathbf{Q}$ is *positive definite*, then there exists a unique positive definite $\mathbf{P}$ satisfying the previous linear matrix equation, if and only if $\mathbf{A}$ is asymptotically stable or the real part (Re) of $\lambda\{\mathbf{A}\} < 0$. Then (B.2.1) is called the *Lyapunov equation*, the solution of which is given by

$$\mathbf{P} = \int_0^\infty e^{\mathbf{A'}t} \mathbf{Q} e^{\mathbf{A}t} dt. \tag{B.2.2}$$

B.3   State Space Form for Discrete-Time Systems

A linear time-invariant (LTI), discrete-time, dynamical system is described by

$$\mathbf{x}(k+1) = \mathbf{Ax}(k) + \mathbf{Bu}(k), \quad \text{state equation}$$
$$\mathbf{y}(k) = \mathbf{Cx}(k) + \mathbf{Du}(k), \quad \text{output equation} \tag{B.3.1}$$

with initial conditions $\mathbf{x}(k = k_0) = \mathbf{x}(k_0)$. Here, $\mathbf{x}(k)$ is an n-dimensional *state* vector, $\mathbf{u}(k)$ is an r-dimensional *control* vector, and $\mathbf{y}(k)$ is a p-dimensional *output* vector and the various matrices $\mathbf{A}, \mathbf{B}, ...,$ are matrices of appropriate dimensionality. The $\mathcal{Z}$-transform (in terms of the complex variable z) is

$$z\mathbf{X}(z) - \mathbf{x}(k_0) = \mathbf{AX}(z) + \mathbf{BU}(z)$$
$$\mathbf{Y}(z) = \mathbf{CX}(z) + \mathbf{DU}(z) \tag{B.3.2}$$

which becomes

$$\mathbf{X}(z) = [z\mathbf{I} - \mathbf{A}]^{-1} [\mathbf{x}(k_0) + \mathbf{BU}(z)]$$
$$\mathbf{Y}(z) = \mathbf{C} [z\mathbf{I} - \mathbf{A}]^{-1} [\mathbf{x}(k_0) + \mathbf{BU}(z)] + \mathbf{DU}(z). \tag{B.3.3}$$

In terms of the transfer function $\mathbf{G}(z)$ with zero initial conditions $\mathbf{x}(k_0) = 0$, we have

$$\mathbf{G}(z) = \frac{\mathbf{Y}(z)}{\mathbf{U}(z)} = \mathbf{C} [z\mathbf{I} - \mathbf{A}]^{-1} \mathbf{B} + \mathbf{D}. \tag{B.3.4}$$

An LTV, discrete-time, dynamical system is described by

$$\mathbf{x}(k+1) = \mathbf{A}(k)\mathbf{x}(k) + \mathbf{B}(k)\mathbf{u}(k), \quad \text{state equation}$$
$$\mathbf{y}(k) = \mathbf{C}(k)\mathbf{x}(k) + \mathbf{D}(k)\mathbf{u}(k), \quad \text{output equation} \quad \text{(B.3.5)}$$

with initial conditions $\mathbf{x}(k = k_0) = \mathbf{x}(k_0)$. The solution of the LTI discrete-time system (B.3.1) is given by

$$\mathbf{x}(k) = \mathbf{\Phi}(k, k_0)\mathbf{x}(k_0) + \sum_{m=k_0}^{k-1} \mathbf{\Phi}(k, m+1)\mathbf{B}\mathbf{u}(m)$$

$$\mathbf{y}(k) = \mathbf{C}\mathbf{\Phi}(k, k_0)\mathbf{x}(k_0) + \mathbf{C}\sum_{m=k_0}^{k-1} \mathbf{\Phi}(k, m+1)\mathbf{B}\mathbf{u}(m) + \mathbf{D}\mathbf{u}(k)$$

$$\text{(B.3.6)}$$

where, $\mathbf{\Phi}(k, k_0)$, called the *state transition matrix* of the discrete-time system (B.3.1), is given by

$$\mathbf{\Phi}(k, k_0) = \mathbf{A}^k \quad \text{(B.3.7)}$$

having the properties

$$\mathbf{\Phi}(k_0, k_0) = \mathbf{I}, \quad \mathbf{\Phi}(k_2, k_1)\mathbf{\Phi}(k_1, k_0) = \mathbf{\Phi}(k_2, k_0). \quad \text{(B.3.8)}$$

Similarly, the solution of the LTV, discrete-time system (B.3.5) is given by

$$\mathbf{x}(k) = \mathbf{\Phi}(k, k_0)\mathbf{x}(k_0) + \sum_{m=k_0}^{k-1} \mathbf{\Phi}(k, m+1)\mathbf{B}(m)\mathbf{u}(m)$$

$$\mathbf{y}(k) = \mathbf{C}(k)\mathbf{\Phi}(k, k_0)\mathbf{x}(k_0) + \mathbf{C}(k)\sum_{m=k_0}^{k-1} \mathbf{\Phi}(k, m+1)\mathbf{B}(m)\mathbf{u}(m) + \mathbf{D}(k)\mathbf{u}(k)$$

$$\text{(B.3.9)}$$

where,

$$\mathbf{\Phi}(k, k_0) = \mathbf{A}(k-1)\mathbf{A}(k-2)\cdots\mathbf{A}(k_0) \quad k \text{ terms}, \quad \text{(B.3.10)}$$

is called the *state transition matrix* of the discrete-time system (B.3.5) satisfying the properties (B.3.8).

B.4   *Controllability and Observability*

Let us first consider the LTI, continuous-time system (B.1.1). Similar results are available for discrete-time systems [35]. The system (B.1.1) with the pair $(\mathbf{A} : nxn, \mathbf{B} : nxr)$ is called *completely state controllable* if any of the following conditions is satisfied:

1. rank of the controllability matrix

$$\mathbf{Q}_c = [\mathbf{B}\ \ \mathbf{AB}\ \ \mathbf{A}^2\mathbf{B}\cdots\mathbf{A}^{n-1}\mathbf{B}] \qquad (B.4.1)$$

   is n (full row rank), or

2. the *controllability Grammian*

$$\mathbf{W}_c(t) = \int_0^t e^{\mathbf{A}\tau}\mathbf{BB}'e^{\mathbf{A}'\tau}d\tau$$

$$= \int_0^t e^{\mathbf{A}(t-\tau)}\mathbf{BB}'e^{\mathbf{A}'(t-\tau)}d\tau \qquad (B.4.2)$$

   is nonsingular for any $t > 0$.

The system (B.1.1) with the pair $(\mathbf{A} : nxn, \mathbf{C} : pxn)$ is *completely observable* if any of the following conditions is satisfied:

1. rank of the observability matrix

$$\mathbf{Q}_o = [\mathbf{C}\ \ \mathbf{CA}\ \ \mathbf{CA}^2\cdots\mathbf{CA}^{n-1}]' \qquad (B.4.3)$$

   has rank n (full column rank).

2. the *observability Grammian*

$$\mathbf{W}_o(t) = \int_0^t e^{\mathbf{A}'\tau}\mathbf{C}'\mathbf{C}e^{\mathbf{A}\tau}d\tau \qquad (B.4.4)$$

   is nonsingular for any $t > 0$.

Other conditions also exist for controllability and observability [35].

B.5   *Stabilizability, Reachability and Detectability*

Stabilizability

A system is *stabilizable* if its uncontrollable states or modes if any, are stable. Its controllable states or modes may be stable or unstable. Thus, the pair $(\mathbf{A}, \mathbf{B})$ is stabilizable if $(\mathbf{A} - \mathbf{BF})$ can be made asymptotically stable for some matrix $\mathbf{F}$.

Reachability

A system is said to be *reachable* if the system can be transferred from initial state to any other specified final state. Thus, a continuous-time system is reachable if and only if the system is controllable and hence reachability is equivalent to controllability.

Detectability

A system is *detectable* if its unobservable states, if any, are stable. Its observable states may be stable or unstable. Thus, the pair $(\mathbf{A}, \mathbf{C})$ is detectable if there is a matrix $\mathbf{L}$ such that $(\mathbf{A} - \mathbf{L}\mathbf{C})$ can be made asymptotically stable. This is equivalent to the observability of the unstable modes of $\mathbf{A}$.

Appendix C

MATLAB Files

This appendix contains MATLAB© files required to run programs used in solving some of the problems discussed in the book. One needs to have the following files in one's working directory before using the MATLAB©.

C.1 MATLAB© for Matrix Differential Riccati Equation

The following is the typical MATLAB© file containing the various given matrices for a problem, such as, Example 3.1 using analytical solution of matrix Riccati differential equation given in Chapter 3. This file, say *example.m* requires the other two files *lqrnss.m* and *lqrnssf.m* given below. The electronic version of ll these files can also be obtained by sending an email to *naiduds@isu.edu*.

```
%%%%%%%%%%%
clear all
A=[0.,1.;-2.,1.];
B=[0.;1.];
Q=[2.,3.;3.,5.];
F=[1.,0.5;0.5,2.];
R=[.25];
tspan=[0 5];
x0=[2.,-3.];
[x,u,K]=lqrnss(A,B,F,Q,R,x0,tspan);
%%%%%%%%%%%
```

C.1.1   MATLAB File lqrnss.m

This MATLAB© file *lqrnss.m* is required along with the other files *example.m* and *lqrnssf.m* to solve the matrix Riccati equation using its analytical solution.

```
%%%%%%%%%%%%
%% The following is lqrnss.m
function [x,u,K]=lqrnss(As,Bs,Fs,Qs,Rs,x0,tspan)
%Revision Date 11/14/01
%%
% This m-file calculates and plots the outputs for a
% Linear Quadratic Regulator (LQR) system based on given
% state space matrices A and B and performance index
% matrices F, Q and R.  This function takes these inputs,
% and using the analytical solution to the
%% matrix Riccati equation,
% and then computing optimal states and controls.
%
%
%       SYNTAX:   [x,u,K]=lqrnss(A,B,F,Q,R,x0,tspan)
%
%       INPUTS (All numeric):
%       A,B           Matrices from xdot=Ax+Bu
%       F,Q,R         Performance Index Parameters;
%       x0            State variable initial condition
%       tspan         Vector containing time span [t0 tf]
%
%       OUTPUTS:
%       x         is the state variable vector
%       u         is the input vector
%       K         is the steady-state matrix inv(R)*B'*P
%
%       The system plots Riccati coefficients, x vector,
% and u vector
%
%Define variables to use in external functions
%
global A E F Md tf W11 W12 W21 W22 n
%
%Check for correct number of inputs
```

```
%
if nargin<7
    error('Incorrect number of inputs specified')
    return
end
%
%Convert Variables to normal symbology to prevent
% problems with global statement
%
A=As;
B=Bs;
F=Fs;
Q=Qs;
R=Rs;
plotflag=0;    %set plotflag to 1 to avoid plotting of
% data on figures
%
%Define secondary variables for global passing to
% ode-related functions and determine matrice size
%
[n,m]=size(A);            %Find dimensions of A
[nb,mb]=size(B);          %Find dimensions of B
[nq,mq]=size(Q);          %Find dimensions of Q
[nr,mr]=size(R);          %Find Dimensions of R
[nf,mf]=size(F);          %Find Dimensions of F
if n~=m                   %Verify A is square
    error('A must be square')
else
    [n,n]=size(A);
end
%
%Data Checks for proper setup
if length(A)>rank(ctrb(A,B))
%Check for controllability
    error('System Not Controllable')
    return
end
if (n ~= nq) | (n ~= mq)
%Check that A and Q are the same size
```

```
      error('A and Q must be the same size');
      return
   end
   if ~(mf==1&nf==1)
      if (nq ~= nf) | (mq ~= mf)
%Check that Q and F are the same size
         error('Q and F must be the same size');
         return
      end
   end
   if ~(mr==1&nr==1)
      if (mr ~= nr) | (mb ~= nr)
         error('R must be consistent with B');
         return
      end
   end
mq = norm(Q,1);
% Check if Q is positive semi-definite and symmetric
if any(eig(Q) < -eps*mq) | (norm(Q'-Q,1)/mq > eps)
   disp('Warning: Q is not symmetric and positive ...
semi-definite');
end
mr = norm(R,1);
% Check if R is positive definite and symmetric
if any(eig(R) <= -eps*mr) | (norm(R'-R,1)/mr > eps)
disp('Warning: R is not symmetric and positive ...
definite');
end
%
%Define Initial Conditions for
%numerical solution of x states
%
t0=tspan(1);
tf=tspan(2);
tspan=[tf t0];
%
%Define Calculated Matrices and Vectors
%
E=B*inv(R)*B';          %E Matrix E=B*(1/R)*B'
```

```
%
%Find Hamiltonian matrix needed to use
% analytical solution to
% matrix Riccati differential equation
%
Z=[A,-E;-Q,-A'];
%
%Find Eigenvectors
%
[W,D]=eig(Z);
%
%Find the diagonals from D and pick the
% negative diagonals to create
% a new matrix M
%
j=n;
[m1,index1]=sort(real(diag(D)));
   for i=1:1:n
       m2(i)=m1(j);
       index2(i)=index1(j);
       index2(i+n)=index1(i+n);
       j=j-1;
   end
Md=-diag(m2);
%
%Rearrange W so that it corresponds to the sort
% of the eigenvalues
%
for i=1:2*n
   w2(:,i)=W(:,index2(i));
end
W=w2;
%
%Define the Modal Matrix for D and Split it into Parts
%
W11=zeros(n);
W12=zeros(n);
W21=zeros(n);
W22=zeros(n);
```

```
j=1;
  for i=1:2*n:(2*n*n-2*n+1)
    W11(j:j+n-1)=W(i:i+n-1);
    W21(j:j+n-1)=W(i+n:i+2*n-1);
    W12(j:j+n-1)=W(2*n*n+i:2*n*n+i+n-1);
    W22(j:j+n-1)=W(2*n*n+i+n:2*n*n+i+2*n-1);
    j=j+n;
  end
%
%Define other initial conditions for
% calculation of P, g, x and u
%
t1=0.;
tx=0.;                          %time array for x
tu=0.;                          %time array for u
x=0.;                           %state vector
%
%Calculation of optimized x
%
[tx,x]=ode45('lqrnssf',fliplr(tspan),x0,...
odeset('refine',2,'RelTol',1e-4,'AbsTol',1e-6));
%
%Find u vector
%
j=1;
us=0.;          %Initialize computational variable
for i=1:1:mb
  for tua=t0:.1:tf
    Tt=-inv(W22-F*W12)*(W21-F*W11);
    P=(W21+W22*expm(-Md*(tf-tua))*Tt*...
    expm(-Md*(tf-tua)))*inv(W11+W12*expm(-Md*(tf-tua))...
*Tt*expm(-Md*(tf-tua)));
    K=inv(R)*B'*P;
    xs=interp1(tx,x,tua);
    us1=real(-K*xs');
    us(j)=us1(i);
    tu(j)=tua;
    j=j+1;
  end
```

```
  u(:,i)=us';
  us=0;
   j=1;
end
%
%Provide final steady-state K
%
P=W21/W11;
K=real(inv(R)*B'*P);
%
%Plotting Section, if desired
%
if plotflag~=1
%
%Plot diagonal Riccati coefficients using a
% flag variable to hold and change colors
%
fig=1;            %Figure number
cflag=1;          %Variable used to change plot color
j=1;
Ps=0.;            %Initialize P matrix plot variable
for i=1:1:n*n
  for t1a=t0:.1:tf
 Tt=-inv(W22-F*W12)*(W21-F*W11);
 P=real((W21+W22*expm(-Md*(tf-t1a))*Tt*expm(-Md*...
(tf-t1a)))*inv(W11+W12*expm(-Md*(tf-t1a))*Tt...
*expm(-Md*(tf-t1a))));
    Ps(j)=P(i);
    t1(j)=t1a;
    j=j+1;
  end
  if cflag==1;
    figure(fig)
    plot(t1,Ps,'b')
    title('Plot of Riccati Coefficients')
    xlabel('t')
    ylabel('P Matrix')
    hold
    cflag=2;
```

```
  elseif cflag==2
     plot(t1,Ps,'m:')
     cflag=3;
  elseif cflag==3
     plot(t1,Ps,'g-.')
     cflag=4;
  elseif cflag==4
     plot(t1,Ps,'r--')
     cflag=1;
     fig=fig+1;
  end
 Ps=0.;
 j=1;
end
if cflag==2|cflag==3|cflag==4
   hold
   fig=fig+1;
end
%
%Plot Optimized x
%
if n>2
   for i=1:3:(3*fix((n-3)/3)+1)
      figure(fig);
      plot(tx,real(x(:,i)),'b',tx,real(x(:,i+1)),'m:',tx,...
real(x(:,i+2)),'g-.')
      title('Plot of Optimized x')
      xlabel('t')
      ylabel('x  vectors')
      fig=fig+1;
   end
end
if (n-3*fix(n/3))==1
   figure(fig);
   plot(tx,real(x(:,n)),'b')
elseif (n-3*fix(n/3))==2
   figure(fig);
   plot(tx,real(x(:,n-1)),'b',tx,real(x(:,n)),'m:')
end
```

```
title('Plot of Optimized x')
xlabel('t')
ylabel('x  vectors')
fig=fig+1;
%
%Plot Optimized u
%
if mb>2
   for i=1:3:(3*fix((mb-3)/3)+1)
       figure(fig);
       plot(tu,real(u(:,i)),'b',tu,real(u(:,i+1)),'m:',...
tu,real(u(:,i+2)),'g-.')
       title('Plot of Optimized u')
       xlabel('t')
       ylabel('u  vectors')
       fig=fig+1;
 %
   end
end
if (mb-3*fix(mb/3))==1
   figure(fig);
   plot(tu,real(u(:,mb)),'b')
elseif (mb-3*fix(mb/3))==2
   figure(fig);
   plot(tu,real(u(:,mb-1)),'b',tu,real(u(:,mb)),'m:')
end
title('Plot of Optimized u')
xlabel('t')
ylabel('u  vectors')
%
end
%%
%%%%%%%%%%%%%%%
```

C.1.2   MATLAB File lqrnssf.m

This file *lqrnssf.m* is used along with the other two files *example.m* and *lqrnss.m* given above.

```
%%%%%%%%%%%%%%%%%
```

```
%% The following is lqrnssf.m
%%
function dx = lqrnssf(t,x)
% Function for x
%
global A E F Md tf W11 W12 W21 W22 n
%Calculation of P, Riccati Analytical Solution
Tt=-inv(W22-F*W12)*(W21-F*W11);
P=(W21+W22*expm(-Md*(tf-t))*Tt*expm(-Md*(tf-t)))*...
inv(W11+W12*expm(-Md*(tf-t))*Tt*expm(-Md*(tf-t)));
%
xa=[A-E*P];
%
%Definition of differential equations
%
dx=[xa*x];
%%%%%%%%
```

C.2   MATLAB© for Continuous-Time Tracking System

The following MATLAB© files are used to solve the Example 4.1. The main file is Example4.1(example4_1.m) which requires the files: Example 4.1 (example4_1p.m), Example 4.1(example4_1g.m), and Example 4.1 (example4_1x.m). The file Example 4.1(example4_1.m) is for solving the set of first order Riccati differential equations; Example 4.1 (example4_1g.m) is the set of first order g vector differential equations; and Example 4.1 (example4_1x.m) is the set of state differential equations.

C.2.1   MATLAB File for Example 4.1(example4_1.m)

```
clear all
%
%Define variables to use in external functions
global tp p tg g
%
%Define Initial Conditions for numerical solution
```

```
% of g and x states
%
tf=20;
tspan=[tf 0];
tp=0.;
tg=0.;
tx=0.;
pf=[2.,0.,0.];
gf=[2.,0.];
x0=[-0.5,0.];
%
%Calculation of P
%
[tp,p]=ode45('example4_1p',tspan,pf,odeset('refine',2,...
'RelTol',1e-4,'AbsTol',1e-6));
%
%Calculation of g
%
[tg,g]=ode45('example4_1g',tp,gf,odeset('refine',2,...
'RelTol',1e-4,'AbsTol',1e-6));
%
%Calculation of optimized x
%
[tx,x]=ode45('example4_1x',flipud(tg),x0,...
odeset('refine',2,'RelTol',1e-4,'AbsTol',1e-6));
%
%Plot  Riccati coefficients
%
fig=1;                    %Figure number
figure(fig)
plot(tp,real(p(:,1)),'k',tp,real(p(:,2)),'k',tp,...
real(p(:,3)),'k')
grid on
xlabel('t')
ylabel('Riccati Coefficients')
hold
%
fig=fig+1;
%
```

```
%Plot g values
%
figure(fig);
plot(tg,real(g(:,1)),'k',tg,real(g(:,2)),'k')
grid on
xlabel('t')
ylabel('g vector')
%%
%
fig=fig+1;
%
%Plot Optimal States x
%
figure(fig);
plot(tx,real(x(:,1)),'k',tx,real(x(:,2)),'k')
grid on
xlabel('t')
ylabel('Optimal States')
%
fig=fig+1;
%
%Plot Optimal Control u
%
[n,m]=size(p);
p12=p(:,2);
p22=p(:,3);
x1=x(:,1);
x2=x(:,2);
g2=flipud(g(:,2));
for i=1:1:n
    u(i) = -250*(p12(i)*x1(i) + p22(i)*x2(i) - g2(i));
end
figure(fig);
plot(tp,real(u),'k')
grid on
xlabel('t')
ylabel('Optimal Control')
```

C.2.2   MATLAB File for Example 4.1(example4_1p.m)

```
function dp = example4_1p(t,p)
% Function for P
%
%Define variables to use in external functions
%
%Definition of differential equations
%
dp=[250*p(2)^2+4*p(2)-2
   250*p(2)*p(3)-p(1)+3*p(2)+2*p(3)
   250*p(3)^2-2*p(2)+6*p(3)];
%
```

C.2.3   MATLAB File for Example 4.1(example4_1g.m)

```
function dg = example4_1g(t,g)
% Function for g
%
%Define variables to use in external functions
%
global tp p
%
%Definition of differential equations
%
dg=[(250*interp1(tp,p(:,2),t)+2)*g(2)-2
   -g(1)+(250*interp1(tp,p(:,3),t)+3)*g(2)];
```

C.2.4   MATLAB File for Example 4.1(example4_1x.m)

```
function dx = example4_1x(t,x)
% Function for x
%
%Define variables to use in external functions
global tp p tg g
%
%Definition of differential equations
%
dx=[x(2)
   -2*x(1)-3*x(2)-250*(interp1(tp,p(:,2),t)*x(1)+...
```

```
interp1(tp,p(:,3),t)*x(2)-interp1(tg,g(:,2),t))];
%
```

C.2.5   MATLAB File for Example 4.2(example4_1.m)

```
clear all
%
%Define variables to use in external functions
global tp p tg g
%
%Define Initial Conditions for numerical solution of
% g and x states
%
tf=20;
tspan=[tf 0];
tp=0.;
tg=0.;
tx=0.;
pf=[0.,0.,0.];
gf=[0.,0.];
x0=[-1.,0.];
%
%Calculation of P

    [tp,p]=ode45('example4_2p',tspan,pf,...
odeset('refine',2,'RelTol',1e-4,'AbsTol',1e-6));
%
%Calculation of g
%
    [tg,g]=ode45('example4_2g',tp,gf,...
odeset('refine',2,'RelTol',1e-4,'AbsTol',1e-6));
%
%Calculation of optimized x
%
    [tx,x]=ode45('example4_2x',flipud(tg),x0,...
odeset('refine',2,'RelTol',1e-4,'AbsTol',1e-6));
%
fig=1;                      %Figure number
figure(fig)
plot(tp,real(p(:,1)),'k',tp,real(p(:,2)),'k',tp,...
```

```
real(p(:,3)),'k')
grid on
title('Plot of P')
xlabel('t')
ylabel('Riccati Coefficients')
hold
%
fig=fig+1;
%
%Plot g values
%
figure(fig);
plot(tg,real(g(:,1)),'k',tg,real(g(:,2)),'k')
grid on
title('Plot of g Vector')
xlabel('t')
ylabel('g vector')
%
fig=fig+1;
%
%Plot Optimized x
%
figure(fig);
plot(tx,real(x(:,1)),'k',tx,real(x(:,2)),'k')
grid on
title('Plot of Optimal States')
xlabel('t')
ylabel('Optimal States')
%
fig=fig+1;
%
%Calculate and Plot Optimized u
%
[n,m]=size(p);
p12=flipud(p(:,2));
p22=flipud(p(:,3));
x1=x(:,1);
x2=x(:,2);
g2=flipud(g(:,2));
```

```
for i=1:1:n
   u(i) = -25*(p12(i)*x1(i) + p22(i)*x2(i) + g2(i));
end
figure(fig);
plot(tx,real(u),'b')
grid on
title('Plot of Optimal Control')
xlabel('t')
ylabel('Optimal Control')
%%%%%%%%%%%%%%%%%%%
```

C.2.6 *MATLAB File for Example 4.2(example4_2p.m)*

```
function dp = example4_2p(t,p)
% Function for P
%
%Define variables to use in external functions
%
%Definition of differential equations
%
dp=[25*p(2)^2+4*p(2)-2
    25*p(2)*p(3)-p(1)+3*p(2)+2*p(3)
    25*p(3)^2-2*p(2)+6*p(3)];
%%
```

C.2.7 *MATLAB File for Example 4.2(example4_2g.m)*

```
function dg = example4_2g(t,g)
% Function for g
%
%Define variables to use in external functions
%
global tp p
%
%Definition of differential equations
%
dg=[(25*interp1(tp,p(:,2),t)+2)*g(2)-4*t
    -g(1)+(25*interp1(tp,p(:,3),t)+3)*g(2)];
%%
```

C.2.8 MATLAB File for Example 4.2(example4_2x.m)

```
function dx = example4_2x(t,x)
% Function for x
%
%Define variables to use in external functions
global tp p tg g
%
%Definition of differential equations
%
dx=[x(2)
    -2*x(1)-3*x(2)-25*(interp1(tp,p(:,2),t)*x(1)+...
interp1(tp,p(:,3),t)*x(2)-interp1(tg,g(:,2),t))];
%%
```

C.3   MATLAB© for Matrix Difference Riccati Equation

The following is the typical MATLAB© file containing the various given matrices for a problem, such as Example 5.5, using analytical solution of matrix Riccati difference equation given in Chapter 5. This file, say *example.m* requires the other file *lqrdnss.m* given below.

```
%%%%%%%%%%%%%%%%
clear all
A=[.8,1;0,.5];
B=[1;.5];
F=[2,0;0,4];
Q=[1,0;0,1];
R=1;
kspan=[0 10];
x0(:,1)=[5.;3.];
[x,u]=lqrdnss(A,B,F,Q,R,x0,kspan);
%%%%%%%%%%%%%%%%%%%%%%%
```

C.3.1 ˜MATLAB File lqrdnss.m

This MATLAB© file *lqrdnss.m* is required along with the other file *example.m* to solve the matrix Riccati difference equation using its analytical solution.

```
%%%%%%%%%%%%
function [x,u]=lqrdnss(As,Bs,Fs,Qs,Rs,x0,kspan)
%
%This m-file calculates and plots the outputs for a
% discrete Linear Quadratic Regulator system
%Based on provided linear state space matrices
% for A and B and Performance Index matrices
%  for F, Q and R.
%This function takes these inputs, and using the
% analytical solution to the matrix Riccati equation,
% formulates the optimal states and inputs.
%
%
%       SYNTAX:   [x,u]=lqrdnss(A,B,F,Q,R,x0,tspan)
%
%   INPUTS (All numeric):
%  A,B     Matrices from xdot=Ax+Bu
%  F,Q,R  Performance Index Parameters; terminal cost,
%   error and control weighting
%  x0  State variable initial condition.  Must be a
%         column vector [x10;x20;x30...]
%  kspan    Vector containing sample span [k0 kf]
%
%   OUTPUTS:
%       x              is the state variable vector
%       u              is the input vector
%
%       The system plots the Riccati coefficients in
% combinations of 4,
%              and the x vector, and u vector in
%       combinations of 3.
%


%Check for correct number of inputs

if nargin<7
   error('Incorrect number of inputs specified')
   return
```

```
end

%Convert Variables to normal symbology to
% prevent problems with global statement

A=As;
B=Bs;
F=Fs;
Q=Qs;
R=Rs;
plotflag=0;    %set plotflag to a 1 to avoid plotting
% of data on figures

%Define secondary variables for global passing to
% ode-related functions and determine matrice size

[n,m]=size(A);              %Find dimensions of A
[nb,mb]=size(B);            %Find dimensions of B
[nq,mq]=size(Q);            %Find dimensions of Q
[nr,mr]=size(R);            %Find Dimensions of R
[nf,mf]=size(F);            %Find Dimensions of F
if n~=m                     %Verify A is square
   error('A must be square')
else
   [n,n]=size(A);
end

%Data Checks for proper setup
%Check for controllability
if length(A)>rank(ctrb(A,B))
   error('System Not Controllable')
   return
end
if (n ~= nq) | (n ~= mq)
%Check that A and Q are the same size
   error('A and Q must be the same size');
   return
end
if ~(mf==1&nf==1)
```

```matlab
    if (nq ~= nf) | (mq ~= mf)
%Check that Q and F are the same size
        error('Q and F must be the same size');
        return
    end
end
if ~(mr==1&nr==1)
    if (mr ~= nr) | (mb ~= nr)
        error('R must be consistent with B');
        return
    end
end
mq = norm(Q,1);
% Check if Q is positive semi-definite and symmetric
if any(eig(Q) < -eps*mq) | (norm(Q'-Q,1)/mq > eps)
disp('Warning: Q is not symmetric and ...
positive semi-definite');
end
mr = norm(R,1);
% Check if R is positive definite and symmetric
if any(eig(R) <= -eps*mr) | (norm(R'-R,1)/mr > eps)
disp('Warning: R is not symmetric and ...
positive definite');
end

%Define Calculated Matrix

E=B*inv(R)*B';

%Find matrix needed to calculate Analytical Solution
%    to Riccati Equation

H=[inv(A),inv(A)*E;Q*inv(A),A'+Q*inv(A)*E];

%Find Eigenvectors

[W,D]=eig(H);

%Find the diagonals from D and pick the negative
```

```
%    diagonals to create a new matrix M

j=n;
[m1,index1]=sort(real(diag(D)));
   for i=1:1:n
       m2(i)=m1(j);
       index2(i)=index1(j);
       index2(i+n)=index1(i+n);
       j=j-1;
   end
Md=diag(m2);

%Rearrange W so that it corresponds to the
% sort of the eigenvalues

for i=1:2*n
   w2(:,i)=W(:,index2(i));
end
W=w2;

%Define the Modal Matrix for D and split it into parts

W11=zeros(n);
W12=zeros(n);
W21=zeros(n);
W22=zeros(n);
j=1;
  for i=1:2*n:(2*n*n-2*n+1)
    W11(j:j+n-1)=W(i:i+n-1);
    W21(j:j+n-1)=W(i+n:i+2*n-1);
    W12(j:j+n-1)=W(2*n*n+i:2*n*n+i+n-1);
    W22(j:j+n-1)=W(2*n*n+i+n:2*n*n+i+2*n-1);
    j=j+n;
  end

%Find M
M=zeros(n);
j=1;
  for i=1:2*n:(2*n*n-2*n+1)
```

```
    M(j:j+n-1)=D(i:i+n-1);
    j=j+n;
  end

%Zero Vectors
x=zeros(n,1);

%Define Loop Variables (l=lambda)
k0=kspan(1);
kf=kspan(2);

%x and P Conditions
x(:,1)=x0(:,1);
Tt=-inv(W22-F*W12)*(W21-F*W11);
P=real((W21+W22*((Md^-(kf-0))*Tt*(Md^-(kf-0))))...
*inv(W11+W12*((Md^-(kf-0))*Tt*(Md^-(kf-0)))));
L=inv(R)*B'*(inv(A))'*(P-Q);
u(:,1)=-L*x0(:,1);
k1(1)=0;

for k=(k0+1):1:(kf)
   Tt=-inv(W22-F*W12)*(W21-F*W11);
   P=real((W21+W22*((Md^-(kf-k))*Tt*(Md^-(kf-k))))...
*inv(W11+W12*((Md^-(kf-k))*Tt*(Md^-(kf-k)))));
   L=inv(R)*B'*(inv(A))'*(P-Q);
   x(:,k+1)=(A-B*L)*x(:,k);
   u(:,k+1)=-L*x(:,k+1);
   k1(k+1)=k;
end

%Plotting Section, if desired

if plotflag~=1

%Plot Riccati coefficients using flag variables
% to hold and  change colors
%Variables are plotted one at a time and the plot held

fig=1;          %Figure number
```

```matlab
cflag=1;          %Variable used to change plot color
j=1;
Ps=0.;            %Initialize P Matrix plot variable
for i=1:1:n*n
  for k=(k0):1:(kf)
    Tt=-inv(W22-F*W12)*(W21-F*W11);
    P=real((W21+W22*((Md^-(kf-k))*Tt*(Md^-(kf-k))))...
*inv(W11+W12*((Md^-(kf-k))*Tt*(Md^-(kf-k)))));
    Ps(j)=P(i);
    k2(j)=k;
    j=j+1;
  end
  if cflag==1;
     figure(fig);
     plot(k2,Ps,'b')
     title('Plot of Riccati Coefficients')
     grid on
     xlabel('k')
     ylabel('P Matrix')
     hold
     cflag=2;
  elseif cflag==2
     plot(k2,Ps,'b')
     cflag=3;
  elseif cflag==3
     plot(k2,Ps,'b')
     cflag=4;
  elseif cflag==4
     plot(k2,Ps,'b')
     cflag=1;
     fig=fig+1;

  end
 Ps=0.;
 j=1;
end
if cflag==2|cflag==3|cflag==4
   hold
   fig=fig+1;
```

```
end

%Plot Optimized x

x=x';
if n>2
   for i=1:3:(3*fix((n-3)/3)+1)
      figure(fig);
      plot(k1,real(x(:,i)),'b',k1,real(x(:,i+1)),'b',k1,...
         real(x(:,i+2)),'b')
      grid on
      title('Plot of Optimal States')
      xlabel('k')
      ylabel('Optimal States')
      fig=fig+1;
   %
   end
end
if (n-3*fix(n/3))==1
   figure(fig);
   plot(k1,real(x(:,n)),'b')
elseif (n-3*fix(n/3))==2
   figure(fig);
   plot(k1,real(x(:,n-1)),'b',k1,real(x(:,n)),'b')
end
grid on
title('Plot of Optimal States')
xlabel('k')
ylabel('Optimal States')
fig=fig+1;
  %
%Plot Optimized u
%
u=u';
if mb>2
   for i=1:3:(3*fix((mb-3)/3)+1)
      figure(fig);
      plot(k1,real(u(:,i)),'b',k1,real(u(:,i+1)),...
'm:',k1,real(u(:,i+2)),'g-.')
```

```
        grid on
        title('Plot of Optimal Control')
        xlabel('k')
        ylabel('Optimal Control')
        fig=fig+1;
    end
end
if (mb-3*fix(mb/3))==1
   figure(fig);
   plot(k1,real(u(:,mb)),'b')
elseif (mb-3*fix(mb/3))==2
   figure(fig);
   plot(k1,real(u(:,mb-1)),'b',k1,real(u(:,mb)),'m:')
end
grid on
title('Plot of Optimal Control')
xlabel('k')
ylabel('Optimal Control')
  gtext('u')
end
%%%%%%%%
```

C.4 MATLAB© for Discrete-Time Tracking System

This MATLAB© file for tracking Example 5.6 is given below.

```
% Solution Using Control System Toolbox (STB) in
%  MATLAB Version 6
clear
A=[0.8 1;0,0.6]; %% system matrix A
B=[1;0.5]; %% system matrix B
C=[1 0;0 1]; %% system matrix C
Q=[1 0;0 0]; %% performance index
%% state weighting matrix Q
R=[0.01]; %% performance index control
%% weighting matrix R
F=[1,0;0,0]; %% performance index weighting matrix F
%
x1(1)=5; %% initial condition on state x1
```

```
x2(1)=3; %% initial condition on state x2
xk=[x1(1);x2(1)];
zk=[2;0];
zkf=[2;0];
% note that if kf = 10 then
% k = [k0,kf] = [0 1 2,...,10],
% then we have 11 points and an array x1 should
% have subscript
% x1(N) with N=1 to 11. This is because x(o) is
% illegal in array
% definition in MATLAB. Let us use N = kf+1
k0=0; % the initial instant k_0
kf=10; % the final instant k_f
N=kf+1; %
[n,n]=size(A); % fixing the order of the system matrix A
I=eye(n); % identity matrix I
E=B*inv(R)*B'; % the matrix E = BR^{-1}B'
V=C'*Q*C;
W=C'*Q;
%
% solve matrix difference Riccati equation
%   backwards starting from kf to k0
% use the form P(k) = A'P(k+1)[I + EP(k+1)]^{-1}A + V
% first fix the final conditionS P(k_f) = F;
% g(k_f) = C'Fz(k_f)
% note that P, Q, R, F are all symmatric ij = ji
Pkplus1=C'*F*C;
gkplus1=C'*F*zkf;
p11(N)=F(1);
p12(N)=F(2);
p21(N)=F(3);
p22(N)=F(4);
%
g1(N)=gkplus1(1);
g2(N)=gkplus1(2);
%
Pk=0;
gk=0;
for k=N-1:-1:1,
```

```
      Pk = A'*Pkplus1*inv(I+E*Pkplus1)*A+V;
      Lk = inv(R+B'*Pkplus1*B)*B'*Pkplus1*A;
      gk=(A-B*Lk)'*gkplus1+W*zk;
      p11(k) = Pk(1,1);
      p12(k) = Pk(1,2);
      p21(k) = Pk(2,1);
      p22(k) = Pk(2,2);
      pkplu1 = Pk;
%
      g1(k) = gk(1);
      g2(k) = gk(2);
      gkplus1 = gk;
end
%
% calcuate the feedback coefficients L and Lg(k)
% L(k) = (R+B'P(k+1)B)^{-1}BP(k+1)A
% Lg(k) = [R + B'P(k+1)B]^{-1}B'
%
for k = N:-1:1,
   Pk=[p11(k),p12(k);p21(k),p22(k)];
   gk=[g1(k);g2(k)];
   Lk = inv(R+B'*Pkplus1*B)*B'*Pkplus1*A;
   Lgk= inv(R+B'*Pkplus1*B)*B';
   l1(k) = Lk(1);
   l2(k) = Lk(2);
   lg1(k) = Lgk(1);
   lg2(k) = Lgk(2);
end
%
% solve the optimal states
% x(k+1) = [A-B*L]x(k) + BLg(k+1)g(k+1) given x(0)
%
xk=0.0;
for k=1:N-1,
   Lk = [l1(k),l2(k)];
   Lgk = [lg1(k),lg2(k)];
   Lgkplus1=[lg1(k+1),lg2(k+1)];
   xk = [x1(k);x2(k)];
   xkplus1 = (A-B*Lk)*xk + B*Lgkplus1*gk;
```

```
    x1(k+1) = xkplus1(1);
    x2(k+1) = xkplus1(2);
end
%
% solve for optimal control
% u(k) = - L(k)x(k) + Lg(k)g(k+1)
%
xk=0.0;
 %    for k=1:N,
   for k=1:N-1,
   Lk = [l1(k),l2(k)];
   Lgk = [lg1(k),lg2(k)];
   gkplus1=[g1(k+1);g2(k+1)];
   xk = [x1(k);x2(k)];
   u(k) = - Lk*xk + Lgk*gkplus1;
end
%
% plot various values: P(k), g(k), x(k), u(k)
% let us first reorder the values of k = 0 to kf
%
% first plot P(k)
%
k = 0:1:kf;
figure(1)
plot(k,p11,'k:o',k,p12,'k:+',k,p22,'k:*')
grid
xlabel('k')
ylabel('Riccati coefficients')
gtext('p_{11}(k)')
gtext('p_{12}(k)')
gtext('p_{22}(k)')
%
% Plot g(k)
%
k = 0:1:kf;
figure(2)
plot(k,g1,'k:o',k,g2,'k:+')
grid
xlabel('k')
```

```
ylabel('Vector coefficients')
gtext('g_{1}(k)')
gtext('g_{2}(k)')
%
k=0:1:kf;
figure(3)
plot(k,x1,'k:o',k,x2,'k:+')
grid
xlabel('k')
ylabel('Optimal States')
gtext('x_1(k)')
gtext('x_2(k)')
%
figure(4)
k=0:1:kf-1;
plot(k,u,'k:*')
grid
xlabel('k')
ylabel('Optimal Control')
gtext('u(k)')
%
% end of the program
%
```

References

[1] N. I. Akhiezer. *The Calculus of Variations*. Blaisdell Publishing Company, Boston, MA, 1962.

[2] B. D. O. Anderson and J. B. Moore. Linear system optimization with prescribed degree of stability. *Proceedings of the IEE*, 116(12):2083–2087, 1969.

[3] B. D. O. Anderson and J. B. Moore. *Optimal Control: Linear Quadratic Methods*. Prentice Hall, Englewood Cliffs, NJ, 1990.

[4] P. J. Antsaklis and A. N. Michel. *Linear Systems*. The McGraw Hill Companies, Inc., New York, NY, 1997.

[5] K. Asatani. Sub-optimal control of fixed-end-point minimum energy problem via singular perturbation theory. *Journal of Mathematical Analysis and Applications*, 45:684–697, 1974.

[6] M. Athans and P. Falb. *Optima Control: An Introduction to The Theory and Its Applications*. McGraw Hill Book Company, New York, NY, 1966.

[7] T. Başar and P. Bernhard. H^∞-*Optimal Control and Related Minimax Design Problems: Second Edition*. Birkhäuser, Boston, MA, 1995.

[8] A. V. Balakrishnan. *Control Theory and the Calculus of Variations*. Academic Press, New York, NY, 1969.

[9] M. Bardi and I. C.-Dolcetta. *Optimal Control and Viscosity Solutions of Hamilton-Jacobi-Bellman Equations*. Birkhäuser, Boston, MA, 1997.

[10] S. Barnett. *Matrices in Control Theory*. Van Nostrand Reinhold, London, UK, 1971.

[11] J. S. Bay. *Fundamentals of Linear State Space Systems*. The McGraw Hill Companies, Inc., New York, NY, 1999.

[12] R. E. Bellman. *Dynamic Programming*. Princeton University Press, Princeton, NJ, 1957.

[13] R. E. Bellman. *Introduction to Matrix Analysis*. Mc-Graw Hill Book Company, New York, NY, second edition, 1971.

[14] R. E. Bellman and S. E. Dreyfus. *Applied Dynamic Programming.* Princeton University Press, Princeton, NJ, 1962.

[15] R. E. Bellman and R. E. Kalaba. *Dynamic Programming and Modern Control Theory.* Academic Press, New York, NY, 1965.

[16] A. Bensoussen, E. Hurst, and B. Naslund. *Management Applications of Modern Control Theory.* North-Holland Publishing Company, New York, NY, 1974.

[17] L. D. Berkovitz. *Optimal Control Theory.* Springer-Verlag, New York, NY, 1974.

[18] D. P. Bertsekas. *Dynamic Programming and Optimal Control: Volume I.* Athena Scientific, Belmont, MA, 1995.

[19] D. P. Bertsekas. *Dynamic Programming and Optimal Control: Volume II.* Athena Scientific, Belmont, MA, 1995.

[20] J. T. Betts. *Practical Methods for Optimal Control Using Nonlinear Programming.* SIAM, Philadelphia, PA, 2001.

[21] S. Bittanti. History and prehistory of the Riccati equation. In *Proceedings of the 35th Conference on Decision and Control,* page 15991604, Kobe, Japan, December 1996.

[22] S. Bittanti, A. J. Laub, and J. C. Willems, editors. *The Riccati Equation.* Springer-Verlag, New York, NY, 1991.

[23] G. A. Bliss. *Lectures on the Calculus of Variations.* University of Chicago Press, Chicago, IL, 1946.

[24] V. G. Boltyanskii. *Mathematical Methods of Optimal Control.* Rinehart and Winston, New York, NY, 1971.

[25] V. G. Boltyanskii, R. V. Gamkrelidze, and L. S. Pontryagin. On the theory of optimal processes. *Dokl. Akad. Nauk SSSR,* 110:7–10, 1956. (in Russian).

[26] O. Bolza. *Lectures on the Calculus of Variations.* Chelsea Publishing Company, New York, NY, Third edition, 1973.

[27] A. E. Bryson, Jr. , W. F. Denham, and S. E. Dreyfus. Optimal programming problems with inequality constraints, I: Necessary conditions for extremal solutions. *AIAA Journal,* 1(3):2544–2550, 1963.

[28] A. E. Bryson, Jr. Optimal control-1950 to 1985. *IEEE Control Systems,* 16(3):26–33, June 1996.

[29] A. E. Bryson, Jr. *Dynamic Optimization.* Addison Wesley Longman, Inc., Menlo Park, CA, 1999.

[30] A. E. Bryson, Jr. and Y. C. Ho. *Applied Optimal Control: Optimization, Estimation and Control.* Hemisphere Publishing Com-

pany, New York, NY, 1975. Revised Printing.

[31] R. S. Bucy. *Lectures on Discrete Time Filtering*. Springer-Verlag, New York, NY, 1994.

[32] J. B. Burl. *Linear Optimal Control: H_2 and H_∞ Methods*. Addison-Wesley Longman Inc., Menlo Park, CA, 1999.

[33] J. A. Cadzow and H. R. Martens. *Discrete-Time and Computer Control Systems*. Prentice Hall, Englewood Cliffs, NJ, 1970.

[34] B. M. Chen. *Robust and H_∞ Control*. Springer-Verlag, London, UK, 2000.

[35] C. T. Chen. *Linear System Theory and Design*. Oxford University Press, New York, NY, Third edition, 1999.

[36] G. S. Christensen, M. E. El-Hawary, and S. A. Soliman. *Optimal Control Applications in Electric Power Systems*. Plenum Publishing Company, New York, NY, 1987.

[37] P. Cicala. *An Engineering Approach to the Calculus of Variations*. Levrotto and Bella, Torino, 1957.

[38] S. J. Citron. *Elements of Optimal Control*. Rinehart and Winston, New York, NY, 1969.

[39] P. Colaneri, J. C. Geromel, and A. Locatelli. *Control Theory and Design: An RH_2 and RH_∞ Viewpoint*. Academic Press, San Diego, CA, 1997.

[40] W. F. Denham and A. E. Bryson, Jr. Optimal programming problems with inequality constraints II: Solution by steepest ascent. *AIAA Journal*, 2(1):25–34, 1964.

[41] P. M. DeRusso, R. J. Roy, C. M. Close, and A. A. Desrochers. *State Variables for Engineers*. John Wiley & Sons, New York, NY, Second edition, 1998.

[42] P. Dorato, C. Abdallah, and V. Cerone. *Linear-Quadratic Control: An Introduction*. Prentice Hall, Englewood Cliffs, NY, 1995.

[43] J. C. Doyle, B. A. Francis, and A. R. Tannenbaum. *Feedback Control Theory*. Macmillan Publishing, New York, NY, 1992.

[44] J. C. Doyle, K. Glover, P. P. Khargonekar, and B. A. Francis. State-space solutions to standard H_2 and H_∞ control problems. *IEEE Transactions on Automatic Control*, 34:831–847, 1989.

[45] S. E. Dreyfus. *Dynamic Programming and the Calculus of Variations*. Academic Press, New York, NY, 1966.

[46] L. Elsgolts. *Differential Equations and Calculus of Variations*. Mir Publishers, Moscow, Russia, 1970.

[47] L. E. Elsgolts. *Calculus of Variations*. Addison-Wesley, Rading,

MA, 1962.

[48] G. M. Ewing. *Calculus of Variations with Applications.* Dover Publications, Inc., New York, NY, 1985.

[49] W.F. Feehery and P.I. Barton. Dynamic optimization with state variable path constraints. *Computers in Chemical Engineering*, 22:1241–1256, 1998.

[50] M. J. Forray. *Variational Calculus in Science and Engineering.* McGraw-Hill Book Company, New York, NY, 1968.

[51] B. A. Francis. *A Course in H_∞ Optimal Control Theory*, volume 88 of *Lecture Notes in Control and Information Sciences.* Springer-Verlag, Berlin, 1987.

[52] R. V. Gamkrelidze. Discovery of the Maximum Principle. *Journal of Dynamical and Control Systems*, 5(4):437–451, 1999.

[53] R. V. Gamkrelidze. Mathematical Works of L. S. Pontryagin. (personal communication), August 2001.

[54] F. R. Gantmacher. *The Theory of Matrices, Vols. 1 and 2.* Chelsea Publishing, New York, NY, 1959.

[55] I. M. Gelfand and S. V. Fomin. *Calculus of Variations.* John Wiley & Sons, New York, NY, 1988.

[56] M. Giaquinta and S. Hildebrandt. *Calculus of Variations: Volume I: The Lagrangian Formalism.* Springer-Verlag, New York, NY, 1996.

[57] M. Giaquinta and S. Hildebrandt. *Calculus of Variations: Volume II: The Hamiltonian Formalism.* Springer-Verlag, New York, NY, 1996.

[58] H. H. Goldstine. *A History of the Calculus of Variations: From the 17th through the 19th Century.* Springer-Verlag, New York, NY, 1980.

[59] R. J. Gran. Fly me to the Moon - then and now. *MATLAB News & Notes*, Summer:4–9, 1999.

[60] M. Green and D. Limebeer. *Linear Robust Control.* Prentice-Hall, Englewood Cliffs, NJ, 1995.

[61] J. Gregory and C. Lin. *Constrained Optimization in the Calculus of Variations and Optimal Control Theory.* Van Nostrand Reinhold, New York, NY, 1992.

[62] R. F. Hartl, S. P. Sethi, and R. G. Vickson. A survey of the maximum principles for optimal control problems with state constraints. *SIAM Review*, 37(2):181–218, June 1995.

[63] A. Heim and O. V. Stryk. *Documentation of PAREST - A Mul-*

tiple Shooting Code for Optimization Problems in Differential - Algebraic Equations, November 1996.

[64] M. R. Hestenes. A general problem in the calculus of variations with applications to paths of least time. Technical Report RM-100, RAND Corporation, 1950.

[65] M. R. Hestenes. *Calculus of Variations and Optimal Control*. John Wiley & Sons, New York, NY, 1966.

[66] L. M. Hocking. *Optimal Control: An Introduction to the Theory and Applications*. Oxford University Press, New York, NY, 1991.

[67] J. C. Hsu and A. U. Meyer. *Modern Control Principles and Applications*. McGraw Hill, New York, NY, 1968.

[68] D. H. Jacobson and M. M. Lee. A transformation technique for optimal control problems with state inequality constraints. *IEEE Transactions on Automatic Control*, AC-14:457–464, 1969.

[69] T. Kailath. *Linear Systems*. Prentice Hall, Englewood Cliffs, NJ, 1980.

[70] R. E. Kalman. Contribution to the theory of optimal control. *Bol. Soc. Matem. Mex.*, 5:102–119, 1960.

[71] R. E. Kalman. A new approach to linear filtering in prediction problems. *ASME Journal of Basic Engineering*, 82:34–45, March 1960.

[72] R. E. Kalman. Canonical structure of linear dynamical systems. *Proc. Natl. Acad. Sci.*, 148(4):596–600, April 1962.

[73] R. E. Kalman. Mathematical description of linear dynamical systems. *J. Soc. for Industrial and Applied Mathematics*, 1:152–192, 1963.

[74] R. E. Kalman. New methods in wiener filtering theory. In *Proceedings of the Symposium on Engineering Applications of Random Function Theory and Probability*, New York, 1963. Wiley.

[75] R. E. Kalman. The theory of optimal control and the calculus of variations. In R. Bellman, editor, *Mathematical Optimization Techniques*, chapter 16. University of California Press, 1963.

[76] R. E. Kalman and R. S. Bucy. New results in linear filtering and prediction theory. *Transactions ASME J. Basic Eng.*, 83:95–107, 1961.

[77] R. E. Kalman, Y. Ho, and K. S. Narendra. Controllability of linear dynamical systems. *Contributions to Differential Equations*, 1(2):189–213, 1963.

[78] M. I. Kamien and N. L. Schwartz. *Dynamic Optimization: The*

Calculus of Variations and Optimal Control in Economics and Management, Second Edition. Elsvier Science Publishing Company, New York, NY, 1991.

[79] D. E. Kirk. *Optimal Control Theory.* Prentice Hall, Englewood Cliffs, NJ, 1970.

[80] G. E. Kolosov. *Optimal Design of Control Systems: Stochastic and Deterministic Problems.* Marcel Dekker, Inc., New York, NY, 1999.

[81] M. L. Krasnov, G. I. Makarenko, and A. I. Iselev. *Problems and Exercises in Calculus of Variations.* Mir Publishers, Moscow, Russia, 1975.

[82] B. C. Kuo. *Digital Control Systems, Second Edition.* Holt, Rinehart, and Winston, New York, NY, 1980.

[83] B. C. Kuo. *Automatic Control Systems, Seventh Edition.* Prentice Hall, Englewood Cliffs, NJ, 1995.

[84] H. Kwakernaak and R. Sivan. *Linear Optimal Control Systems.* Wiley-Interscience, New York, NY, 1972.

[85] J. L. Lagrange. *Mechanique Analytique, 2 Volumes.* Paris, France, 1788.

[86] E. B. Lee and L. Markus. *Foundations of Optimal Control Theory.* John Wiley & Sons, New York, NY, 1967.

[87] G. Leitmann. *An Introduction to Optimal Control.* McGraw Hill, New York, NY, 1964.

[88] G. Leitmann. *The Calculus of Variations and Optimal Control: An Introduction.* Plenum Publishing Co., New York, NY, 1981.

[89] F. L. Lewis. *Optimal Control.* John Wiley & Sons, New York, NY, 1986.

[90] F. L. Lewis. *Applied Optimal Control and Estimation: Digital Design and Implementation.* Prentice Hall, Englewood Cliffs, NJ, 1992.

[91] F. L. Lewis and V.L. Syrmos. *Optimal Control, Second Edition.* John Wiley & Sons, New York, NY, 1995.

[92] A. Locatelli. *Optimal Control: An Introduction.* Birkhäuser, Boston, MA, 2001.

[93] D. G. Luenberger. *Optimization by Vector Space Methods.* John Wiley and Sons, New York, NY, 1969.

[94] M. A. Lyapunov. The general problem of motion stability. *Comm. Soc. Math. Kharkov*, 1892. Original paper in Russian. Translated in French, *Ann. Fac. Sci. Toulouse*, 9, pp. 203-474, (1907),

Reprinted in *Ann. Math. Study*, No. 17, Princeton University Press, Princeton, NJ, (1949).

[95] A. G. J. MacFarlane and I. Postlethwaite. The generalized Nyquist stability criterion and multivariable root loci. *International Journal Control*, 25:81–127, 1977.

[96] J. M. Maciejowski. *Multivariable Feedback Design*. Addison-Wesley Publishing Company, Reading, MA, 1989.

[97] J. Macki and A. Strauss. *Introduction to Optimal Control Theory*. Springer-Verlag, New York, NY, 1982.

[98] E. J. McShane. On multipliers for Lagrange problems. *American Journal of Mathematics*, LXI:809–818, 1939.

[99] E. J. McShane. The calculus of variations from the beginning through optimal control theory. *SIAM Journal of Control and Optimization*, 27:916–939, September 1989.

[100] J. S. Meditch. On the problem of optimal thrust programming for a lunar soft landing. *IEEE Transactions on Automatic Control*, AC-9:477–484, 1864.

[101] L. Meirovitch. *Dynamcis and Control of Structures*. John Wiley & Sons, New York, NY, 1990.

[102] G. H. Meyer. *Initial Value Methods for Boundary Value Problems*. Academic Press, New York, NY, 1973.

[103] A. A. Milyutin and N. P. Osmolovskii. *Calculus of Variations and Optimal Control*. American Mathematical Society, Providence, RI, 1997. Translations of Mathematical Monographs.

[104] I. H. Mufti, C. K. Chow, and F. T. Stock. Solution of ill-conditioned lienar two-point, boundary value problem by Riccati transformation. *SIAM Review*, 11:616–619, 1969.

[105] D. S. Naidu and D. B. Price. Singular perturbations and time scales in the design of digital flight control systems. Technical Paper 2844, NASA Langley Research Center, Hampton, VA, December 1988.

[106] I. P. Petrov. *Variational Methods in Optimal Control Theory*. Academic Press, New York, NY, 1968.

[107] D. A. Pierre. *Optimization Theory with Applications*. John Wiley & Sons, New York, NY, 1969.

[108] E. R. Pinch. *Optimal Control and Calculus of Variations*. Oxford University Press, New York, NY, 1993.

[109] L. S. Pontryagin, V. G. Boltyanskii, R. V. Gamkrelidze, and E. F. Mishchenko. *The Mathematical Theory of Optimal Processes*.

Wiley-Interscience, New York, NY, 1962. (Translated from Russian).

[110] L. Pun. *Introduction to Optimization Practice.* John Wiley, New York, NY, 1969.

[111] R. Pytlak. *Numerical Methods for Optimal Control Problems with State Constraints*, volume 1707 of *Lecture Notes in Mathematics*. Springer-Verlag, Berlin, Germany, 1999.

[112] W. F. Ramirez. *Process Control and Identification.* Academic Press, San Diego, CA, 1994.

[113] W. T. Reid. *Riccati Differential Equations.* Academic Press, New York, NY, 1972.

[114] C. J. Riccati. Animadversiones in aequations differentiales secundi gradus. *Acta Eruditorum Lipsiae*, 8:67–23, 1724.

[115] H. H. Rosenbrock. *Computer-Aided Control System Design.* Academic Press, New York, 1974.

[116] E. P. Ryan. *Optimal Relay and Saturating Control System Synthesis.* Peter Peregrinus Ltd., Stevenage, UK, 1982.

[117] A. Saberi, P. Sannuti, and B. M. Chen. H_2 *Optimal Control.* Prentice Hall International (UK) Limited, London, UK, 1995.

[118] A. Sagan. *Introduction to the Calculus of Variations.* Dover Publishers, Mineola, NY, 1992.

[119] A. P. Sage. *Optimum Systems Control.* Prentice Hall, Englewood Cliffs, NJ, 1968.

[120] A. P. Sage and C. C. White III. *Optimum Systems Control, Second Edition.* Prentice Hall, Englewood Cliffs, NJ, 1977.

[121] D. G. Schultz and J. L. Melsa. *State Functions and Linear Control Systems.* McGraw Hill, New York, NY, 1967.

[122] A. Seierstad and K. Sydsaeter. *Optimal Control Theory with Economic Applications.* Elsevier Science Publishing Co., New York, NY, 1987.

[123] S. P. Sethi and G. L. Thompson. *Optimal Control Theory: Applications to Management Science and Economics: Second Edition.* Kluwer Academic Publishers, Hingham, MA, 2000.

[124] V. Sima. *Algorithms for Linear-Quadratic Optimization.* Marcel Dekker, Inc., New York, NY, 1996.

[125] G.M. Siouris. *An Engineering Approach to Optimal Control and Estimation Theory.* John Wiley & Sons, New York, NY, 1996.

[126] D. R. Smith. *Variational Methods in Optimization.* Prentice Hall, Englewood Cliffs, NJ, 1974.

[127] R. F. Stengel. *Stochastic Optimal Control: Theory and Application.* Wiley-Interscience, New York, NY, 1986.

[128] A. Stoorvogel. *The H_∞ Control Problem: A State Space Approach.* Prentice Hall, Englewood Cliffs, NJ, 1992.

[129] O. V. Stryk. Numerical solution of optimal control problems by direct collocation. *International Series of Numerical Mathematics*, 111:129–143, 1993.

[130] O. V. Stryk. *User's Guide for DIRCOL, A Direct Collocation Method for the Numerical Solution of Optimal Control Problems*, 1999.

[131] M. B. Subrahmanyam. *Finite Horizon H_∞ and Related Control Problems.* Birkhäuser, Boston, MA, 1995.

[132] H. J. Sussmann and J. C. Willems. 300 years of optimal control: from the brachistochrone to the maximum principle. *IEEE Control Systems Magazine*, 17:32–44, June 1997.

[133] K. L. Teo, C. J. Goh, and K. H. Wong. *A Unified Computational Approach to Optimal Control Problems.* Longman Scientific and Technical, Harlow, UK, 1991.

[134] I. Todhunter. *A History of Progress of the Calculus of Variations in the Nineteenth Century.* Chelsea Publishing Company, New York, NY, 1962.

[135] J. Tou. *Modern Control Theory.* McGraw Hill, New York, NY, 1964.

[136] J. L. Troutman. *Variational Calculus and Optimal Control, Second Edition.* Springer-Verlag, New York, NY, 1996.

[137] F. A. Valentine. The problem of Lagrange with differential inequalities as added side conditions. In *Contributions to the Calculus of Variations*, pages 407–408. Chicago University Press, Chicago, IL, 1937.

[138] D. R. Vaughan. A nonrecursive algebraic solution for the discrete Riccati equation. *IEEE Transactions Automatic Control*, AC-15:597–599, October 1970.

[139] T. L. Vincent and W. J. Grantham. *Nonlinear and Optimal Control Systems.* John Wiley & Sons, New York, NY, 1997.

[140] R. Vinter. *Optimal Control.* Birkhäuser, Boston, MA, 2000.

[141] F. Y. M. Wan. *Introduction to the Calculus of Variations and its Applications.* Chapman and Hall, London, 1994.

[142] J. Warga. *Optimal Control of Differential and Functional Equations.* Academic Press, New York, NY, 1972.

[143] R. Weinstock. *Calculus of Variations with Applications to Physics and Engineering*. Dover Publishing, Inc., New York, NY, 1974.

[144] N. Wiener. *Cybernetics*. Wiley, New York, NY, 1948.

[145] N. Wiener. *Extrapolation, Interpolation, and Smoothing of Stationary Time Series*. Technology Press, Cambridge, MA, 1949.

[146] L. C. Young. *Lectures on the Calculus of Variations and Optimal Control Theory*. W. B. Saunders Company, Philadelphia, PA, 1969.

[147] L. A. Zadeh and C. A. Desoer. *Linear System Theory*. McGraw-Hill Book Company, New York, 1963.

[148] G. Zames. Feedback and optimal sensitivity: model reference transformation, multiplicative seminorms and approximate inverses. *IEEE Transactions Automatic Control*, 26:301–320, 1981.

[149] M. I. Zelikin. *Control Theory and Optimization I: Homogeneous Spaces and the Riccati Equation in the Calculus of Variations*. Springer-Verlag, Berlin, Germany, 2000.

[150] K. Zhou, J. C. Doyle, and K. Glover. *Robust and Optimal Control*. Prentice Hall, Upper Saddle River, NJ, 1996.

Index